76.00
7SC

Galaxy Distances and
Deviations from Universal Expansion

NATO ASI Series

Advanced Science Institutes Series

A series presenting the results of activities sponsored by the NATO Science Committee, which aims at the dissemination of advanced scientific and technological knowledge, with a view to strengthening links between scientific communities.

The series is published by an international board of publishers in conjunction with the NATO Scientific Affairs Division

A	Life Sciences	Plenum Publishing Corporation
B	Physics	London and New York
C	Mathematical and Physical Sciences	D. Reidel Publishing Company Dordrecht, Boston, Lancaster and Tokyo
D	Behavioural and Social Sciences	Martinus Nijhoff Publishers
E	Engineering and Materials Sciences	The Hague, Boston and Lancaster
F	Computer and Systems Sciences	Springer-Verlag
G	Ecological Sciences	Berlin, Heidelberg, New York and Tokyo

Series C: Mathematical and Physical Sciences Vol. 180

Galaxy Distances and Deviations from Universal Expansion

edited by

Barry F. Madore

Department of Astronomy, David Dunlap Observatory,
University of Toronto, Canada and
California Institute of Technology, Pasadena, California, U.S.A.

and

R. Brent Tully

Institute for Astronomy, University of Hawaii,
Honolulu, Hawaii, U.S.A.

D. Reidel Publishing Company

Dordrecht / Boston / Lancaster / Tokyo

Published in cooperation with NATO Scientific Affairs Division

Proceedings of the NATO Advanced Research Workshop on
Galaxy Distances and Deviations from Universal Expansion
Kona, Hawaii, U.S.A.
January 13-17, 1986

Library of Congress Cataloging in Publication Data

NATO Advanced Research Workshop on Galaxy Distances and Deviations from Universal
    Expansion (1986 : Kailua Kona, Hawaii)
    Galaxy distances and deviations from universal expansion.

    (NATO ASI series. Series C, Mathematical and physical sciences; vol. 180)
    "Proceedings of the NATO Advanced Research Workshop on Galaxy Distances and
Deviations from Universal Expansion, Kona, Hawaii, U.S.A., January 13–17, 1986"—T.p. verso.
    "Published in cooperation with NATO Scientific Affairs Division."

    Includes index.
    1. Extragalactic distances—Congresses. 2. Expanding universe—Congresses.
I. Madore, B. II. Tully, R. Brent. III. North Atlantic Treaty Organization. Scientific Affairs
Division. IV. Title. V. Series: NATO ASI series. Series C, Mathematical and physical
sciences; vol. 180.
QB857.N366   1986   523.1'8   86–10157
ISBN 90–277–2277–3

Published by D. Reidel Publishing Company
P.O. Box 17, 3300 AA Dordrecht, Holland

Sold and distributed in the U.S.A. and Canada
by Kluwer Academic Publishers,
101 Philip Drive, Assinippi Park, Norwell, MA 02061, U.S.A.

In all other countries, sold and distributed
by Kluwer Academic Publishers Group,
P.O. Box 322, 3300 AH Dordrecht, Holland

D. Reidel Publishing Company is a member of the Kluwer Academic Publishers Group

---

All Rights Reserved
© 1986 by D. Reidel Publishing Company, Dordrecht, Holland.
No part of the material protected by this copyright notice may be reproduced or utilized
in any form or by any means, electronic or mechanical, including photocopying, recording
or by any information storage and retrieval system, without written permission from the
copyright owner.

Printed in The Netherlands

TABLE OF CONTENTS

Preface	ix
List of Participants	xi
New Results on the Distance Scale and the Hubble Constant G. de Vaucouleurs	1
The Magellanic Clouds and the Distance Scale M. W. Feast	7
Main Sequence Fitting and the Distance to the Large Magellanic Cloud R. A. Schommer	15
Recent CCD Observations of Cepheids in Nearby Galaxies: A Progress Report W. L. Freedman	21
Distances Using A Supergiant Stars R. B. Tully and J. M. Nakashima	25
An Infrared Calibration of the Cepheid Distance Scale B. F. Madore	29
RR Lyrae Stars in the Halo of M31 — A Preliminary Report C. J. Pritchet	35
Two Stepping Stones to the Hubble Constant S. van den Bergh	41
Physical Models of Supernovae and the Distance Scale J. C. Wheeler and R. P. Harkness	45
Calibration of the Infrared Tully-Fisher Relation and the Global Value of the Hubble Constant M. Aaronson	55
Distances from HI Observations of Nearby Groups and Clusters W. K. Huchtmeier	63
21cm Line Widths and Distances of Spiral Galaxies R. C. Kraan-Korteweg, L. Cameron, and G. A. Tammann	65
Malmquist Bias and Virgocentric Model in Tully-Fisher Relation L. Bottinelli, P. Fouqué, L. Gouguenheim, G. Paturel, and P. Teerikorpi	73

The Luminosity Index as a Distance Indicator and the Structure
of the Virgo Cluster
G. de Vaucouleurs 77

Infrared Colour-Luminosity Relations for Field Galaxies
R. S. Ellis, B. Mobasher, and R. M. Sharples 81

Minimizing the Scatter in the Tully-Fisher Relation
G. D. Bothun 87

I versus (B-I) Color-Magnitude Relation for Galaxies and I-Band Distances
M. J. Pierce 93

Photometry of Galaxies and the Local Peculiar Motion
N. Visvanathan 99

The Motion of the Sun and the Local Group and the Velocity Dispersion of
"Field" Galaxies
O.-G. Richter 105

The Motion of the Local Group relative to the Nearest Clusters of Galaxies
J. Mould 111

Distances and Motions of 2000 Galaxies in the Pisces-Perseus Supercluster
M. P. Haynes and R. Giovanelli 117

Elliptical Galaxies and Non-Uniformities in the Hubble Flow
D. Burstein, R. L. Davies, A. Dressler, S. M. Faber,
D. Lynden-Bell, R. Terlevich, and G. Wegner 123

Large-Scale Anisotropy in the Hubble Flow
C. A. Collins, R. D. Joseph, and N. A. Robertson 131

An Improved Distance Indicator for Elliptical Galaxies
S. Djorgovski and M. Davis 135

A Quadrupolar Component in the Velocity Field of the Local Supercluster
P. B. Lilje, A. Yahil, and B. J. T. Jones 139

Is the Criticism of the Virgocentric Flow Model Justified?
A. Yahil 143

Anisotropy of the Galaxies Detected by IRAS
A. Meiksin and M. Davis 147

The Local Gravitational Field
A. Yahil 151

Mass Segregation, HI Deficiency and Dwarf Irregular Galaxies
in the Virgo Cluster
E. E. Salpeter 159

TABLE OF CONTENTS

Effects of Mass Segregation in Rotating Clusters of Galaxies
   A. Kashlinsky 163

Measurements of the Cosmic Background Radiation
   P. Lubin and T. Villela 169

On the Measure of Cosmological Times
   A. Renzini 177

Ages from Nuclear Cosmochronology
   F.-K. Thielemann and J. W. Truran 185

Constraints on the Ages of Distant Giant Elliptical Radio Galaxies
   R. A. Windhorst, D. C. Koo, and H. Spinrad 197

Extragalactic Gas at High Redshift:
A Chronograph of Nonlinear Departures from Hubble Flow
   P. R. Shapiro 203

Microwave and X-ray Constraints on Local Large-Scale Overdensities
   L. Danese and G. De Zotti 215

Supercluster Infall Models
   J. V. Villumsen and M. Davis 219

Virgo Infall and the Mass Density of the Universe
   A. L. Melott 225

A Determination of Ω from the Dynamics of the Local Supercluster
   E. Shaya 229

The Dynamics of Superclusters and Ω_0
   Y. Hoffman 237

The Distribution of Galaxies versus Dark Matter
   A. Dekel 243

Peculiar Velocities and Galaxy Formation
   N. Vittorio 251

Large Scale Structure in Universes Dominated by Cold Dark Matter
   J. R. Bond 255

Dark Matter, Cosmic Strings, and Large Scale Structure
   J. R. Primack, G. R. Blumenthal, and A. Dekel 265

Clustering in Real Space vs. Redshift Space
   N. Kaiser 271

Cosmological Shells and Blast Waves
   J. P. Ostriker 273

More on Structure Involving $10^{18} M_\odot$
   R. B. Tully 279

Galaxy Distances and Deviations from the Hubble Flow; Summary Remarks
    J. P. Ostriker   283
Galaxy Surface Brightness and Clustering:
    A Test for Biased Galaxy Formation Scenarios?
        S. Djorgovski and M. Davis   291
Index   295

Preface

It was a general feeling among those who attended the NATO/ARW meeting on the *Galaxy Distances and Deviations from Universal Expansion*, that during the week in Hawaii a milestone had been passed in work on the distance scale. While not until the last minute did most of the participants know who else would be attending, no one was displeased with the showing. As it turned out, scarcely a single active worker in the field of the distance scale missed the event. Few knew all of the outstanding work that was to be revealed, and/or the long-term programs that were to be encapsulated in the first few days. Areas of general agreement were pinpointed with candid speed, and most of the discussion moved on quickly to new data, and areas deserving special new attention. As quickly as one project was reported as being brought successfully to a close, a different group would report on new discoveries with new directions to go. New data, new phenomena; but the sentiment was that we were building on a much safer foundation, even if the Universe was unfolding in a much more complex and unexpected way than was previously anticipated.

In editing these proceedings a decision was made well in advance of the Meeting that no attempt would made to record the discussion. This was done for many reasons. First and foremost, the Meeting was designed to be a workshop, from beginning to end. The logistics of documenting the discussion in such an environment, it was felt, would only work toward defeating the atmosphere of uninhibited, free discussion that was a primary goal of the organizers. Secondly, because the field is known to be moving so fast, the Proceedings were deemed to be of greatest value if they could be distributed as quickly as possible. Editing, correcting, and verifying the discussion in whatever form it *might* have been captured in, would have considerably slowed the publication. Other editors might have chosen a different course, we chose the route to the fastest publication.

Still that is not to say that the flavor of the Meeting is entirely lost in the publication. The order of presentation of the bulk of the papers given at the Meeting is preserved in the text; so the reader can still appreciate the flow of the Workshop. Most of the papers were written within a few weeks either side of their presentation, and so they must capture the spirit of the participant to some extent. In fact, many of the early contributions were immediately revised in light of the discussion and/or new data that came to light in the Meeting. In addition, several extra papers, based on impromptu, extended discussions of new results, were sent to the Editor for inclusion; and they are duly published here. Only two or three of the presented papers failed to make it into the published proceedings; apparently a few authors did not have the time to meet the strictly enforced deadlines. We regret their loss, but we held to our stated deadline out of respect and allegiance to those who met theirs. We hope that the book is not too much the worse for our severity.

It must, of course be stated, with proper emphasis, that none of this, neither the Meeting, nor the Publication, would have been possible without appreciable support from the *North Atlantic Treaty Organization*. the *National Aeronautics and Space Administration* and the *National Science Foundation*, and the *University of Hawaii*. To the administrators of these programs our thanks; all of the participants (and now, we hope, the readers) have benefited from their generosity. The Scientific Organizing Committee are to be commended for taking the initiative and setting up the necesssary ground-work for the Meeting, while Jaye Nakashima and David Block deserve special thanks for their energy in making the Meeting happen in a smooth and efficient way. Finally, our thanks to the many people at the *Institute for Astronomy*, at the *Canada-France-Hawaii Telescope Corporation*, and at the *United Kingdom Infrared Telescope* for helping to make the Meeting a social as well as a scientific success. May this be the first in a continuing series.

List of Participants

Dr. Marc Aaronson
Steward Observatory
University of Arizona
Tucson, AZ 85721

Dr. Neta Bahcall
Space Telescope Science Institute
Homewood Campus
Baltimore, MA 21218

Dr. J. Bland
Institute for Astronomy
University of Hawaii
2680 Woodlawn Dr.
Honolulu, HI 96822

Dr. J. R. Bond
C. I. T. A.
60. St. George St.
University of Toronto
Toronto, Ontario
Canada M5S 1A1

Dr. Greg Bothun
Robinson Lab. 105-24
California Institute of Technology
Pasadena, CA 91125

Dr. Lucette Bottinelli
Dept. de Radioastronomie
Observatoire de Meudon
92190 Meudon
France

Dr. David Burstein
Physics Department
Arizona State
Tempe, AZ 85281

Dr. Guido Chincarini
Department of Physics
Univ. of Oklahoma
Norman, OK 73069

Dr. Carol Christian
C. F. H. T.
P. O. Box 1597
Kamuela, HI 96743

Dr. C. Collins
Blackett Laboratory
Imperial College
London SW7 2BZ
United Kingdom

Dr. Avishai Dekel
Department of Physics
Weizmann Institute of Science
Rehovot 76100
Israel

Dr. Gerard de Vaucouleurs
Department of Astronomy
RLM 15.212
University of Texas
Austin, TX 78712

Dr. N. Devereux
Institute for Astronomy
2680 Woodlawn Dr.
University of Hawaii
Honolulu, HI 26822

Dr. G. De Zotti
Instituto di Astronomia
Vicolo Observatorio 5
35100 Padova
Italy

Dr. S. Djorgovski
Center for Astrophysics
OIR Division
60 Garden Street
Cambridge, MA 02138

Dr. S. Eales
Institute for Astronomy
2680 Woodlawn Dr.
University of Hawaii
Honolulu, HI 26822

Dr. Richard Ellis
Department of Physics
University of Durham
Durham DH1 E1E
United Kingdom

Dr. Michael Feast
S. A. A. O.
P.O. Box 9
Observatory 7935, Cape
South Africa

Dr. Wendy L. Freedman
Las Campanas Observatories
813 Santa Barbara St.
Pasadena, CA 91101

Dr. Lucienne Gouguenheim
Lab. d'Astronomie
Bat 426
Univertite de Paris Sud
91405 Orsay Cedex
France

Dr. David Hanes
Department of Physics
Queens University
Kingston, Ontario
Canada K7L 3N6

Dr. David Hartwick
Department of Astronomy
University of Victoria
Victoria, British Columbia
Canada V8W 2Y2

Dr. Martha Haynes
Department of Astronomy
Space Sciences Building
Cornell University
Ithaca, NY 14853

Dr. J. Hill
Institute for Astronomy
University of Hawaii
2680 Woodlawn Dr.
Honolulu, HI 96822

Dr. Yehuda Hoffman
Physics Department
University of Pennsylvannia
Philadelphia, PA 19104

Dr. Walter Huchtmeier
Max-Planck Institut fur Radioastronomie
Auf dem Hugel 69
D-5300 Bonn
Federal Republic of Germany

LIST OF PARTICIPANTS

Dr. Bob Joseph
Astronomy Group
Imperial College
London SW7 2BZ
United Kingdom

Dr. A. Kashlinsky
N. R. A. O.
Edgemont Road
Chrlottesville, VA 22903-2475

Dr. Nick Kaiser
Dept. of Applied Mathematics
Queen Mary College
Mile End Road
London E1 4NS
United Kingdom

Dr. R. Lavery
Institute for Astronomy
University of Hawaii
2680 Woodlawn Dr.
Honolulu, HI 96822

Dr. Per Lilje
Institute of Astronomy
Madingley Road
Cambridge CB3 OHA
United Kingdom

Dr. Phil Lubin
Space Science Lab.
University of California
Berkeley, CA 94720

Dr. Barry F. Madore
David Dunlap Observatory
P.O. Box 360
Richmond Hill, Ontario
Canada L4C 4Y6

Dr. R. A. McLaren
Canada-France-Hawaii Telescope
P.O. Box 1597
Kamuela, HI 96743

Dr. B. Marano
Dipartimento di Astronomia
Università di Balogna
I-40126 Balogna
Italy

Dr. A. **Meiksin**
Physics Department
University of California
Berkeley, CA 94720

Dr. Adrian Melott
Astronomy and Astrophysics Center
5640 S. Ellis Avenue
Chicago, IL 60637

Dr. Jeremy Mould
Robinson Lab
California Institute of Technology
Pasadena, CA 91125

Dr. Jaye Nakashima
Institute for Astronomy
2680 Woodlawn Dr.
University of Hawaii
Honolulu, HI 26822

Dr. Jerry Ostriker
Astrophysical Sciences, Peyton **Hall**
Princeton University
Princeton, NJ 08544

Dr. M. Pierce
Institute for Astronomy
2680 Woodlawn Dr.
University of Hawaii
Honolulu, HA 96822

Dr. Joel Primack
Stanford Linear Accelerator Center
Stanford University
Stanford, CA 94035

Dr. Chris Pritchet
Department of Physics
University of Victoria
Victoria, BC
Canada V8W 2Y2

Dr. Alvio Renzini
Dipartimento de Astronomia
Università di Bologna, CP 596
I-40100 Bologna
Italy

Dr. Otto-Georg Richter
Space Telescope Science Institute
Homewood Campus
Johns Hopkins University
Baltimore, MD 21218

Dr. Jim Rose
Institute for Astronomy
2680 Woodlawn Dr.
University of Hawaii
Honolulu, HI 26822

Dr. Edwin Salpeter
Newman Lab.
Cornell University
Ithaca, NY 14853

Dr. Allan Sandage
Center for Astrophysics
University of California, San Diego
La Jolla, CA 92093

Dr. Robert Schommer
Dept. of Physics and Astronomy
Serin Physics Lab
Rutgers University
P.O. Box 849
Piscataway, NJ 08854

Dr. Paul Shapiro
Astronomy Department
University of Texas
Austin, TX 78712

Dr. Ed Shaya
Robinson Lab. 105-24
California Institute of Technology
Pasadena, CA 91125

Dr. Alex Szalay
Department of Astronomy
University of Chicago
5640 S. Ellis Avenue
Chicago, IL 60637

Dr. Gustav Tammann
St. Alban-Ring 172
CH-4052 Basel
Switzerland

Dr. F.-K. Thielemann
Department of Astronomy
University of Illinois
Urbana, IL 61801

Dr. Tripicco
Institute for Astronomy
2680 Woodlawn Dr.
University of Hawaii
Honolulu, HI 26822

LIST OF PARTICIPANTS

Dr. Brent Tully
Institute for Astronomy
2680 Woodlawn Dr.
University of Hawaii
Honolulu, HI 26822

Dr. Sidney van den Bergh
Herzberg Institute
National Research Council
Dominion Astrophysical Obs.
5071 W. Saanich Road
Victoria, BC V8X 4M6
Canada

Dr. Harry van der Laan
Sterrewacht
University of Leiden
P.O. Box 9513
2300 RA Leiden
The Netherlands

Dr. G. Vettolani
Inst. de Radioastronomia CNR
Via Irnerio 46
I-40126 Bologna
Italy

Dr. T. Villela
Space Sciences Lab.
University of California
Berkeley, CA 94720

Dr. Jens Villumsen
Robinson Lab. 105-24
California Institute of Technology
Pasadena, CA 91125

Dr. N. Visvanathan
Mt. Stromlo Observatory
Private Bag
P.O. Woden
ACT 2606, Australia

Dr. Nicola Vittorio
Department of Astronomy
University of California
Berkely, CA 94720

Dr. D. Ward
Institute for Astronomy
University of Hawaii
2680 Woodlawn Dr.
Honolulu, HI 96822

Dr. Craig Wheeler
Astronomy Department
University of Texas
Austin, TX 78712

Dr. James Wright
National Science Foundation
1800 G Street N.W.
Washington, D.C. 20550

Dr. Amos Yahil
Astronomy Program
State University of New York
ESS Building
Stony Brook, NY 11794

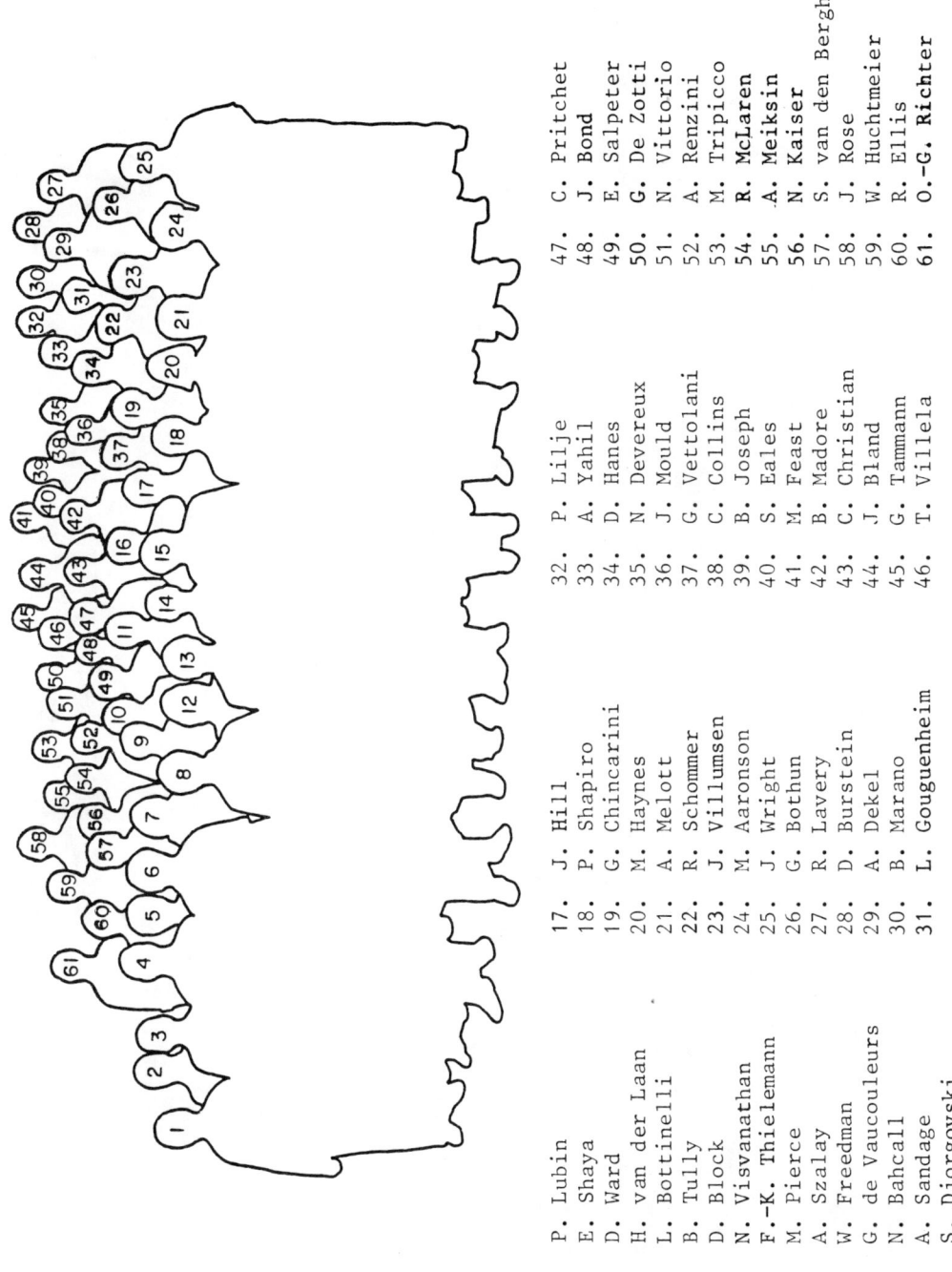

1. P. Lubin
2. E. Shaya
3. D. Ward
4. H. van der Laan
5. L. Bottinelli
6. B. Tully
7. D. Block
8. N. Visvanathan
9. F.-K. Thielemann
10. M. Pierce
11. A. Szalay
12. W. Freedman
13. G. de Vaucouleurs
14. N. Bahcall
15. A. Sandage
16. S. Djorgovski
17. J. Hill
18. P. Shapiro
19. G. Chincarini
20. M. Haynes
21. A. Melott
22. R. Schommer
23. J. Villumsen
24. M. Aaronson
25. J. Wright
26. G. Bothun
27. R. Lavery
28. D. Burstein
29. A. Dekel
30. B. Marano
31. L. Gouguenheim
32. P. Lilje
33. A. Yahil
34. D. Hanes
35. N. Devereux
36. J. Mould
37. G. Vettolani
38. C. Collins
39. B. Joseph
40. S. Eales
41. M. Feast
42. B. Madore
43. C. Christian
44. J. Bland
45. G. Tammann
46. T. Villela
47. C. Pritchet
48. J. Bond
49. E. Salpeter
50. G. De Zotti
51. N. Vittorio
52. A. Renzini
53. M. Tripicco
54. R. McLaren
55. A. Meiksin
56. N. Kaiser
57. S. van den Bergh
58. J. Rose
59. W. Huchtmeier
60. R. Ellis
61. O.-G. Richter

NEW RESULTS ON THE DISTANCE SCALE AND THE HUBBLE CONSTANT

Gérard de Vaucouleurs
Department of Astronomy
University of Texas
Austin, TX 78712

ABSTRACT. The following topics are discussed: galactic extinction, primary distance indicators, distance moduli in the Local and Sculptor groups, luminosity index, HI line width, velocity dispersion, sosies, supernovae, Malmquist bias, distance to the Hercules supercluster and Hubble constant. The best estimate is still $90 \leq H_0 \leq 100$.

1. GALACTIC EXTINCTION MODELS. Studies of five tracers of galactic extinction (galaxy counts, cluster counts, galaxy surface brightness, HI index, and color indexes) give no support to the concept of large dust-free polar windows and confirm the statistical validity of the cosec b law up to both poles with a B-band extinction coefficient $A_1(B)$ = 0.20±0.02 mag (dV. and Buta 1983) in agreement with RC2 and the IRAS data. This does not exclude the presence of small windows, not centered at the poles, as in the model of Burstein and Heiles (1978, 1982). This model predicts smaller extinction than the RC2 formula at high latitudes, larger values at low latitudes, and in parts of the southern sky ($\delta < -23°$) where it rests on HI column densities only. The maximum systematic effect on the zero point of the extragalactic distance scale of a change from the RC2 to the B-H model would be < 0.2 mag (i.e. increase distances, reduce H_0 by less than 10 %).

2. PRIMARY INDICATORS. At least four (possibly six) independent primary distance indicators are now available: novae, cepheids, RR Lyr and HB stars, and long-period variables (possibly AB supergiants and eclipsing binaries). Type I supernovae are also approaching this status.
(a) <u>Novae</u>: the original m_{15} rule (Buscombe and dV. 1955, dV. 1978a) was good to ±0.25 mag; with an empirical fit of the Shara (1981) second-order correction to the Magellanic novae it is now good to ±0.1. The new determinations of expansion parallaxes of galactic novae (Cohen 1985) confirm within 0.1 mag the zero points of M_0 and M_{15} adopted in 1978. Including the Shara correction the revised zero point for log t_3 = 1.5 is $M_{15}(pg)$ = -5.78±0.10 (8 best novae). Applications to the Magellanic and M 31 novae give apparent moduli $\mu(pg)$ = 18.83±0.11 (LMC,

8 novae), 19.0±0.15: (SMC, 2 novae), and 24.28±0.14 (M31, 17 novae).
(b) <u>Cepheids</u>: New IR photometry (Freedman et al. 1985) shows a way out of the inconclusive debates on the effects of color, metallicity, amplitude on the visible light curves. Particularly valuable are comparisons of multicolor photometry in different galaxies which allow simultaneous determination of the extinction and modulus difference of individual stars (Freedman 1984); an application to M 33 and the LMC yields an extinction-free differential modulus of 5.6 and a V-band total (galactic + internal) extinction of 0.6 mag. This may be compared with 5.2 and ≈ 0.75 (Humphreys 1984), 6.0 and 0.35 (dV. 1978b) and 6.76 and 0.1 (Sandage 1983). The adopted zero point, $\langle M_V \rangle$ = - 3.60±0.15 (for log P_0 = 0.8), of the P-L relation (dV. 1978a) has been confirmed within ±0.13 by more recent reviews (-3.47, Balona and Shobbrook 1984, -3.50, Schmidt 1984; -3.62, Stothers 1983; -3.70, Feast 1984; - 3.73, Caldwell 1983). Because more galactic cepheids of longer periods have now independent calibrations than a decade ago, giving a better overlap with extragalactic cepheids, it is advisable to change the reference period from 6.3 d (log P_0 = 0.8) to 10 d (log P_0 = 1.0). With the mean slopes (-2.9; -3.8,+2.7) of the P-L and P-L-C relations in this range, the new zero point is $\langle M_V(1) \rangle$ = -4.2±0.1 with $\langle B_0 - V_0 \rangle$ = +0.75 for log P_0 = 1.0.
(c) <u>RR Lyrae variables</u>: New determinations of the zero point, $\langle M_V(RR) \rangle$, of ab-type RR Lyr variables continue to fluctuate between extrema of +0.4 and +1.0, with the majority in the range +0.6 to +0.8 (0.61±0.15, Stothers 1983; 0.76±0.14, Hawley et al. 1986), but tend to average perhaps 0.1 mag brighter than the previously adopted value (+0.8±0.15, dV. 1978a). The large scatter may reflect population, age and metallicity effects, that have not yet been disentangled. Until such second order effects have been clarified the zero point may be revised to $\langle M_V(RR) \rangle$ = +0.7±0.1:.
(d) <u>Long period variables</u>: The discovery of a bolometric P-L relation for Mira-type variables (Robertson and Feast 1981, Glass and Feast 1982) introduced an important new primary distance indicator whose zero point can be calibrated independently in the Galaxy (+0.76±0.11, Glass and Lloyd Evans 1981). This indicator, using IR light curves, has the advantage of reduced sensitivity for galactic extinction. The distances of the galactic center (R_0 = 7.95 ±0.6 kpc) and of the LMC (μ_0 = 18.49±0.13) derived from this indicator are in excellent agreement with previous determinations (dV. 1983a; Glass and Lloyd-Evans 1981).
Altogether the 4 to 6 primary indicators define the zero point of the scale with a mean error ≤0.1 mag (all agree within ±0.1; dV. 1978b).

3. DISTANCE MODULI IN THE LOCAL GROUP. The uncertainty in extinction model does not affect the distances of the primary calibrating galaxies in the Local Group whose extinction corrections are determined directly from observed color excesses, not from an extinction model. With the revised zero points and measured extinction corrections the distance moduli of the four primary calibrators are: LMC, μ_V = 18.6, E(B - V)(total) = 0.11, μ_0 = 18.3 (18.31); SMC, 18.9, 0.09, 18.6 (18.62); M 31, 24.6, 0.15:,24.1 (24.07); M 33, 24.5, 0.2:, 23.9 (24.30) (dV. 1978b values are given in parenthesis).

4. DISTANCE MODULI IN THE SCULPTOR GROUP. The mean modulus of 27.0 previously adopted (dV. 1978d) depended on a small number of precariously calibrated secondary indicators (dV. 1978c). Recent determinations by primary and recalibrated secondary indicators give more reliable estimates: N55, μ_0 = 26.57 (methods 1,2,3); N247, 27.21 (4,5,6); N253, 27.42 (5,6); N300, 26.43 (1,4,5,7,8); N7793, 27.74 (1,2,4,5,6) [Methods: 1, brightest blue stars; 2, brightest superassociation; 3, brightest cluster; 4, luminosity index; 5, HI line width (B-band); 6, HI line width (H-band); 7, largest HII ring; 8, cepheids]. The group is now clearly resolved in distance, with a depth ($\approx$1 Mpc, allowing for errors in the distance moduli) comparable to the projected width (20° = 0.9 Mpc) on the sky. The mean modulus, $\langle\mu_0\rangle$ = 27.07 (27.01 with B-H model) is practically unchanged.

5. LUMINOSITY INDEX. The small dispersion ($\sigma \approx 0.4$ mag) and slope $dM/d\Lambda_c$ = 3.0 of the calibrating relation (dV. 1979a) have been confirmed by several tests: (1) calibrating galaxies, absolute magnitude vs. Λ_c; (2) volume limited samples in clusters (Virgo, Hercules), apparent magnitude vs. Λ_c; (3) Type I supernovae, magnitude difference between galaxy and SNI vs. Λ_c. The different conclusions of Sandage et al. (SBT)(1985) (that $\sigma \approx 1.0$ mag and $dM/d\Lambda$ = 4.86) based on a magnitude - Λ plot for galaxies claimed to be in the "Virgo cluster" apply only to their own data. A detailed analysis shows that their morphological types T have large systematic and accidental errors, their L classes are not on the van den Bergh system, they neglect important inclination corrections to both magitudes and luminosity classes, introduce objects which are not members of the Virgo S-cloud, but are in the more distant S', M, W and X clouds, use eyeball magnitudes subject to larger errors at m > 14, etc, which all contribute to the bias and excessive scatter of the SBT plot.

6. HI LINE WIDTH INDICATOR. In spite of some technical differences the B-band and H-band versions of the Tully-Fisher relation applied to disk-dominated galaxies define distance scales which are in excellent agreement. Both lead to values of H_0 in the 90 to 100 range, depending only on zero point and extinction corrections. Both give distance moduli in the Virgo-S cloud which are independent of inclination (dV. 1983a), while the Sandage-Tammann (1976) version causes a rapid dependence of μ_0 on i. Comparisons of distance moduli derived from the B-band T-F relation and from the corrected luminosity index shows that the latter have larger accidental errors ($\sigma \approx 0.4$-0.5 mag versus 0.2-0.3) at all i > 35°; the reverse is true for near face on galaxies for which the inclination corrections to line widths become large and uncertain (Bottinelli et al. 1984). Thus, Λ_c and $\log V_M$ are complementary distance indicators and give the best results when used jointly with appropriate weights depending on i.

7. VELOCITY DISPERSION. The Faber-Jackson relation for the spheroidal components of early-type galaxies (T < 0), has a slope smaller than 10 in the visible (Tonry 1981, dV. and Olson 1982). It is applicable to both ellipticals and lenticulars with the same zero point when a color

term is included. It may also apply to the bulges of spirals, but with a zero point varying with morphological type (dV. 1983b). There is good systematic agreement between the distance scales independently derived (on a common zero point) from the F-J and T-F relations for groups including both spheroidal and disk galaxies, but the distance moduli from the F-J relation have larger errors, ≈ 0.6 mag, than those from the T-F relation, ≈ 0.3, because of the lower precision with which σ_v can be measured and the unfavorable slope, $dM/d(\log \sigma_v) \approx 8-9$ vs. $dM/d(\log V_M) \approx 5-6$.

8. SOSIES. The method of "sosie" (or look-alike) galaxies (Paturel 1984, Bottinelli et al. 1985), requiring only an equality of observable parameters, is more direct than methods depending on the calibration of a functional relation between absolute magnitude (or linear diameter) and luminosity indicator. Applications have been made (i) to Sbc sosies of our Galaxy (dV. 1983a,b) to tie directly the extragalactic distance scale to the galactocentric distance R_o of the Sun. [A suggestion by van der Kruit (1985) that the scale length of the galactic disk model is 5.5 kpc, rather than 3.7 (dV. and Pence 1978, dV. 1983a) is in conflict with a recent dynamical determination, 3.9 kpc, by Freeman (1985) from the trend of σ_z velocity dispersion of old disk K giants versus galactocentric distance]; (ii) to sosies of a small sample of nearby galaxies (Paturel 1984) and to all RC2 sosies of a score of calibrating galaxies (Bottinelli et al. 1985); (iii) to sosies of M 31 and of our Galaxy in the Virgo cluster and in the Hercules supercluster (dV. and Corwin 1985b, 1986) with remarkably uniform results for the derived value of the Hubble ratio (sect. 12).

9. INNER RINGS. The well-defined inner rings in barred disk-galaxies which are associated with resonances near co-rotation (Buta 1984) are useful metric distance indicators after suitable calibration (Buta and dV. 1982). Applications have been made to a large number of field galaxies (Buta and dV. 1983a), to several groups, clusters and the Hercules supercluster (Buta and dV. 1983b; Buta and Corwin 1986). The absolute calibration, however, must for the present rest on other tertiary indicators because none of the nearby calibrating galaxies is of this type (except, perhaps, our Galaxy; see dV. 1970, 1983a,b), and the dispersion of the derived moduli ($\sigma \approx 0.6-0.7$ mag) is larger than in the case of photometric indicators. Nevertheless, applications to large volume-limited samples (sect. 12) lead to valid results, substantially independent of galactic extinction model (except for zero point).

10. SUPERNOVAE. The evidence for or against the proposals that maximum luminosity is, either, a constant independent of rate of decay (or that the rate of decay is constant, Cadonau et al. 1985) or, alternatively, depends on the rate on decay (Rust 1974, dV. and Pence 1976, Pskovskii 1977) needs to be re-examined. In this respect V-band light curves (Younger and van den Bergh 1985) which do not exhibit the 30-day "kink" in the B (or pg) light curves (reflecting the rapid rise to maximum of the color index) allow a better measure of the decay rate by application

of the interpolation formula $V = V_o + a[\log(t - t_o)]^2$, recently found to give a very good representation of the definitive visual light curve of SN 1885 (S And) in M 31 (dV. and Corwin 1985a). Preliminary calculations for ten well-observed Type I supernovae indicate a range of decay rate parameter, $0.7 \leq a \leq 1.65$ (or 1.35, excluding S And) with $\langle a \rangle = 1.11 \pm 0.08$ ($\sigma = 0.26$). Whether this range correlates with the absolute magnitude at maximum or with the different sub-types recently identified by infra-red photometry remains to be seen. A redetermination of the absolute magnitude of SN 1572 (B Cas) by means of improved estimates of the distance of its radio remnant (3C 10), gives $M_V = -18.55 \pm 0.3$ (dV. 1985), in excellent agreement with the SN I mean value (-18.5 ± 0.2) on the short distance scale (dV. 1979b).

11. MALMQUIST BIAS. Recent tests show that the short scale is not grossly affected by the Malmquist bias: (1) the all-sky average of the Hubble ratio, H^*, determined on the short scale by a score of different approaches is rigorously independent of distance in the range $10 \leq \Delta \leq 100$ Mpc; (2) different indicators having intrinsic variances ranging all the way from $\sigma^2 = 0.05-0.10$ (B-band T-F relation) to $0.2-0.3$ (Λ_c index), $0.3-0.4$ (Type I SN), $0.4-0.5$ (F-J relation), and $0.5-0.6$ (inner rings) are in complete agreement (Buta and dV. 1983a; Bottinelli et al. 1984; dV. and Olson 1984). There is no indication of systematic departures increasing with σ^2; (3) a model-independent mapping of the velocity field near the supergalactic plane within ≈ 30 Mpc from the Local Group (dV. and Peters 1985) shows that while H^* <u>increases</u> with distance in the general direction of the south galactic pole (in excess of reflex Local Group motion), it <u>decreases</u> in the general direction of the north galactic pole, and is substantially independent of distance in intermediate directions. Malmquist biases cannot be positive in some directions, negative in others, absent elsewhere.

Finally, the Malmquist bias can be eliminated altogether by fitting (almost) complete luminosity (or diameter) functions in a volume-limited sample, such as the Hercules supercluster (sect. 12), with the corresponding functions for nearby calibrating galaxies.

12. DISTANCE OF HERCULES SUPERCLUSTER AND HUBBLE CONSTANT. The Hercules supercluster ($\langle z \rangle = 0.0370$, $\langle b \rangle = +33°$) is probably the nearest large system at a distance, $\Delta > 100$ Mpc, sufficient to be well outside the sphere of influence of the Local supercluster, but still near enough to allow proper classification and measurement of galaxian types, diameters, luminosities, HI line widths, etc. A large data base has been assembled and used to estimate its distance by means of a variety of metric and photometric indicators and, in particular, by fitting (almost) complete distribution functions of (1) metric diameters of inner ring structures in SB(r) galaxies (Buta and dV. 1983b), and (2) isophotal diameters and total magnitudes of galaxies of known Λ_c and/or HI line widths (Buta and Corwin 1986). The results are in excellent agreement with distances from Sosies of M 31 and of our Galaxy (sect.8) and from Type I supernovae (sect.10) (dV. and Corwin 1985, 1986).

With the adopted mean modulus $\mu_0 = 35.25 \pm 0.2$ (RC2 extinction) and the mean redshift corrected to the background radiation reference frame, $V =$

11,078 ± 108 km s^{-1}, the mean Hubble ratio is $H^* = 99 \pm 10$ km s^{-1} Mpc^{-1} (or ≈ 90 with the B-H extinction model). The Hubble constant may differ by a few percent, depending on cosmological model.

REFERENCES

Balona, L.A. and Shobbrook, R.R. 1984, M.N.R.A.S., 211, 375.
Bottinelli, L., Gouguenheim, L., Paturel, G., de Vaucouleurs, G.
    1984, Astr.Ap. Suppl., 56, 381; 1985, Ap.J. Suppl., 59, 293.
Burstein, D. and Heiles, C. 1978, Ap.J., 225, 40; 1982, A.J., 87, 1165.
Buscombe, W. and de Vaucouleurs, G. 1955, Observatory, 75, 170.
Buta, R.J. 1984, Univ. of Texas Publ. in Astr. No. 23.
Buta, R.J. and Corwin, H.G. 1986, Ap.J. submitted.
Buta, R.J. and de Vaucouleurs, G. 1982, Ap.J. Suppl., 48, 219;
    1983a, Ap.J. Suppl., 51, 419; 1983b, Ap.J., 226, 1.
Cadonau, R., Sandage, A., Tammann, G. 1985, Lect Notes Phys. 224, 151.
Caldwell, J.A.R. 1983, Observatory, 103, 244.
Cohen, J. 1985, Ap.J., 292, 90.
de Vaucouleurs, G. 1970, IAU Symp. 38, Spiral Structure of Galaxy, 18.
    1978a,b Ap.J., 223, 351,730; 1978c,d 224, 14, 710;
    1979a,b 227, 380,729; 1983a,b 268, 461,468; 1985, 289, 5.
-------- and Buta, R.J. 1983, A.J., 88, 939.
-------- and Corwin, H.G. 1985a, Ap.J., 295, 287; 1985b,
    Ap.J., 297, 23; 1986, Ap. J., submitted.
-------- and Olson, D.W. 1982, Ap.J., 256,346; 1984, Ap.J. Suppl. 56,91.
-------- and Pence, W.D. 1976, Ap. J. 209, 687; 1978, A.J., 83, 1163.
-------- and Peters, W. 1985, Ap. J., 297, 27.
Feast, M.W. 1984, IAU Symp. 108, Struct. & Evo. of Mag. Clouds, p.157.
Freedman, W.L. 1984, Dissert., D.Dunlap Obs., Univ. of Toronto.
Freedman, W.L., Grieve, G.R., Madore, B.F. 1985, Ap.J. Suppl., 59, 311.
Freeman, K.C. 1985, private communication.
Glass, I.S. and Feast, M.W. 1982, M.N.R.A.S., 198, 199.
Glass, I.S. and Lloyd Evans, T. 1981, Nature, 291, 303.
Hawley, S.L., Jefferys, W.S., Barnes, T.J., Lai W. 1986, Ap.J., ...
Humphreys, R.M., Jones, T.J., Sitko, M.L. 1984, A.J., 89, 1155.
Paturel, G. 1984, Ap.J., 282, 382.
Pskovskii, Yu.P. 1977, Soviet Ast., 21, 675.
Robertson, B.S.C. and Feast, M.W. 1981, M.N.R.A.S., 196, 111.
Rust, B.W. 1974, Oak Ridge Nat. Lab. Publ. No.4953.
Sandage, A. 1983, A.J., 88, 1108.
Sandage, A., Binggeli, B. and Tammann, G.A. 1985, A.J., 90, 395.
Sandage, A. and Tammann, G.A. 1976, Ap.J., 210, 7.
Schmidt, E.G. 1984, Ap.J., 285, 501.
Shara, M.M. 1981, Ap.J., 243, 268.
Stothers, R.B. 1983, Ap.J., 274, 20.
Tonry, J.L. 1981, Ap.J. Lett., 251, L1.
van der Kruit, P. 1985, Astr. Ap., in press.
Younger, P.F. and van den Bergh, S. 1985, Astr. Ap. Suppl., 61, 365.

THE MAGELLANIC CLOUDS AND THE DISTANCE SCALE

M.W. Feast
South African Astronomical Observatory
P O Box 9
Observatory 7935 Cape
South Africa

1. INTRODUCTION

The Magellanic Clouds are of crucial importance in establishing suitable distance indicators and in checking their mutual consistency. The SMC contains young objects which are considerably more metal deficient than those in either the Galaxy or the LMC. It is therefore of importance in checking metallicity effects on (young) distance indicators.
However because of its great depth (~18 kpc) in the line of sight it is less suitable than the LMC for precise studies of distance indicators.

2. CEPHEIDS

In the past some of the uncertainties surrounding the distance of the LMC have been concerned with the proper interstellar absorption corrections to adopt. Much of this uncertainty has been removed in the case of Cepheids, at least, since for them reddenings of individual objects can be derived from BVI photometry. The method is somewhat model dependent since the intrinsic line in the B-V/V-I diagram depends on abundance. However recent work by Caldwell and Coulson (1985) shows that this dependence is slight. Their mean LMC Cepheid value is E_{B-V} = 0.074. In any case if we use a P-L-C relation the derived modulus is very insensitive to the adopted absorption. The differential reddenings between LMC Cepheids determined from the BVI photometry are not significantly model dependent and allow one to establish a P-L-C relation entirely empirically. The scatter about such a relation is very small (~0.07) and can be accounted for by observational error. It is apparent that Cepheids are very good distance indicators provided the zero point of the relation can be determined. The latest P-L-C formulation (Table 1) Caldwell and Coulson 1986) does not predict significantly different absolute magnitudes in the mean from that of Martin, Warren and Feast (1979) when the zero points are suitably adjusted and I shall use the latter since it has been applied by a number of workers.

Table 1

P-L-C Relation
Caldwell and Coulson (1986) $\langle M_v \rangle = -3.53 \log P + 2.13 (\langle B_0 \rangle - \langle V_0 \rangle) + \phi_1$
Martin, Warren & Feast (1979) $\langle M_v \rangle = -3.80 \log P + 2.70 (\langle B_0 \rangle - \langle V_0 \rangle + \phi_2$

Adopt
Hyades Modulus	3.28
Hyades Metallicity Correction	-0.22
Pleiades Modulus	5.57
LMC Cepheid Metallicity Corrections	-0.16

Cluster Fits (Adjusted to above Hyades/Pleiades scale)

		ϕ_2	$(m-M)_{0}$LMC
1.	ZAMS (Caldwell 1983)	-2.39	18.64 ± 0.15
2.	β-Index (Balona, Shobbrook 1984)	-2.26	18.51 ± 0.10
3.	As 2 but unit weight per cluster	-2.31	18.56 ± 0.10

	ϕ_1	
Collected Data (Caldwell & Coulson 1986)	-2.25	18.62 ± 0.10

Adopt 18.60 for Cepheids

The zero point of the P-L-C relation can be obtained from galactic Cepheids in open clusters. There have been suggestions in recent times that zero points differing by ~0^m.5 are obtained depending on whether the cluster distances are derived from ZAMS fitting or Strömgren β Photometry (the latter method being pioneered by Schmidt (1984)). This problem is resolved when one refers all the distances to a consistent Hyades/Pleiades scale and uses the new β calibration by Balona and Shobbrook (1984) which is directly tied to the Pleiades. Adopting a Hyades modulus of 3.28 with a metallicity correction of 0.22 and a Pleiades modulus of 5.57 (which form a consistent set (cf. Pel (1985) and references there) and with a metallicity correction of -0.16 to the LMC P-L-C relation (Caldwell, see Feast 1984a) leads to the values shown in Table 1. The main current disadvantage of the β method is that it has so far been applied to a limited sample of clusters only. This makes the method somewhat sensitive to the way one weights the clusters. Balona and Shobbrook discuss six clusters of which one (NGC 7790) has three Cepheids. They adopt unit weight per Cepheid. It might be best however to give unit weight per cluster (since most of the errors are in the cluster distances) in which case the revised value listed would result. Caldwell and Coulson (1986) have recently recalibrated the P-L-C relation using the currently available data and obtain an LMC modulus of 18.62.

More, accurate work is needed on the galactic calibrators. A case in point is NGC 7790 containing 3 Cepheids for which the published β and ZAMS distances differ by $0\overset{m}{.}4$. Pedreros et al. (1985) have recently obtained new photographic photometry and find a ZAMS fit which agrees in modulus with the β photometry. However this is at the expense of adopting an E_{B-V} which is $0\overset{m}{.}10$ higher than the Schmidt (1984) result so the discrepancy cannot be said to have been resolved. A Walker (1985a, b) has obtained significantly improved c-m diagrams of Lynga 6 and NGC 6067 which lead to lower P-L-C zero points and this will tend to reduce the mean value slightly, lowering the LMC distance. He is also carrying out BVI photometry of the 3-day Cepheids in the LMC cluster NGC 1866. Preliminary indications are that these short period Cepheids may give an LMC modulus $\sim 0\overset{m}{.}1$ nearer to us (Walker 1986).

The results currently available suggest that the LMC modulus from Cepheids is 18.60 ± 0.10.

Infrared (JHK) photometry of Cepheids is of importance primarily because a narrow P-L relation, free of abundance effects is expected. The work of Laney and Stobie (1985) shows that there is a significant infrared colour term but this will have a minor effect on distance determinations. Unfortunately the calibration of the infrared P-L relation is sensitive to the reddenings adopted for the calibrating clusters and to the law of reddening (the general cluster stars are measured in the optical where absorption is important, the cluster Cepheids in the infrared where it is not). The revised list of galactic calibrators by the Toronto group (Welch et al. 1985) applied to the K band P-L relation of Laney and Stobie gives 18.52 for the LMC modulus (reduced to the our Hyades/Pleiades zero point). However using a number of calibrators Laney and Stobie derive a preliminary distance of 18.74. Present indications are therefore that an infrared modulus in agreement with the BVI value will be found. Perhaps the most significant point to notice about the infrared work at present is that the difference in moduli between the two Clouds is the same as in the BVI work (defining the centroids in the same way).

JHK Δ Mod 0.32 ± 0.04 (Laney and Stobie 1985)
BVI Δ Mod 0.30 ± 0.06 (Caldwell and Coulson 1986)

The agreement is important because both the P-L-C relation used to determine the luminosities, and the intrinsic colours used to determine interstellar reddenings have had to be corrected for metallicity differences between the Clouds. These effects are more important in BVI than in JHK and the agreement of the results suggests that the way this is being done is sound.

A number of recent papers have investigated "short cut" methods of determining Cepheid mean magnitudes etc. These are of possible significance in extending Cepheid work to more distant galaxies (but see Moffett and Barnes 1986 for a caution). They are of less importance in fixing the LMC distance. The work of Visvanathan (1985) on 1.05 μm photometry falls mainly in this category. His LMC distance modulus is 18.57 on our scale (a much larger distance is given in the paper but this has not been corrected for Hyades metallicity).

Whilst the above results all seem reasonably consistent there is one very discrepant recent result. It should be possible to use early type companions to derive Cepheid luminosities. Bohm-Vitense (1985) has attempted to do this using IUE observations of 5 Cepheids with hot companions. Kurutz models are used to derive effective temperatures, and luminosities come from assuming the stars are on the ZAMS. If one scales Bohm-Vitense ZAMS to the one I have adopted one finds that the luminosity scale is 1.05 mags fainter than in the previous discussion - an LMC modulus of 17.6. This result is very difficult to accept at its face value. It is perhaps trying to tell us something about the model adopted or possibly that the companions are not on the ZAMS. That the latter might be the case can be seen from the CCD c-m diagram derived by A Walker (1985b) for NGC 6067 which contains an 11 day Cepheid. The companions discussed by Bohm-Vitense all have $(B-V)_0$ ~0.0. In NGC 6067 stars of this colour are evolved and spread over ~1 mag as the sequence tends to become vertical. The Cepheids discussed by Bohm-Vitense are of shorter period than 11 days (and hence older) and there is even more opportunity for the companions to be evolved.

3. RR LYRAE VARIABLES

Accurate CCD magnitudes of RR Lyraes in NGC 2210 (LMC) measured by A Walker (1985a) can be used to derive a distance provided we know the absolute magnitudes of RR Lyraes. Table 2 contains some recent estimates. Almost any reasonable way of averaging the determinations gives $\langle M_V(RR) \rangle$ ~0.6 which I adopt. The uncertainty should probably be put at about 0.1. There is no direct evidence for an $[Fe/H]$ -$\langle M_V(RR) \rangle$ relation such as was originally envisaged by Sandage (1982) to explain the Oosterhoff effect. The RR Lyraes in the galactic centre also do not appear to show any significant effect of this kind (Walker and Mack 1986). However the evidence in ω Cen (cf. Feast 1985) suggests that RR Lyraes with $[Fe/H] > -1.0$ are ~$0^m.15$ fainter than those of lower abundance. (The RR Lyrae variable of greatest measured metallicity in ω Cen is anomolously bright. Walker and Mack (1985) interpret this as indicating that it has a companion since its amplitude is low).

4. MIRA VARIABLES

Miras in the LMC show a good infrared P-L relation (Glass & Lloyd Evans 1981, Feast 1984, Glass & Reid 1986) and this can be calibrated by Miras in globular clusters. Adopting horizontal branches at $0^m.6$ leads to a modulus of 18.5. However Menzies and Whitelock (1985) show that the globular cluster Miras gives some evidence for a dependence of $\langle M_V(RR) \rangle$ on $[Fe/H]$. Within the error the results would be satisfied by assuming the HB's of cluster with $[Fe/H] > -1.0$ are $0^m.15$ fainter than the others as the direct evidence in ω Cen suggests. In that case we can calibrate the Mira zero point from metal rich clusters using $\langle M_V(RR) \rangle = +0.75$ and we obtain 18.42 for the LMC modulus.

Table 2

RR Lyrae Absolute Magnitudes

		$\langle M_V(RR)\rangle$
(1)	Statistical Parallaxes (Strugnell et al. 1986)	0.75 ± .2
(2)	Model fit to M5 HB in u.v. (Bohlin et al. 1985)	0.63 (±0.1?)
(3)	Baade-Wesselink, X Ari (Manduca et al 1981)	0.59 ± 0.25
(4)	Baade-Wesselink, RR Lyrae (Manduca et al. 1981)	0.61 ± 0.35
(5)	IR Baade-Wesselink VY Ser (Longmore et al. 1985)	0.63 ± 0.12
(6)	NGC 6752 m.s. fit (Penny & Dickens see Penny 1984)	0.6 (±0.1)
(7)	M15 m.s. fit with Lutz-Kelker correction (Fahlman et al. 1985)	0.32 ± 0.21)
(8)	M15 m.s. fit without Lutz-Kelker correction (Fahlman et al. 1985)	0.43 ± 0.18)
(9)	M15 m.s. fit and model (adopted value) (Fahlman et al. 1985)	0.55)
	Adopt	0.6

Adopting $\langle M_V(RR)\rangle$ = 0.6 leads through NGC 2210 to an LMC modulus of 18.42.

5. CLUSTERS

One can also use the HB's of old clusters (adopting $\langle M_V(HB)\rangle$ = 0.6) to get distances. Andersen et al. (1985) have done this recently for NGC 2210 using electronography and getting a modulus of 18.3. However A Walker (1984) has CCD observations of this cluster which place the HB ~0.2 mag fainter and it seems best at present to stay with the RR Lyrae distance for this cluster. For Hodge 11, another old cluster, Andersen et al. (1984) used electronography to obtain a modulus of 18.4 with an internal error of 0.1.

The above results suggest that old objects give 18.4 or possibly a somewhat greater modulus.

Schommer et al. (1984) obtained c-m diagrams of the intermediate age clusters NGC 2162/NGC 2190. Comparison with theory gave them a LMC modulus of 18.1* (though Chiosi and Pigatto (1986) derived 18.6 by incorporating convective overshooting in the models). It seems best to stay at present with a direct comparison to the Hyades which gives 18.32 ± 0.2 (cf. Feast 1984).

---

*Footnote A similar result is obtained by Mateo and Hodge for H4 (preprint).

Table 3

LMC Modulus

	$(m-M)_0$
Cepheids (Table 1)	18.60
RR Lyraes in NGC 2210 ($<Mv(RR)> = 0.6$) (A Walker 1985a)	18.42 (± 0.10 int)
Miras (calibrated by galactic globular clusters (Menzies, Whitelock 1985, Feast 1984)	
(1) $<Mv(RR)> = 0.6$	18.50 (± 0.06 int)
(2) $<Mv(RR)> = 0.75$ for $[Fe/H] > -1$	18.42 (± 0.05 int)
Old Clusters ($<Mv(HB)> = 0.6$) NGC 2210 (Andersen et al. 1985, A Walker 1984) (See text)	18.3 → 18.42
Hodge 11 (Andersen et al. 1984)	18.4 (± 0.1 int)
Intermediate Age Clusters NGC 2162/NGC 2190 (Schommer et al. 1984)	
(1) Theory	18.1?
(2) Hyades fit	18.32 $\pm$ 0.2
Young Clusters NGC 1866 (A Walker 1985a) (fit to Pleiades)	
(1) $Y = 0.25$ $Z = 0.012$ (Young Clusters in LMC (Cohen 1982))	18.42 $\pm$ 0.2
(2) $Y = 0.273$ $Z = 0.016$ (Fits of Model to NGC 1866 (Becker, Mathews 1983))	18.52 $\pm$ 0.2

Best overall current value 18.5

One can also attempt to use young clusters. Some years ago Merle Walker (1974) gave a modulus of 18.05 for NGC 1866 in the LMC from electronography. However a new CCD c-m diagram by A Walker (1985a) fits the Pleiades with an LMC modulus of 18.42 or 18.52 depending on the metallicity adopted for NGC 1866.

6. CONCLUSION

There are a number of other distance indicators but the ones I have discussed probably deserve most of the weight.
In view of the uncertainties involved it would seem unwise to overemphasise the differences between the various determinations. Whilst it is possible that future work will reveal clear discrepancies between different indicators, none is definitely established by the data so far and it would seem best to adopt a value of the LMC modulus near

18.5 which would be consistent with all the data, as the best currently available value.

I would like to thank my colleagues at SAAO whose work is mentioned in the text for preprints and helpful discussions.

REFERENCES

Andersen, J., Blecha, A. & Walker, M.F., 1984. Mon. Not. R. astr. Soc., **211**, 695.
Andersen, J., Blecha, A. & Walker, M.F., 1985. Astr. Astrophys., **150**, L12.
Balona, L.A. & Shobbrook, R.R., 1984. Mon. Not. R. astr. Soc., **211**, 375.
Becker, S.A. & Mathews, G.J., 1983. Astrophys. J., **270**, 155.
Bohlin, R.C., Cornett, R.H., Hill, J.K., Smith, A.M. & Stecker, T.P., 1985. Astrophys. J., **292**, 687.
Böhm-Vitense, E., 1985. Astrophys. J., **296**, 169.
Caldwell, J.A.R., 1983. Observatory, **103**, 244.
Caldwell, J.A.R. & Coulson, I.M., 1985. Mon. Not. R. astr. Soc., **212**, 879. (Errata **214**, 639).
Caldwell, J.A.R. & Coulson, I.M., 1986. Mon. Not. R. astr. Soc., in press.
Chiosi, C. & Pigatto, L., 1986. In press.
Cohen, J.G., 1982. Astrophys. J., **258**, 143.
Fahlman, G.G., Richer, H.B. & VandenBerg, D.A., 1985. Astrophys. J. Suppl., **58**, 225.
Feast, M.W., 1984. Structure and Evolution of the Magellanic Clouds, IAU Symp. 108, p. 157, eds. van den Bergh, S. & de Boer, K.S., Reidel.
Feast, M.W., 1984b. Mon. Not. R. astr. Soc., **211**, 51p.
Feast, M.W., 1985. Mem. Soc. astr. Ital., **56**, 213.
Glass, I.S. & Lloyd Evans, T., 1981. Nature, **291**, 303.
Glass, I.S. & Reid, I.N., 1986. Mon. Not. R. astr. Soc., in press.
Laney, C.D. & Stobie, R.S., 1985. Mon. Not. R. astr. Soc., in press.
Longmore, A.J., Fernley, J.A., Jameson, R.F., Sherrington, M.R. & Frank, J., 1985. Mon. Not. R. astr. Soc., **216**, 873.
Manduca, A., Bell, R.A., Barnes, T.G., Moffett, T.J. & Evans, D.S., 1981. Astrophys. J., **250**, 312.
Martin, W.L., Warren, P.R. & Feast, M.W., 1979. Mon. Not. R. astr. Soc., **188**, 139.
Menzies, J.W. & Whitelock, P.A., 1985. Mon. Not. R. astr. Soc., **212**, 783.
Moffett, T.J. & Barnes, T.G., 1986. Astrophys. J., in press.
Pedreros, M., Madore, B.F. & Freedman, W.L., 1984. Astrophys. J., **286**, 563.
Pel, J.W., 1985. Cepheids: Theory and Observations, IAU Coll. 82, p. 1, ed. by Madore, B.F., Cambridge Univ. Press.
Penny, A.J., 1984. Gemini (Royal Greenwich Observatory), No. 12, 1.
Sandage, A., 1982. Astrophys. J., **252**, 553.
Schmidt, E.G., 1984. Astrophys. J., **285**, 501.

Schommer, R.A., Olszewski, E.W. & Aaronson, M., 1984. Astrophys. J., **285**, L53.
Strugnell, P.R., Reid, I.N. & Murray, C.A., 1986. Mon. Not. R. astr. Soc., in press.
Visvanathan, N., 1985. Astrophys. J., **288**, 182.
Walker, M.F., 1974. Mon. Not. R. astr. Soc., **169**, 199.
Walker, A.R., 1984. Mon. Notes astr. Soc. S. Afr., **43**, 89.
Walker, A.R. & Mack, P., 1985. SAAO preprint, No. 445.
Walker, A.R. & Mack, P., 1986. Mon. Not. R. astr. Soc., in press.
Walker, A.R., 1985a. Mon. Not. R. astr. Soc., **213**, 889.
Walker, A.R., 1985b. Mon. Not. R. astr. Soc., **214**, 45.
Walker, A.R., 1985c. Mon. Not. R. astr. Soc., **212**, 343.
Walker, A.R., 1985d. Mon. Not. R. astr. Soc., **217**, 13p.
Walker, A.R., 1986. IAU Symposium 118, in press.
Welch, D.L., McAlary, C.W., Madore, B.F., McLaren, R.A. & Neugebauer, G., 1985. Astrophys. J. **292**, 217.

MAIN SEQUENCE FITTING AND THE DISTANCE TO THE LARGE MAGELLANIC CLOUD

R. A. Schommer
Rutgers University
Dept. of Physics and Astronomy
P.O. Box 849
Piscataway, NJ 08854

ABSTRACT. The development of new photometric techniques has permitted accurate measurements of the stellar main sequences in the Magellanic Clouds. Reasonable choices of the relevant parameters (i.e., reddening and abundances) have led to a "short" distance modulus, $(m-M)_o = 18.2$, for the Large Cloud. A consistent short value for the distance is derived from ZAMS fits, isochrone fits from the Yale and VandenBerg models, and comparisons with Galactic clusters. The sources of random and systematic error are briefly discussed, and the effects due to convective overshoot are estimated. The possibility of extending this technique to other galaxies is mentioned.

1. INTRODUCTION

The Magellanic Clouds have played a vital role in the determination of galactic distances, most notably through the discovery of the Cepheid P-L relation (Leavitt 1908). The Clouds have provided a vital resource for astronomers studying individual stars and stellar evolution, and also the structure and evolution of these nearest extragalactic systems. The richness of the cluster system in the clouds has received considerable attention in the past twenty years. Recent attempts to study the age and enrichment history of the Cloud clusters has led, rather surprisingly, to a new, accurate, and short determination of the distance to the LMC by direct measurement of the unevolved main sequences (MS).

2. CLUSTERS, PHOTOMETRY, AND COLOR-MAGNITUDE DIAGRAMS

Two technical developments, one in detector hardware and one in reduction software, have enabled us to measure magnitudes fainter than the vertical portions of the unevolved MS in the Clouds. The development of high quantum efficient, linear detectors (CCDs) have provided the necessary data. In particular, the PFCCD system on the CTIO 4 m has been a tremendously successful implementation; it has the required photometric linearity and stability, on a large aperture at an excellent photometric site. To reduce these data, especially in the crowded cluster fields, requires point spread function fitting techniques. Our first efforts used RICHFLD, developed at KPNO by D.Tody, but for the past two years we have relied on DAOPHOT, developed by Stetson and Olszewski, because of its multiple star fitting algorithm.

Figure 1 shows a resulting color-magnitude (C-M) diagram for NGC 2162, in BV colors (Schommer, Olszewski, and Aaronson 1984; SOA), with more than three magnitudes of main sequence evident. It is a typical, although high quality example, of the billion year old Cloud clusters (Hodge 1983). The observed width of the MS, internal error estimates, comparison with data taken on several nights, and tests adding artificial stars to the frames, all indicate that the photometric errors are ±0.05 in B-V down to V~ 23rd. These are fairly stringent photometric requirements, and of the 15 clusters we have analyzed to date, only 3-4 satisfy these requirements; poor seeing, more crowded fields, or inappropriate clusters (older than 2 billion years), compromise the main sequence determination. The main source of scatter is the crowding effects on the evaluation of sky levels; the central 10-20 arcseconds of most of these clusters are not measureable, due to the inability to determine a nearby sky value for each star.

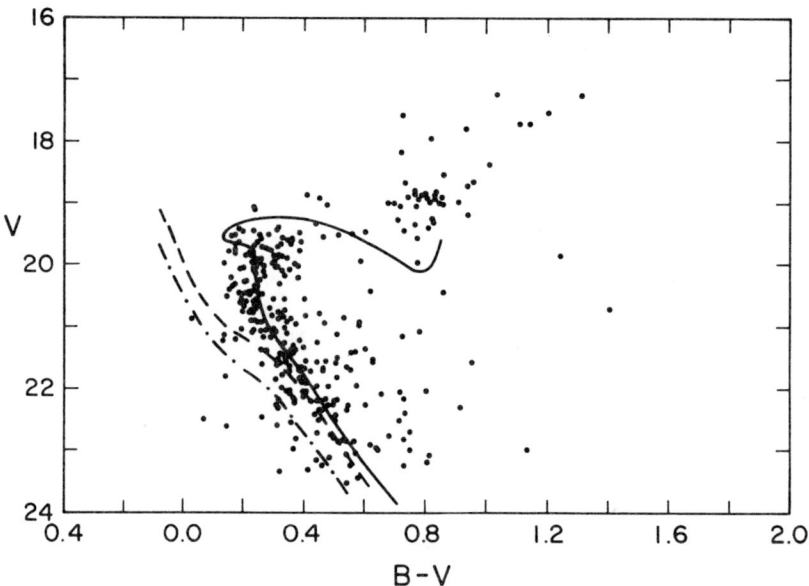

Figure 1. Color-magnitude diagram of NGC 2162 in the LMC. Superposed ZAMS for two distances (18.1, 18.6) and a billion yr. isochrone (18.1).

The locus of cluster points on the C-M diagram is determined by the distance modulus, the reddening, the cluster age, and the abundance parameters (both Y and Z). We will adopt a foreground reddening of E(B-V)=0.06 (McNamara and Feltz 1980). This is consistent with the foreground reddening maps of Bustein and Heiles (1982). Abundances are more uncertain and difficult to evaluate; they undoubtedly represent the most serious source of possible systematic error. The available measures of [Fe/H] for intermediate age objects (Cohen 1982; Harris 1982) indicate -0.2 < [Fe/H]< -1.0. The clusters clearly contain less than solar abundances, for the giant branches are considerably bluer than their galactic counterparts.

Figure 1 shows the ZAMS for [Fe/H]=-0.46 from VandenBerg (1985) at two distance moduli: 18.1 (---) and 18.6 (-··-). The evolutionary brightening away from the MS limits the fitting accuracy to about ±0.2, but the short value is clearly preferable. The long modulus only becomes acceptable at very high abundance (solar or greater), contrary to all external measures of Cloud abundances and grossly inconsistent with the colors of the giants in these clusters. The discussion by Mateo and Hodge (1985) on this matter is noteworthy; only very convoluted parameters allow the long modulus to fit.

Also shown in Figure 1 is an isochrone (solid line) by VandenBerg, with [Fe/H]=-0.23 for an age of 1 billion years (SOA includes Yale models), which appears to be a reasonable fit. For the long modulus, a younger isochrone (about 0.6 billion yrs) gives the closest fit, but the observed MS is redder than the models unless [Fe/H] is solar or greater. While such a metal rich MS can fit, the resultant giant branch will be at least 0.2 mag redder in B-V than the observed giants.

Worries about the reliability of the models led us to fit galactic open clusters to these LMC C-M diagrams. This is made rather difficult by the lack of suitable galactic objects; only a few have the proper ages; the reddening is high, uncertain or variable in many cases; and the abundances in the Galaxy are larger than the Clouds. Possible objects include: the Hyades; NGC 7789; NGC 2477 (but high and probable variable reddening); NGC 2158 (reddening problems); NGC 2420 and NGC 2506 (2-3 billion yrs old). Proper compensation for these effects leads repeatedly to the short modulus (see SOA).

While the short modulus is ~0.5 mag less than the long scale traditionally used for Cepheids (Sandage and Tammann 1974; Martin, Warren, and Feast 1979), there have been earlier short determinations, most notably by Eggen (1977), using Cepheids. DeVaucouleurs (1980b) has long championed the short scale using a variety of techniques, with a somewhat larger reddening. Some recent studies indicate possible problems with the Galactic Cepheid zero point (Schmidt 1984).

Other investigators working on the LMC clusters who find the short value preferable include Walker (1974), but especially the recent work on NGC 2213 by DaCosta, Mould and Crawford (1985) and on H4 by Mateo and Hodge (1985); these last two works include data of similar quality to that presented here.

The RR Lyrae stars also point toward the shorter distance. The photographic values of Graham (1984), augmented by new CCD measures by Walker (1985), derive a mean V=19.2 for the LMC RR Lyrae. With the adopted reddening, the long modulus requires M_v < 0.4, unacceptably bright. Our modulus implies M_v ~ 0.8, entirely appropriate for the mean Cloud abundance, according to the precepts of Sandage (1982).

Chiosi and Pigotti (1985) have stressed the neglect of convective overshoot in the isochrone calculations, due to the incomplete theory of convective energy transport. While these effects are noticeable when fitting the isochrones to the cluster (see the discussion by DaCosta et al. 1985), our distance modulus is not affected because:

1) we fit the unevolved MS, which is unaffected by the convective parameters; it is the evolution near core hydrogen exhaustion (the hook region in the isochrones of Fig.1) which is strongly affected by evolutionary brightening, changing the derived ages, but not the MS or distance;

2) we fit to Galactic clusters, because of similar worries with the models; to first order effects of convection drop out;

3) their arguement would imply different moduli for different age clusters, whereas all clusters are consistent with the short modulus (Andersen, et al. 1985; Jones 1985) and, as argued above, the RR Lyrae results favor our value; and

4) the luminosity function technique they espouse cannot uniquely determine age and distance, which both affect the magnitudes in similar ways; they are measuring the tip of the evolved main sequence, which is most sensitive to evolutionary brightening, and even given a distance, this technique requires accurate compensation for uncertain convective effects.

3. OTHER GALAXIES

While the controversy over the Cloud distance will continue for a time, such debates have seldom stopped the pursuit of the distance grail. In that spirit, I mention the interesting possibilty of determining MS distances to other local group galaxies. The main sequences of several nearby dwarf spheroidals can be reached with current techniques, but these systems are too old to allow accurate distance determination, and while interesting for other reasons, the absence of Cepheids eliminates their use in the distance ladder. The next likliest galaxies are the spirals M31 and M33, and the irregular NGC 6822. The latter object unfortunately suffers from severe foreground contamination.

To estimate the photometric requirements necessary to measure main sequences in M31 or M33, we can extrapolate the data from Fig.1. Thus we need photometry to V=29-30th mag., with errors of ~0.05 down to 28-29th. The crucial problem will be crowding; we need a 10x improvement in resolution over the ~1 arcsecond available from the ground. These are parameters that should push the limit of the Hubble Space Telescope, due for launch this year.

Measurement of field stars in these galaxies are complicated by the combined effects of depth, reddening, age and abundance variations, so a cluster population would be desireable. The best LMC clusters are ~ 1 billion yrs, with integrated $(B-V)_o$ = 0.4-0.68. M33 appears to offer a variety of similar clusters (Christian and Schommer 1983), and the attempt should be challenging.

I would like to acknowledge the help and advice of many colleagues, including M.Aaronson, G.DaCosta, P.Flower, J.Graham, P.Hodge, M. Mateo, J. Jones, P.Papenhausen and D.VandenBerg. These results would not have been possible without the collaboration of E. Olszewski, and the telescope allocations of the CTIO TAC. I acknowledge partial support from NSF grant AST84-12515.

REFERENCES

Andersen,J.,Blecha,A., and Walker, M.F. 1985, Astron. Astrophys., 150, L12.
Burstein,D., and Heiles,C. 1982, A. J., 87, 1165.
Chiosi,C. and Pigatto,L. 1985, preprint.
Christian, C.A., and Schommer,R.A. 1983, Ap.J., 275, 92.
Cohen,J. 1982, Ap.J., 258, 143.
DaCosta,G.S., Mould,J.R., and Crawford,M.D.1985, Ap.J., 297, 582.
deVaucouleurs,G. 1980a, P.A.S.P., 92, 576.
deVaucouleurs,G. 1980b, P.A.S.P., 92, 579.
Eggen,O.J. 1977, Ap.J. Supple., 34, 1.
Graham, J.A. 1984, in IAU Symposium 108, Structure and Evolution of the Magellanic Clouds, ed. S. van den Bergh and K.S. de Boer (Dordrecht: Reidel), p.207.
Harris,H.C. 1982, A.J., 88, 507.
Hodge,P.W. 1983, Ap.J., 264, 470.
Jones, J. 1985, thesis, Clemson University.
Leavitt,H.S. 1908, Harvard Ann., 60, 87.
Martin,W.L., Warren,P.R., and Feast,M.W. 1979, M.N.R.A.S., 188, 139.
Mateo, M., and Hodge, P.W. 1985, preprint.
McNamara,P.H., and Feltz,K.A. 1980, P.A.S.P., 92, 587.
Sandage,A. 1982, Ap.J., 252, 553.
Sandage,A., and Tammann,G.A. 1974, Ap.J., 190, 525.
Schmidt, E.G. 1984, Ap.J., 285, 501.
Schommer, R.A., Olszewski,E.W., and Aaronson, M. 1984, Ap.J., 285, L53.
VandenBerg, D.A. 1985, preprint.
Walker, A. 1985, M.N.R.A.S., 212, 343.
Walker, M.F. 1974, M.N.R.A.S., 169, 199.

Quips and Quotes from Bob Schommer's Notebook

"But I want to emphasize, there *is* a Hubble constant."

"This work has been done by Martha Haynes and her husband."

"Assuming velocity equals distance. Yes, of course."

"This result will make Dave Burstein's day."

"You're a nice guy, so you wind up getting $H_0 = 70$."

"I'll show you a little twiddle that will broaden that line out."

"Nobody has done it so far, but we can invent a model."

"You're seeing bubble motions of the whole sheebang."

"So *somebody* in the Universe knows about Arecibo."

"I'll accept $H_0 = 50$ ..., if $c = 600,000$"

"My measuring errors are a few percent, unless I talk to David."

"The observers in the Northern Hemispere were stabilizing."

"It goes right off and falls off the edge of what I am talking about."

"This is *my* kind of theory."

"Let's stop here. I think Jerry can convince everbody later."

"Here is Virgo; but it's not focussed.
 It doesn't matter, you can't see Virgo anyway."

"Here's to Malmquist."

"*Surf's up!*"

RECENT CCD OBSERVATIONS OF CEPHEIDS IN NEARBY GALAXIES: A PROGRESS REPORT

Wendy L. Freedman
Mount Wilson and Las Campanas Observatories
Pasadena, CA 91101-1292

ABSTRACT New *BVRI* CCD data for previously known Cepheids in the the nearby galaxies IC 1613, NGC 6822, M31, M33, WLM, NGC 300, Sextans A and NGC 2403 have been obtained. A method for determining distances and mean reddenings, utilizing the data at four wavelength bands, is discussed.

1. INTRODUCTION

Due to their high intrinsic luminosities, and characteristic light curves, Cepheids have been detected in many nearby galaxies, and over the last 60 years, these variables have proven to be very useful distance indicators. Once Cepheids have been discovered and monitored, a straightforward period-luminosity relation can, in principle, yield a distance modulus to the galaxy.

However, the calibration and application of the period-luminosity (P-L) or the period-luminosity-color (P-L-C) relations are still subject to some uncertainty. This is because both the observed magnitudes and colors will be affected by the amount of obscuration (anywhere along the line of sight to the observed Cepheids), and also presumably by the metallicity of the individual Cepheids. The magnitude of the latter effect is very uncertain, and has not been well-treated thus far in the application of the Cepheid distance scale. It is complicated by the fact that theoretical models predict effects due to metallicity for both stellar interiors and atmospheres, but the combined effects have not been addressed in much detail, nor have they been unequivocally demonstrated empirically. The combined impact of reddening and possible metallicity effects is nontrivial. For instance, if one is trying to solve for the color coefficient in the P-L-C relation, then one first needs to determine individual reddenings to each of the Cepheids. However, since the magnitudes and colors are affected by both reddening and metallicity, without strictly independent determinations, there is an inherent circularity present in the problem.

In this paper, a method is described for obtaining the mean reddening for the Cepheids in each galaxy. The method has the advantage that the reddenings are directly applicable to the Cepheids (and are not derived, say, from nearby young stars, where the extinction may differ). This method has been previously described with reference to the Cepheids in M33 (Freedman 1985). The effects of metallicity are not treated in the present work; however, a brief summary of a program in collaboration with Madore and Aaronson to test empirically for a metallicity dependence of the zero point of the Cepheids P-L relation is included in the final section.

2. THE CCD PROGRAM

A long-term program is currently underway to obtain charge coupled device (CCD) data at four wavelengths, (B V R and I), for galaxies in which Cepheids have already been discovered. The program galaxies include IC 1613, NGC 6822, WLM, M31, M33, NGC 300, Sextans A and NGC 2403. The Cepheids in these galaxies have been discovered and discussed previously by Sandage (1971), Kayser(1967), Sandage and Carlson (1985), Hubble (1926), Sandage and Carlson (1983), Hubble (1929), Baade and Swope (1963, 1965), Graham (1984), Sandage and Carlson (1982) and, Tammann and Sandage (1968).

At present, single-epoch CCD data have been obtained for several of the above-mentioned galaxies, including IC 1613, NGC 6822 (both discussed briefly in Freedman 1984), as well as the Wolf-Lundmark-Mellotte system, and Sextans A. For M31, NGC 300, NGC 2403 and M33 (for which preliminary results from single-epoch data were discussed in Freedman 1985), some further phase information has now been obtained. Still, it is found (e.g., Freedman 1985) that even for single-epoch observations, the CCD period-luminosity relations are very narrow. For instance, the full width of the P-L relation at I is 0.6 magnitudes, which is already about half that of time-averaged P-L relation at B. In addition, the multi-wavelength observations permit an estimate of the total extinction of the Cepheids to be obtained. Thus, accurate distance estimates can be obtained even from these single-epoch measurements. It is hoped that eventually the scatter in the P-L relations can be further improved with additional phase information, but such results are several years in the future.

The observations so far obtained were made at the Canada-France-Hawaii 3.6m, the Kitt Peak 4m, and the Cerro Tololo 4m telescopes, using a 320 x 512 pixel RCA CCD in all cases. The scales at the prime foci of each telescope are 0.4, 0.6 and 0.6 arcsec/pixel, respectively. The data reduction is being undertaken using *DAOPHOT*, which is a point-spread-function fitting routine capable of making simultaneous fits to up to 60 crowded stellar images. *DAOPHOT* was developed by Peter Stetson at the Dominion Astrophysical Observatory. Relative distance moduli between the LMC and the program galaxies are obtained by least-squares minimization of the residuals found from slide-fitting the two P-L relations into coincidence. This, of course, can be done independently for $B\ V\ R$ and I. Then, if it is assumed, as for example in Freedman (1985), that the true distance modulus is asymptotically being approached at infinite wavelengths, an estimate of the total extinction and the true distance modulus to the galaxy can be obtained by linearly extrapolating to the origin in a plot of relative distance modulus versus inverse wavelength.

The data reduction for this program is at various stages for each of the individual galaxies. Distance moduli based on the data and methods discussed here will be reported at a later date.

3. ON-GOING AND FUTURE CCD STUDIES

A direct test of the zero-point dependence of the P-L relation on metallicity is being undertaken at the Canada-France-Hawaii Telescope with Madore and Aaronson. Several CCD frames of Cepheids distributed as a function of radius have recently been obtained in M31, with the aim of empirically determining whether the zero point of the P-L relation changes with the measured abundance gradient in this galaxy. The results of this program should provide the first indication of both the magnitude and the sense of the effects of metallicity on observed quantities (for the atmosphere and interior combined). Then if needed, an empirical correction to the Cepheid distances can be applied.

In order to increase the number of Cepheid calibrating galaxies for the IR Tully-Fisher distance scale, a search for Cepheids in Sculptor Group galaxies is being undertaken in collaboration with Madore, Aaronson, Mould, and Graham using the CTIO 4m.

At Palomar, a search for Cepheids in M101 and NGC 4214 is being undertaken by a large consortium using the *Four-Shooter*. Aaronson and Cook are independently searching for Cepheids in M101 at Kitt Peak. The discovery of Cepheids in galaxies more distant than M101 will have to await the *Space Telescope*.

ACKNOWLEDGEMENTS

This work was done while the author was a Carnegie Research Fellow at the Mount Wilson and Las Campanas Observatories. Computing, research and travel support from the Carnegie Institution of Washington is gratefully acknowledged.

REFERENCES

Baade, W., and Swope, H. H. 1963, *A. J.*, **68**, 435.

Baade, W., and Swope, H. H. 1965, *A. J.*, **70**, 212.

Freedman, W. L. 1985, in *IAU Colloquium No. 82, Cepheids: Theory and Observations*, ed. B. F. Madore, (Cambridge: Cambridge University Press), pg. 225.

Freedman, W. L. 1984, *B. A. A. S.*, **16**, 888.

Graham, J. A. 1984, *A. J.*, **89**, 1332.

Hubble, E. E. 1926, *Ap. J.*, **63**, 236.

Hubble, E. E. 1929, *Ap. J.*, **69**, 103.

Kayser, S. 1967, *A. J.*, **72**, 134.

Sandage, A. R. 1971, *Ap. J.*, **166**, 13.

Sandage, A. R. 1983, *A. J.*, **88**, 1108.

Sandage, A. R., and Carlson, G. 1982, *Ap. J.*, **258**, 439.

Sandage, A. R., and Carlson, G. 1983, *Ap. J. (Letters)*, **267**, L25.

Sandage, A. R., and Carlson, G. 1985, *A. J.*, **90**, 1464.

Tammann, G. A., and Sandage, A. R. 1968, *Ap. J.*, **151**, 825.

DISTANCES USING A SUPERGIANT STARS

R. Brent Tully and Jaye M. Nakashima
University of Hawaii, Institute for Astronomy
2680 Woodlawn Drive, Honolulu, Hawaii 96822 USA

INTRODUCTION

With modern detectors it is possible to obtain spectra of individual stars in nearby galaxies. A and B supergiants are among the brightest stars in galaxies with young populations, and it has long been known that these objects have spectral characteristics that are luminosity dependent. There are two physical processes that conspire to weaken the Balmer absorption lines in more luminous stars. (1) Surface gravity is reduced in the brighter stars, leading to narrower photospheric hydrogen absorption lines. (2) Mass loss is greater in the brighter stars, leading to hydrogen line emission in extended atmospheres that tends to fill in the photospheric absorption lines.

The luminosity dependence of the widths of hydrogen Balmer lines in A stars could be used to determine the distances of nearby galaxies. A major hindrance has been the lack of proper calibration at the highest luminosities because there are few extremely bright Galactic A or B supergiants with well-determined distances. However, there are many such stars in the Large Magellanic Cloud, all at approximately the same distance.

We have already published a calibration of $H\alpha$ and $H\beta$ equivalent width-luminosity relationships for samples of stars in the two Magellanic Clouds (Tully and Wolff: Ap. J. 281, 67, 1984). In this article, we will present a relationship between $H\gamma$ and $H\delta$ equivalent widths and luminosities for a very large sample of LMC stars.

THE SAMPLE

Observations were made at the Cerro Tololo Inter-American Observatory with the 4 m R-C spectrograph and 40 mm ultraviolet SIT-vidicon detector. A sample of 127 stars with spectral types in the interval B1-A9 were observed with spectral resolution of 3 Å. Some 25 of the same stars were reobserved with spectral resolution of 1.5 Å. Most stars in the LMC classified B5-A5 and with $V \leq 12$ were observed, although stars in confused regions were avoided. Random samples of stars as faint as V = 14 and of hotter and cooler stars were also included in the program.

Equivalent widths in the Hγ and Hδ lines were measured with a hand planimeter. No systematic differences were discerned between the data at 3 Å and 1.5 Å resolution or between the Hγ and Hδ material. In the subsequent discussions, we use the average of all observations in a given line on a given star and retain separate Hγ and Hδ measurements. Hence, the analysis is based on 254 equivalent width measurements.

TEMPERATURE CORRECTIONS

Spectral classifications are available for all the stars that we observed. Plots of equivalent widths vs. V magnitudes for individual spectral types revealed systematic differences between different types. Figure 1 schematically illustrates what was seen. A0 stars lie in the region of the heavy central line. Cooler stars lie along parallel relationships but offset in zero point, such that at a given magnitude the cooler star has the stronger absorption lines. Hotter stars lie along steeper relationships with a common zero point, such that at fainter magnitudes the hotter star has weaker absorption lines than an A0 star. Whether or not it is physically significant, all spectral classes from B1 to A0 have the common extrapolated zero point (where the Balmer lines would pass from absorption into emission) of $M_V = -9.1$ (assuming a distance modulus uncorrected for reddening of 19.0).

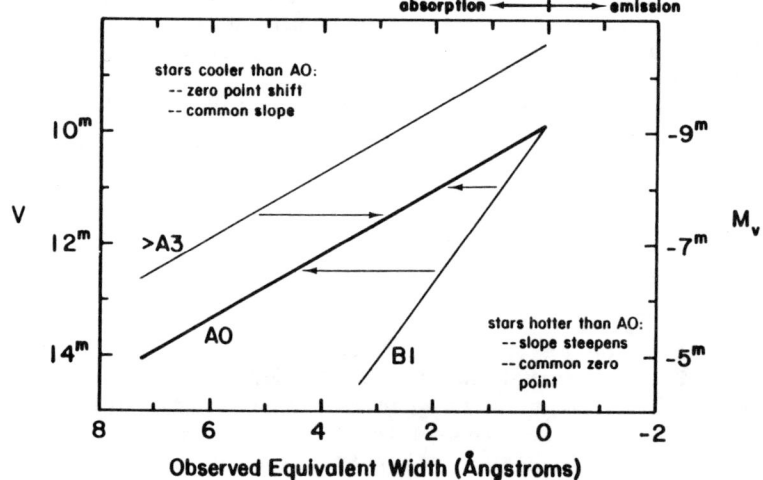

Figure 1

All our observations can be plotted together if equivalent widths are adjusted for the empirically observed temperature effects. The shift is to the A0 relationship. Cooler stars only require a differential shift:

$$W' = W - 0.593 \, \Delta S \qquad 1 \leq \Delta S \leq 2$$
$$W' = W - 1.48 \qquad 3 \leq \Delta S \leq 9$$

where W is the observed and W' the adjusted equivalent widths and ΔS is the difference in spectral type from A0 (units of 0.1 spectral class). Hotter stars require a multiplicative shift:

$$W' = (1 - 0.157 \, \Delta S) \, W \qquad -9 \leq \Delta S \leq -1$$

RESULTS

Temperature adjusted equivalent widths are plotted as a function of magnitude in Figure 2 for our sample of 127 galaxies and 254 measurements. The dispersion about a least-squares fit line is 0.49 mag. This dispersion includes scatter introduced by temperature adjustments and scatter caused by reddening, for which there have been no corrections.

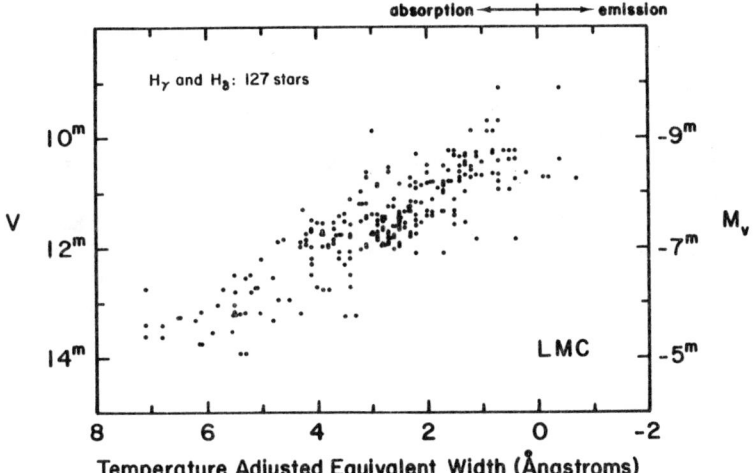

Figure 2

In Figure 3, the data have been binned by intervals of magnitude. For use of the relationship as a distance estimator this is the proper way to bin, since the sample is selected on the basis of magnitudes and subsequent equivalent width measurements are unbiased. The straight line in this plot actually represents the least-squares fit with widths a function of magnitude to all the data of Figure 2

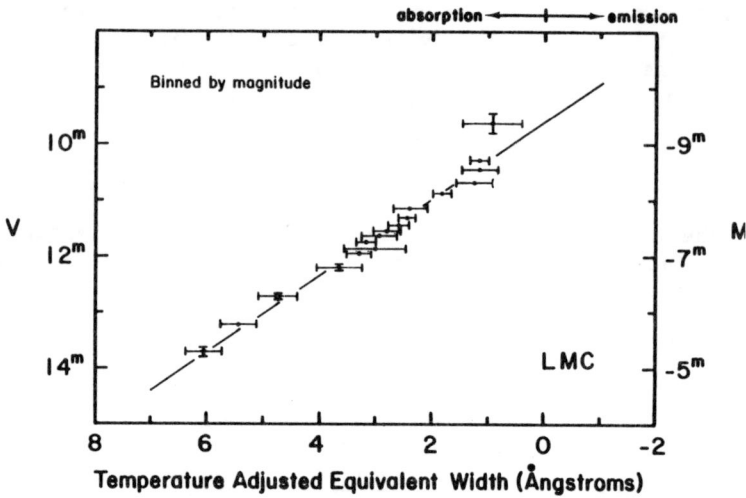

Figure 3

(consistent with magnitude binning). The relationship is seen to be defined to better than 0.1 mag on the magnitude axis.

SUMMARY

The potential of the A star line width-luminosity relationship is confirmed in the magnitude range -9 to -5. The use of A stars as a distance estimator for nearby galaxies looks to be about as attractive as the Cepheid method. A stars are bright enough to be observed to at least distance modulus 29. They are easily identified because of a pronounced Balmer discontinuity. They are plentiful in galaxies with young populations. They tend <u>not</u> to be badly confused. Enough of them can be observed in a single galaxy to define a line width-luminosity relationship that should provide a distance with an uncertainty of order 10%.

AN INFRARED CALIBRATION OF THE CEPHEID DISTANCE SCALE

Barry F. Madore
Department of Astronomy
University of Toronto
Toronto, Canada
   and
Downs Laboratory of Physics
California Institute of Technology
Pasadena, California

ABSTRACT. The results of a five-year campaign to recalibrate the Cepheid period-luminosity relation in the infrared are reviewed. The advantages and limitations of the method as presently used, are highlighted; and finally, new apparent infrared distance moduli relative to the LMC are given for all late-type resolved members of the Local Group with known Cepheids. These include the LMC, SMC, NGC 6822, M33, M31 and IC 1613.

1. INTRODUCTION

When the first near-infrared observations of extragalactic Cepheids were obtained at the Dupont 2.5m reflector of the Las Campanas Observatory in Chile, by Matthew Malkan and myself in June 1980, it was immediately clear that more advantages were to accrue from this new calibration than we had originally hoped. Traditionally, the period-luminosity relation for Cepheids had been calibrated and applied in the blue and visual part of the spectrum; however at these wavelengths the effect of interstellar reddening can be non-negligible, and for Cepheids, problematical. With advances in infrared technology it suddenly became practical to re-observe Cepheids in our own and in other galaxies, repeating a classical path to the distance scale. But this time the path was to be freed to a very large extent from the uncertain effects of reddening and absorption. Had that been the only advantage of going to the infrared those results alone would have been well worth the effort.

Even at the telescope it was apparent that mant additional benefits were in hand. The raw data soon showed that more than just reddening was falling to the infrared. Simultaneously, both the apparent width of the instability strip and the apparent amplitudes of the individual stars themselves were collapsing as projected into the long-wavelength domain. In retrospect, the reasons for this are quite simple; the implications, on the other hand, are still being discussed. At *optical* wavelengths, for an individual Cepheid executing its pulsational cycle, it is well known that the apparent light variations are largely due to variations in the temperature, which give rise to variations in the surface brightness. Physical changes in the radius are finite, (but quite small), amounting to only ten or twenty percent variations in the area. The monochromatic sensitivity of surface brightness to temperature is quite high in the blue; but such is not the case at longer wavelengths. In fact, in the infrared the roles of temperature and radius are nearly reversed, with the temperature variations having such small effects on the surface brightness that the infrared light curves are practically dominated by the small, but persistent, residual radius variations.

The same physical explanation applies to the observed 'collapse' of the instability strip itself, (representing not one star now, but an ensemble). Stars of equal period, but different mean surface temperature find themselves well separated in optical luminosity (that is, up to 1.5 mag at B) due to a primary response of mean surface brightness to mean temperature differences. Again, a secondary contribution to the luminosity width of the P-L relation comes from differences in mean radii across the strip (but according to Fernie 1975 this effect should amount only to 0.08 mag total). In the infrared the mean surface brightness variations from star to star are just as weak a function of temperature as during the cycle of a single star. Accordingly, by swinging out to the infrared the new P-L relation could close down to as little as the geometric limit, and one might expect an equivalent dispersion for the relation of only ±0.02 mag!

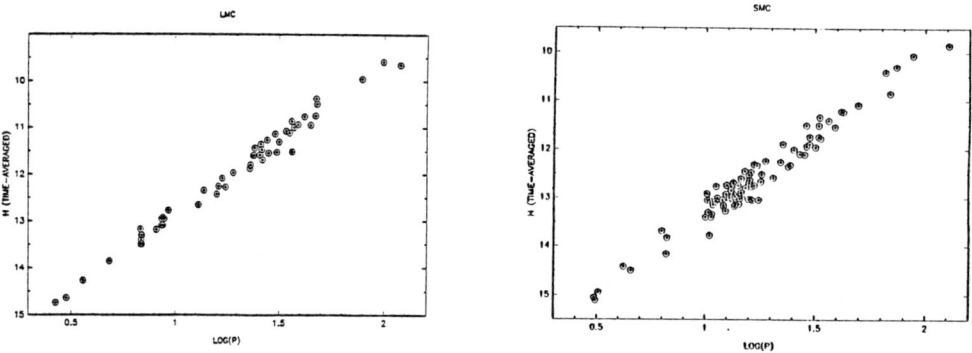

Figure 1. Time-averaged intensity-mean H-band period-luminosity relations for Cepheids in the LMC (left) and SMC (right) based on data fom Welch, McAlary, Madore and McLaren (1986). The SMC relation is more dispersed than the LMC because of line-of-sight depth effects.

While such an unprecedented limit has certainly not been reached yet, the intermediate steps are still very encouraging. With single observations of randomly phase selected Cepheids it is immediately possible to obtain a P-L relation that has a scatter of only ±0.2 mag. And, by obtaining first-order light curves and time-averaging their intensities (see Figure 1) this dispersion has been pushed down to 0.15 mag, for instance, for the Large Magellanic Cloud Cepheids (Welch, McAlary, Madore and McLaren 1986). It is easy to see therefore that an accuracy of ten percent in distance can be readily obtained from a modest number of infrared observations of known Cepheids. And that basically was the whole intent of this campaign.

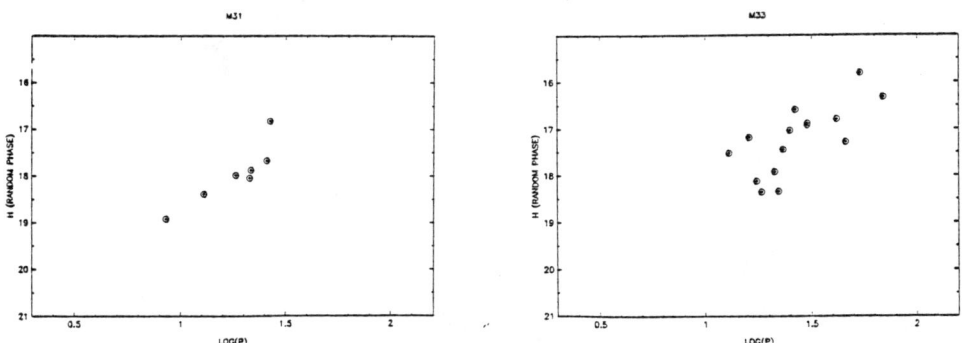

Figure 2. Random-phase H-band period-luminosity relations for Cepheids in M31 (left) and M33 (right). The M31 plot has been purged of probably contaminated data; the M33 plot has not.

2. TECHNICAL LIMITATIONS

Inherent to the method by which present-day near-infrared observations are made there are limitations to the accuracy that can be obtained at the telescope, for the most distant Cepheids. Unlike many limiting observations the primary limitation is not lack of photons, but rather it is source confusion – a situation brought on by the lack of the necessary spatial resolution at the detector. Infrared measurements are made by a rapid succession of comparisons of two (finite-sized) beams, giving an 'on-signal' region and an 'off-signal' region. To balance sensitivities the channels are periodically reversed leading effectively to a final differencing of the

'Cepheid+sky' observation against *two*, different but nearby, 'sky' regions. It is not that such a needed procedure is inherently poor, but it is the large apertures involved that cause the problem. Inside each of the beams (which are typically 5 arcsec in diameter) large numbers of stars in the parent galaxy many be included. Some of these may be optically faint, infrared sources. Their random appearance in one aperture or another would seriously comprome the measurement made on the Cepheid. These crowding and confusion-induced noise sources are almost certainly present (see Figure 2) in the observations of the Cepheids in M33 (Madore et al. 1985) and even more so in the Cepheid sample drawn from M31, where the inclination effects only worsen the problem. Going to more distant systems would inevitably spell greater trouble unless special care were exercised in working on only the absolutely least-crowded stars, far out in the periphery of the galaxy.

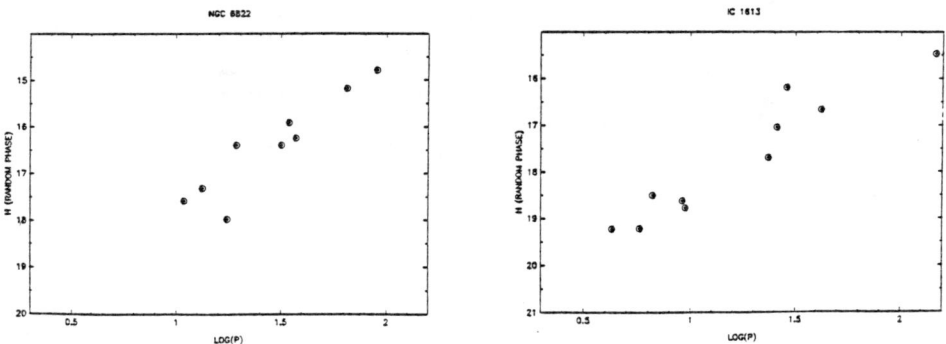

Figure 3. Random-phase H-band period luminosity relations for Cepheids in NGC 6822 (left) and IC 1613 (right).

Indeed, until panoramic infrared detectors become regularly available it is unlikely that much more progress on the distance scale will be made at the near-infrared wavelengths. Admittedly at all wavelengths sky subtraction has to be made quite. This has to be done as close to the star of interest as possible as more and more distant galaxies are studied, because the background disk of the galaxy and structure therein becomes competative with the terrestrial sky as a source of noise. At the present time the only efficient way to proceed is by the use of CCD detectors, as described by Freedman (1986), for instance. There the resolution limit of the seeing can be pressed to the full extent and sky contributions determined in a very local and statistically stable fashion.

3. THE LOCAL GROUP: RELATIVE INFRARED DISTANCE MODULI

Despite the limitations on internal accuracy mentioned above there is nothing known to date that leads us to think that the infrared observations are suffering from *systematic* errors. Certainly all of the major concerns about systematics in the optical are much reduced in the infrared (i.e., reddening, atmospheric metallicity variations, main sequence companion contamination, etc.). However, one fundamental limitation on Cepheids as absolute distance indicators is still with us at all wavelengths. That problem lies in the calibration of the zero-point of all P-L relations using *independent* distance estimates within our Galaxy or anywhere else that it can be done, for that matter (for a recent review, see Schmidt 1984).

Since the *absolute* galactic zero-point for the P-L relation is still being debated at an uncertainty level of about 0.3 mag, it is probably preferable to give the infrared distances in terms of *differential* moduli, derived here from a comparison of each extragalactic infrared data set (as already compiled by Madore 1985) with that of the Large Magellanic Cloud cepheid data (Welch, McAlary, Madore and McLaren 1986) shown in Figure 1. In this way future studies can simply take the differential moduli given in Table 1, input their favorite value of the true distance modulus and reddening to the LMC, as well as their choice of the reddening (foreground and internal) to the parent galaxy, and thereby derive their own absolute distance scale. Hopefully, concensus on those final steps will not be far off as well.

Table 1

Infrared Distance Moduli
Relative to LMC

Galaxy	H modulus
SMC	0.4
NGC 6822	5.0
IC 1613	5.6
M33	5.7
M31	5.9

ACKNOWLEDGEMENTS

It is a pleasure to thank many friends and colleagues who have worked on this project over the last five years. This research and its natural extensions to include other techniques and instruments is being supported by the Natural Sciences and Engineering Research Council of Canada, and by a Killam Research Fellowship made available through the Canada Council. It is an equal pleasure to record my thanks to the many observatories and their staff who provided telescope time and assistance so as to bring this project to a swift and successful completion.

REFERENCES:

Fernie, J. D. 1975, in *I.A.U Symposium No. 67, Variable Stars and Stellar Evolution* ed. V. I. Sherwood and L. Plaut, (Dordrecht: Reidel), pg. 185.

Freedman, W. L. 1986, in *Galaxy Distances and Deviations from Universal Expansion*, ed. B. F. Madore, (Dordrecht: Reidel), pg. 21.

Madore, B. F. 1985, in *IAU Colloquium No. 82, Cepheids: Theory and Observations*, ed. B. F. Madore, (Cambridge University Press: Cambridge), pg. 166.

Madore, B. F., McAlary, C.W., McLaren, R.A., Welch, D.L., Neugebauer, G., and Matthews, K. 1985, *Ap. J.*, **295**, 560.

McAlary, C. W., Madore, B. F., McGonegal, R., McLaren, R. A., and Welch, D. L. 1983, *Ap. J.*, **273**, 539.

McAlary, C. W., Madore, B. F., and Davis, L. E. 1984, *Ap. J.*, **276**, 487.

Schmidt, E. G. 1984, *Ap. J.*, **285**, 511.

Welch, D. L., McAlary, C. W., Madore, B. F., and McLaren, R. A. 1987, *Ap. J. Suppl.*, (to be submitted).

Welch, D. L., McAlary, C. W., McLaren, R. A., and Madore, B. F. 1986, *Ap. J.*, (in press).

RR LYRAE STARS IN THE HALO OF M31 - A PRELIMINARY REPORT

C.J. Pritchet
Physics Department
University of Victoria
P.O. Box 1700
Victoria, B.C., Canada   V8W 2Y2

ABSTRACT. We have detected RR Lyrae variables in a halo field 40' SE of the nucleus of M31. A preliminary analysis of the data for 7 variables yields $(m-M)_0 \cong 24.2$, in good agreement with previous estimates.

1. INTRODUCTION

The first attempt to detect RR Lyrae stars in M31 was made by Baade in the early 1950's using the newly-commissioned 200" telescope at Mt. Palomar. Baade failed; his failure led to the 1.5 mag revision in the zeropoint of the Cepheid P-L relation, together with a factor of two increase in the estimated distance scale and age of the universe. The fascinating story of this work may be traced in the Annual Reports of Mt. Wilson and Palomar Observatories for the years 1951-52 to 1953-54, and is summarized in Baade (1956).
It has been realized for some years that RR Lyrae stars in Local Group Galaxies will be observable with Space Telescope (e.g. Tammann 1979). Such observations are extremely important if one wishes to (i) pin down the zeropoints in the calibration of secondary distance indicators (e.g. novae, IR Tully-Fisher relation, etc.), and (ii) check for population dependence or Oosterhoff group dependence of $< \overline{M_v} >$ for RR Lyrae stars. For several years now Sidney van den Bergh and I have been aware that the time was in fact ripe for a fresh ground-based attack on the problem of RR Lyrae stars in Local Group galaxies. The availability of large telescopes in sites of exceptional seeing (e.g. CFHT) and the development of high quantum efficiency array detectors allow adequate resolution and limiting magnitude for the problem.
In this paper we report on the discovery of RR Lyrae stars in the halo of M31, and on a preliminary determination of the distance modulus of M31.

2. OBSERVATIONS

The observations that we discuss in this paper were obtained in October 1985, and comprise a total of 17 hours of integration (51 × 20 min) of a field 40' SE of the nucleus of M31. The detector used was an RCA 320 × 512 CCD at the prime focus of the CFHT 3.6 m telescope. Seeing for the observations was in the range 0".6 - 1".2 FWHM, with a median seeing of ~0".9. All exposures were sky-noise limited.

The choice of the area observed is of some interest in its own right. From simulations we found that the optimum tradeoff between maximizing the number of RR Lyrae stars per field and the deleterious effects of crowding occurred at μ_B = 27 ± 1 mag arcsec^{-2} (1" FWHM seeing). From our first exposures at the telescope we found that this occurred at approximately r = 40' along the minor axis of M31. This is in remarkable agreement with the luminosity profile of M31 published by de Vaucouleurs (1958).

Preliminary data reduction consisted of the following steps: bias removal, dark subtraction, and flat fielding using dome flats. Individual 20 min exposures were then combined (using a median filter) in groups of three to produce 17 frames, each with 1 hour effective exposure.

Frames were "blinked" independently by C.J. Pritchet and S. van den Bergh. The optimum procedure for visual recognition of variables was empirically found to be a blink comparison of each of the 17-1 hour frames against a mean of all frames. Seeing compensation was effected by convolving one of the two frames being blinked with an appropriate 2D Gaussian.

3. RESULTS

A total of 81 variable objects were found on two or more totally independent pairs of frames. Many of these will turn out to be spurious (cosmetic defects, etc.). To date we have concentrated our attention on 7 objects which were found on 3 or more pairs of frames. An example of one such object is shown in Fig. 1 in high and low states. Photometry of this object (using P. Stetson's DAOPHOT crowded field photometry program) yields magnitudes which, when phased with a $0^d.495$ period, produces a reasonably continuous light curve. (A period of $0^d.595$ produces nearly as continuous a light curve; however, the large [1.5 mag] amplitude of this variable is more consistent with the shorter period.) Data for other variables is shown in Table I.

To use RR Lyrae stars as distance indicators it is necessary to obtain the mean magnitude < B > of the variables. The most resonable method to derive this quantity is take the average of all frames, converting the average intensity of each variable to a magnitude < B > (which is well above the magnitude limit of the combined frame). Such a procedure yields values of < B > which are given in Table I for each variable. A weighted mean of all variables yields < B > = 25.52 ± 0.02 (error in mean)

A recent determination of $<M_V>$ for RR Lyrae variables in the solar neighbourhood yields $<M_V>$ = 0.76 ± 0.14 (Hawley et al. 1986). Taking $<B-V>$ = 0.26 (Hawley, private communication), we obtain $<M_B>$ = 1.02 ± 0.14. Combining this with $<B>$ = 25.52 and A_B = 0.32 (Burstein and Heiles 1984) yields $(m-M)_0$ = 24.18 ± 0.14 or d = 686 kpc for M31. This is in excellent agreement with a number of other M31 distance determination (cf. Table II), in particular that of Welch et al. using IR observations of Cepheids.

It should be emphasized that the above analysis neglects two potentially important biases. 1. The quality $<M_B>$ as defined by Hawley et al. is an average of the light curve in magnitude space, and is computed from a recipe provided by Fitch et al. (1966). This quantity is fainter than an intensity-averaged $<M_B>$, by about 0.1 mag. 2. By initially concentrating on those objects found ≳3 times on independent pairs of frames, we have unwittingly biased our survey to those objects that happen to have their phase coverage concentrated more than average towards maximum light. Monte Carlo simulations suggest that this bias is ~0.15 mag, in the sense that all values of $<B>$ are too bright.

We have not corrected for the above biases (which in any case are of comparable magnitude and opposite sign). Instead we emphasize that the $(m-M)_0$ value for M31 derived above must be considered as preliminary until more variables have been analyzed.

4. CONCLUSIONS

As the first step in a two-pronged attack to improve our estimate of H_0, we have detected RR Lyrae variables in the halo of M31. A preliminary analysis of 7 of the variables yields $(m-M)_0$ = 24.18 ± 0.14, in good agreement with distance moduli derived from a variety of other techniques.

We are grateful to the staff of CFHT for assistance with the observations, and to Peter Stetson for use of his program DAOPHOT. This research was supported in part by the Natural Sciences and Engineering Research Council of Canada.

REFERENCES

de Vaucouleurs, G. (1958). Astrophys. J. **128**, 465.
Burstein, D., and Heiles, G. (1984). Astrophys. J. Suppl., **54**, 33.
Hawley, S., Jeffreys, W., Barnes, T., and Lai, W. (1985). Preprint.
Fitch, W., Wisniewski, W., and Johnson, H. (1966). Publ. Lun. Plan. Lab., **71**, 3.

Baade, W. (1956). P.A.S.P. **68**, 5.
Tammann, G. (1979). In Scientific Research with the Space Telescope (IAU Colloquium 54), ed. M.S. Longair and J.W. Warner, p. 263.
Baade, W., and Swope, H. (1963). Astron. J. **68**, 435.
Welch, D., McAlary, C., McLaren, R., and Madore, B. (1986). Preprint.
van den Bergh, S. (1975). In Stars and Stellar Systems, Vol. **9**, ed. A. Sandage, M. Sandage, and J. Kristian (Chicago: University of Chicago Press), p. 509.
Hartwick, F., and Hutchings, J. (1978). Astrophys. J. **226**, 203.
Cohen, J. (1985). Astrophys. J. **292**, 90.
Mould, J., and Kristian, J. (1985). Preprint.

TABLE I. Data for Variables

Variable	P[d]	B_{max}	B_{min}	$<B>$
160	0.48::	25.1	26.0:	25.63 ± .04
196	0.495	25.0	26.4:	25.51 ± .04
339	0.48:	25.1	26.3:	25.50 ± .03
375	0.535	25.2	26.4:	25.36 ± .04
400	0.595	25.4	26.3:	26.05 ± .10
404	0.56:	25.1	26.4:	25.43 ± .04
428	?	25.5	26.2:	25.53 ± .05

TABLE II. Comparison of $(m-M)_0$ Values for M31

Method	$(m-M)_0$	Reference
Cepheids (blue)	24.20 ± 0.14	Baade & Swope (1963)
Cepheids (IR)	24.26 ± 0.08	Welch et al. (1986)
Novae	24.16 ± 0.12	van den Bergh (1975)
Novae	24.3 ± 0.1	Hartwick & Hutchings (1978)
Novae	24.03 ± 0.20	Cohen (1985)
Pop II giants	24.4 ± 0.2	Mould & Kristian (1985)
RR Lyrae stars	24.18 ± 0.14	This paper

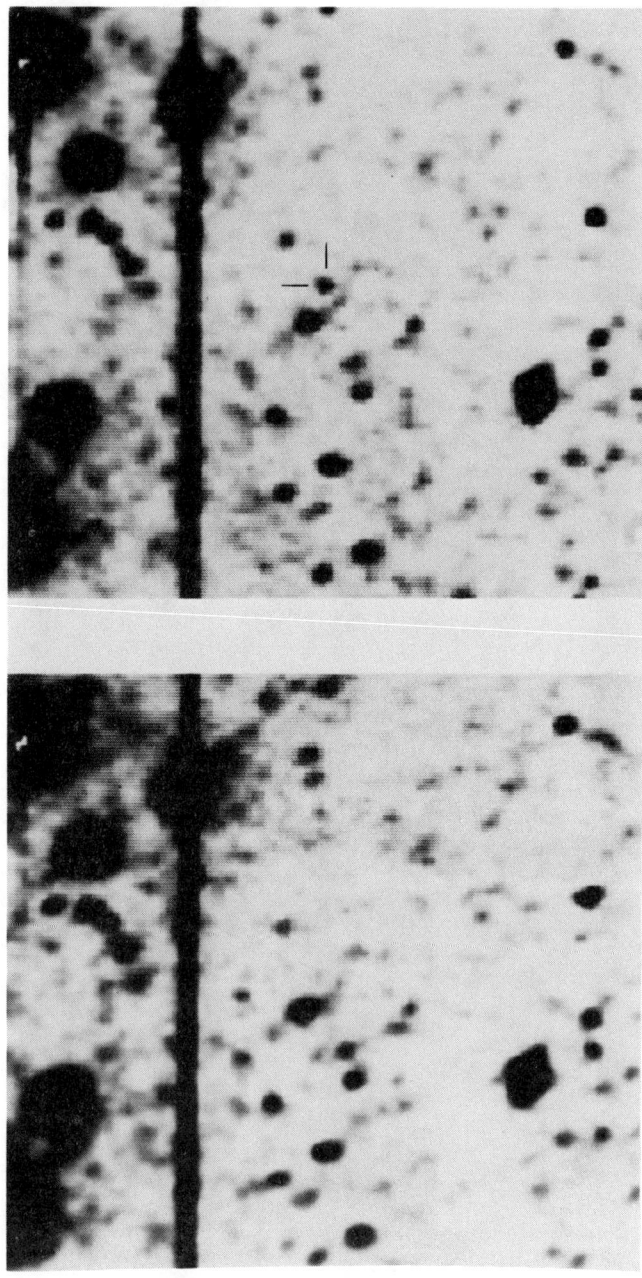

Figure 1. Variable 196 in high (upper panel) and low (lower panel) states. Each frame represents 1 hour of exposure. South is up and east is to the right; the field size is 52" × 52".

TWO STEPPING STONES TO THE HUBBLE CONSTANT

S. van den Bergh
Dominion Astrophysical Observatory
5071 W. Saanich Road
Victoria, B.C. V8X 4M6
Canada

ABSTRACT. Observations of RR Lyrae stars in a halo field in M31 are used to determine the distance to M31. This distance may then be used to calibrate the M(max) versus rate of decline relation for novae. In the spring of 1986 the CFH Telescope will be used to search for novae in Virgo Cluster ellipticals. It is hoped that these observations will result in an improved determination of the Hubble parameter.

1. INTRODUCTION

When Brent Tully and I first discussed his plans for this workshop a year ago I suggested that it might be devoted to the study of the expansion of the Universe. Brent, very wisely I believe, insisted that the time had come to discuss both the global expansion of the Universe and the deviations from the smooth Hubble flow. Whereas this global expansion tells us about the age and the size of the Universe deviations from the Hubble flow contain a great deal of information on both the structure and the content of the Universe. Study of turbulent and messy reality tells us much more about the real world than smooth models ever could!
    I should like to start with a deliberately provocative statement: I believe that the Royal Road to the determination of the Hubble parameter leads via RR Lyrae stars in M31 and novae in the Virgo Cluster and not through the Cepheid swamps. Of course this does not mean that Space Telescope observations of Cepheids in the nearest galaxies might not teach us a great deal about the (presumably messy) kinematics of nearby (D < 10 Mpc) regions of the Universe.

2. WHY NOVAE ARE BEST

There are three reasons why I believe that novae are potentially better calibrators of the extra-Galactic distance scale than Cepheids:
 1. At maximum-light novae are intrinsically more luminous than
      Cepheids.
 2. Novae can be observed in photometrically smooth regions such as

the central bulges of spirals and elliptical galaxies whereas the most luminous Cepheids are concentrated in spiral arms where absorption and crowding problems are severe.
3. The M(max) versus rate of decline relation for novae can be simply calibrated via RR Lyrae stars in M31 (cf. Pritchet in this volume) or using expansion parallaxes (Cohen 1985) in the Galaxy, whereas the calibration of the Cepheid period-luminosity relation still remains bogged down in the endemic problems (Schmidt 1984) associated with main-sequence fitting of clusters containing Cepheids. An additional practical problem is that the presently-known Cepheids in clusters have relatively short periods. Attempts to calibrate the long-period Cepheids - which are the only ones visible in distant galaxies - directly by searching for such objects in young associations (van den Bergh 1985) have so far failed.

3. THE DISTANCE TO M31

From 16 one-hour CCD frames that Chris Pritchet and I have obtained of a field in the halo of M31 we find $\langle B \rangle = 25.50 \pm 0.12$ for 3 RR Lyrae stars. A possible observational bias in this value will be discussed by Pritchet later in this conference. A new and very detailed study of the proper motions of 159 galactic RR Lyrae stars (Hawley et al. 1986) yields $\langle M_B(RR) \rangle = +1.02 \pm 0.14$ so that $(m-M)_B = 24.48 \pm 0.18$ for M31. From the column density of HI (Burstein and Heiles 1984) in the direction of the Andromeda Nebula the Galactic foreground absorption towards M31 is $A_B = 0.32$ mag. The HI map by Cram et al. (1980) shows that the present field lies well beyond the hydrogen gas associated with the disk of M31. We therefore expect no additional absorption along the line of sight that is associated with M31 itself. It therefore follows that the true distance modulus of M31 is $(m-M)_o = 24.48 - 0.32 = 24.16 \pm 0.18$ corresponding to a distance of 679 ± 56 kpc. Our RR Lyrae distance modulus of M31 is in reasonable agreement with the value $(m-M)_o = 24.4 \pm 0.25$ that Mould and Kristian (1985) have obtained by fitting the giant branches of Galactic globular clusters to those of Population II giants in the halo of M31.

4. NOVAE AS DISTANCE INDICATORS

The relation between M(max) and rate of decline of novae (McLaughlin 1945, Schmidt 1957) provides a powerful tool for the calibration of the extra-galactic distance scale. Figure 1 shows a plot of m_{pg}(max) versus log 100 d, in which d is the rate of decline in magnitudes per day, for novae in M31 (Arp 1956, Rosino 1964, 1973). The figure shows a relatively tight correlation between the maximum luminosities of M31 novae and their rates of decline. Nova A4 was both unusually faint and unusually red suggesting that it was dimmed by interstellar absorption within M31. The other objects, which fall about a magnitude above the mean relation between maximum magnitude and rate of decline, may constitute a distinct subclass of novae. Since only two very fast novae have been observed in M31 the shape of the maximum magnitude versus rate of decline relation is not well established at large values of d. It would be particularly important to obtain additional observations of novae in M31, perhaps supplemented by CCD observations of novae in M81, to establish the shape of the M(max) versus rate of decline relation more accurately.

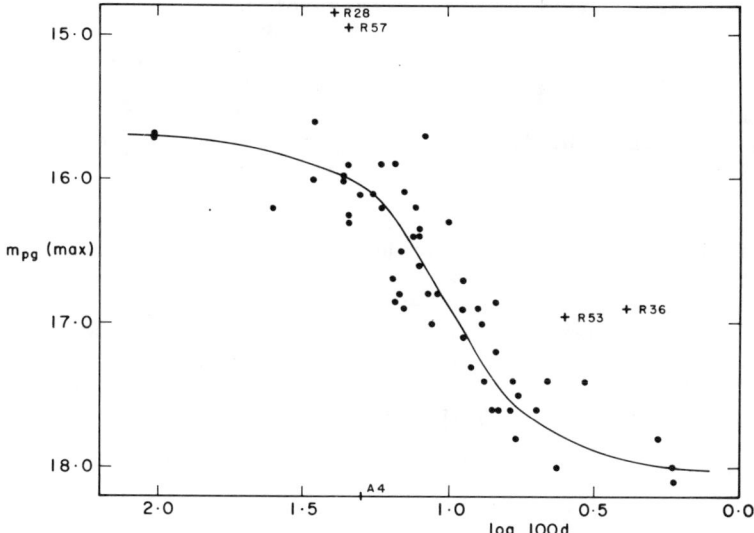

Fig. 1. Period-rate of decline relationship for novae in M31.

Neglecting the 5 deviant objects discussed above the remaining 34 novae have a dispersion of $\sigma = 0.28$ mag about the adopted mean maximum magnitude - rate of decline relation. Because of observational errors the intrinsic dispersion in this relation must be even smaller than this. The fact that T Sco in the relatively metal-poor globular cluster NGC 6093 falls only 0.5 mag (Cohen 1985) above the mean relation between maximum magnitude and rate of decline for Galactic novae shows that the magnitudes of novae at maximum light are not critically dependent on metallicity.

5. THE EXTRA-GALACTIC DISTANCE SCALE

Two years ago Pritchet and van den Bergh (1985) found two novae in the Virgo Cluster elliptical M87. For both of these objects maximum luminosity was missed so that they could not be used to determine the distance of the Virgo Cluster. In a computer simulation, which was based on the frequency of novae per unit luminosity in the bulge of the Andromeda Nebula and the frequency distribution of rates of decline for novae in M31, van den Bergh and Pritchet (1986) calculated that it should be possible to obtain both B(max) and d values for 9 and 14 novae in the Virgo Cluster for assumed distance moduli 31.0 and 32.0 respectively. This calculation assumes that four fields centered on Virgo giant ellipticals are observed on 14 nights in one dark run followed by observations on 4 nights during the next dark run. (This arrangement allows the determination of rates of decline for novae with d > 14 days).

Observations of 10 novae, each of which gives a distance modulus with an error of 0.28 mag, yields an accuracy of $0.28/\sqrt{10} \approx 0.1$ mag in the distance modulus of the Virgo Cluster i.e. a ~5% uncertainty in the distance. To this value should be added the remaining uncertainty in the distance to M31 - which is mainly due to the uncertainty in the absolute magnitude calibration of the RR Lyrae stars. Even including the uncertainty in the infall velocity (250 ± 50 km s^{-1}) of the Local Group into the Virgo Cluster it should be possible to determine the Hubble parameter with an accuracy of $\lesssim 10\%$ from such observations.

Next March and April Chris Pritchet and I have a total of 18 half nights of CFHT time to observe the lightcurves of novae in the Virgo Cluster. If the weather gods are kind to us, and if the demons who cause instrument failures do not rear their ugly heads, we should be able to determine the Hubble parameter before the Space Telescope is launched.

REFERENCES

Arp, H.C., Astron. J., 61, 15 (1956).
Burstein, D., and Heiles, C., Astrophys. J. Suppl., 54, 33 (1984).
Cohen, J.G., Astrophys. J., 292, 90 (1985).
Cram T.R., Roberts, M.S., and Whitehurst, R.N., Astron. Astrophys. Suppl., 40, 215 (1980).
Hawley, S.L., Jefferys, W.H., Barnes, T.G., and Lai, W., preprint (1985).
McLaughlin, D.B., Publ. Astron. Soc. Pac., 57, 69 (1945).
Mould, J., and Kristian, J., preprint (1985).
Pritchet, C., and van den Bergh, S. Astrophys. J. (Letters), 288, L41 (1985).
Pritchet, C.J., and van den Bergh, S., in preparation (1986).
Rosino, L., Ann. Ap., 27, 498 (1964).
Rosino, L., Astron. Astrophys. Suppl., 9, 347 (1973).
Schmidt, E.G., Astrophys. J., 285, 501 (1984).
Schmidt, T., Z. Astrophys., 41, 182 (1957).
van den Bergh, S., "Cepheids: Theory and Observations", Ed. B.F. Madore (Cambridge University Press, 1985) p. 212.
van den Bergh, S., and Pritchet, C.J., Publ. Astron. Soc. Pac., in press (1986).

PHYSICAL MODELS OF SUPERNOVAE AND THE DISTANCE SCALE

J. Craig Wheeler and Robert P. Harkness
Department of Astronomy
University of Texas
Austin, TX 78712
USA

1. INTRODUCTION

There are two basic techniques for employing astronomical phenomena such as supernovae to constrain the cosmic distance scale. One is to use a primarily empirical approach (peak, slope, shape, of light curves) which is intrinsically limited by observational uncertainties. The other is to consider the physical basis, which introduces model dependence. Here we will make the presumption that an understanding of the physics of the explosion is vital to the use of supernovae as distance indicators. In particular, we wish to emphasize that theoretical supernova atmosphere calculations are now available to effect the connection between observations and hydrodynamical models.

In section II we discuss classical Type Ia supernovae (SN Ia) and address the degree to which a carbon deflagration explosion is consistent with the observed spectrum. To the extent that this picture is "correct" it implies a relatively bright event and hence a moderately low value of H_0. The crucial question then concerns uniqueness. In section III we address this question indirectly by reporting recent success in tracing the qualitatively similar, but quantitatively distinct, spectra of Type Ib (SN Ib) to a very different precursor. In section IV we mention briefly the possibility of a third class of hydrogen-deficient supernovae, Type Ic. In section V we summarize the implication of the carbon deflagration model of SN Ia for the distance scale and in section VI present our conclusions and discuss the prognosis for future developments.

II. TYPE Ia SUPERNOVAE AND THE CARBON DEFLAGRATION MODEL

The currently most accepted theory for SN Ia is based on a carbon deflagration explosion. This picture is no better nor worse intrinsically than a variety of others, but it has the special characteristic of starting from viable astrophysical initial conditions and naturally providing an excellent reproduction of the optical spectrum. In particular, we use model W7 of Nomoto, Thielemann, and Yokoi (1984) which assumed a particular prescription for the rate of propagation of the subsonic deflagration, the major parameter of the dynamic model. In this calculation a central portion of the star is converted to nuclear statistical equilibrium such that the dominant component is ^{56}Ni. As the subsonic deflagration dies out, quenched by the expansion, there is a region of incomplete burning which forms elements of intermediate atomic weight, Mg, Si, S, Ca. Finally there is an outer unburned layer of C and O. After the initial hydrodynamic phase, the expansion becomes homologous, $v \propto r$, and velocity can be conveniently used as an independent coordinate. In these terms, the inner ^{56}Ni core falls at v $\lesssim$ 8000 km/s, the partially burned region from 8000 $\lesssim$ v $\lesssim$ 15,000 km/s and the unburned region out to v ~22,000 km/s (limited by the zoning of the dynamical model).

The supernova atmosphere code was developed specifically for this class of problems (Harkness 1985, 1986). LTE is assumed for the Type Ia calculations. With the large abundances of heavy elements, to do otherwise at this time would be too demanding computationally. The agreement between the calculations and the observations gives some ex-post facto justification that the conditions are not far from LTE.

Another parameter is introduced by artificially homogenizing the composition from the outer surface to various depths. This serves primarily to put more of the intermediate mass elements at higher velocity and improves the spectral agreement somewhat (see Branch, Doggett, Nomoto, and Thielemann 1985). The rationalization for this ad hoc procedure is that there may be remnant Rayleigh-Taylor mixing on small or large scales (fingers) that lift some of the deeper material to higher velocities. The basis for this procedure, its degree and necessity must be examined more closely in the future.

Figure 1 gives the spectrum of SN Ia 1981b at maximum light (Branch et al. 1983). This spectrum shows the classical absorption at 6150 Å which demarks this class of Type I supernovae, and the severe ultraviolet deficit which must be successfully reproduced before complete confidence can be had in our understanding of these events. The task facing us is to account self-consistently for the observed ultraviolet deficit as well as the optical spectrum. All relevant continuum processes have been incorporated, including bound-free opacities from excited states (the atomic data is sparse for Fe and

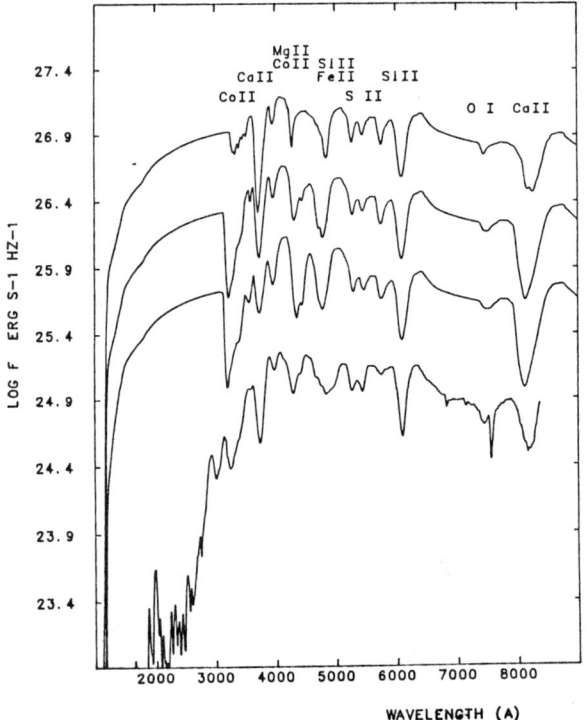

Figure 1 - The observed spectrum of the Type Ia supernova SN 1981b at maximum light (bottom) illustrating the distinctive Si II absorption at 6150 Å and the strong ultraviolet deficit is shown in comparison with three theoretical calculations which indicate the effect of mixing the contents of the outer portions of the model down to 12,000, 10,000, and 8000 km/s (presented from the top down, respectively).

Co so characteristic cross sections have been assumed) and Rayleigh scattering. These continuum processes are insufficient to account for the ultraviolet deficit. We will discuss below various atmosphere models and our preliminary attempts to account for the ultraviolet deficit by resonance line scattering.

Figure 1 also shows a series of theoretical spectra calculated for model W7 of Nomoto, Thielemann and Yokoi (1984) 17 days after explosion. Heating by ^{56}Ni and ^{56}Co decay has been computed during the expansion assuming that all the decay energy is deposited locally to the radioactive matter. This approximation is quite reasonable for epochs prior to maximum light and can be relaxed to produce a somewhat more realistic model. The models given in Figure 1 explore the effect of homogenizing the abundances from v_{max} = 22,000 km/s down to velocities v_{mix} of 17,000, 10,000, and 12,000, km/s. These models use a relatively short line list (30 lines) which includes lines of Co for λ > 3000 Å but only two lines of Fe, 4923 and 5018 Å. Only the continuum is calculated for λ < 3000 Å. Mixing as deeply as 8000 km/s produces far too strong a Co II blend at 3300 Å and the resulting flux redistribution nearly fills in the Ca H and K lines, in contradiction to observation. Mixing down only to 12,000 km/s yields insufficient Co to give the 3300 Å feature, tends to produce too shallow a line at the Ca II infrared triplet at 8300 Å and in general yields lines that are too narrow. Mixing to about 10,000 km/s produces a reasonable compromise and we find the best fit to the eye in a model with mixing to ~ 11,000 km/s.

Figure 2 shows two more models again in comparison with the observed maximum light spectrum of SN 1981b. The models have the abundance homogenized

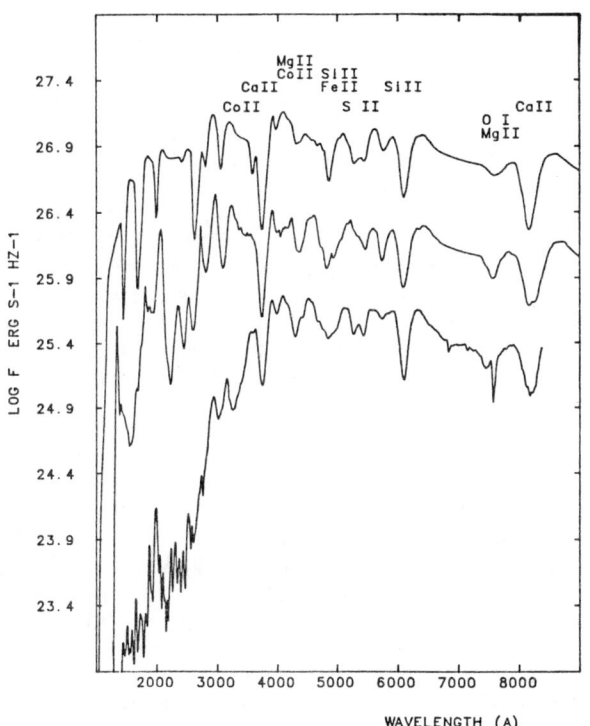

Figure 2 - Two theoretical spectra (upper, middle) are shown in comparison with the maximum light spectrum of SN 1981b (bottom). The top spectrum does not contain Fe or Co lines and is primarily an attempt to reproduce the optical spectrum. The middle spectrum contains an extensive set of lines in the ultraviolet including Fe and Co in an attempt to simulate the ultraviolet deficit. Note the feature at 2000 Å which results from a natural absence of strong scattering lines.

down to $v_{mix} = 10,000$ km/s and incorporate a more extensive line list (~130 lines). The top spectrum has no Fe or Co lines for technical reasons and the feature at 3300 Å is not reproduced. The fit is quite good from 3500 Å to 8500 Å. The "absorption" feature near 4900 Å is too narrow and the dip at 7600 Å is at too long a wavelength in the theoretical model. The few lines at $\lambda < 3000$ Å are clearly insufficient to account for the ultraviolet deficit.

The other theoretical spectrum in Figure 2 (middle) includes lines of Fe and Co, for $\lambda > 1500$ Å. The Fe lines have broadened the feature at 4900 Å to close agreement with the observed spectrum, but at the expense of filling in part of the S II feature at 5300 Å. Higher excitation lines of Si at 3100, 4900 and 5800 Å are too strong relative to the feature at 6150 Å suggesting the temperature may be too high. A more realistic deposition function may help this problem. The line at 3100 Å, in particular, moves flux to fill in the Co line which in other models appears at 3300 Å (see figure 1). Note also the small nip taken out of the peak of the Si II emission feature at 6300 Å. This is caused by C II and is not in the observed spectrum. This allows us in principle to put limits on the carbon abundance but these limits will be a function of the temperature, etc., of the model.

This calculation results in some, but not sufficient, ultraviolet deficit as it stands. There are nevertheless some similarities in the features between the model and the observations. Note especially the peak around 2000 Å. In the model this result is not from emission but from a natural absence of scattering lines in that wavelength range. The magnitude of the ultraviolet flux may be severely affected by the zoning in this model. Many of the ultraviolet lines are very optically thick already in the outer zone, and the failure to properly resolve their contribution may cause an overestimate of the flux. Work is underway to properly resolve the ultraviolet lines.

Although the fastest moving zones of model W7 have an expansion velocity of ~22,000 km s^{-1}, there will be a small amount of mass with considerably higher velocities, possibly becoming mildly relativistic at mass fractions of ~10^{-6} M$_\odot$ (Colgate, private communication). In this case, the strong ultraviolet resonance lines of neutral and singly ionized atoms may have enormous Doppler shifts (widths), and a relatively few strong lines could account for the ultraviolet deficit. Some preliminary calculations have been made to explore this possibility. A small amount of mass, 10^{-4} M$_\odot$, composed of intermediate mass elements has been added to the model with density profiles of $\rho \propto r^{-10}$ and $\rho \propto r^{-7}$ and $v_{max} \sim 0.3$ c. For the $\rho \propto r^{-10}$ case the Mg II $\lambda 2797$ line removes a very significant amount of flux from 2000 to 2700 Å. With a profile $\rho \propto r^{-7}$, the red wing of this line causes interference with some required optical features (e.g., Ca H and K) and hence is unacceptable. These calculations were performed omitting spectral lines of Fe and Co, but they could be significant and will be included in future calculations. The lines of neutral atoms would be highly effective in this scheme, but it appears that the abundance of neutrals is always too low at maximum light.

III. TYPE Ib SUPERNOVAE

Type Ib supernovae were first noted by Bertola (Bertola 1964; Bertola, Mammano, and Perinotto 1965) to be distinct from SN Ia by dint of their lack of the 6150 Å Si II absorption feature. There has been a recent flurry of interest in this class thanks to two bright events, SN 1983n and SN 1984l (Wheeler and Levreault 1985; Panagia et al. 1986; Uomoto and Kirsher 1985; Elias et al. 1985; Graham et al. 1986; Branch 1986). Most of the known SN Ib are associated with H II regions or other Population I environments. Wheeler and Levreault argued that the luminosity is less for SN Ib than SN Ia by about 1.5^m (unless there are still extinction problems) making them about as bright as the average SN II. If this is the case, and nickel decay is the process responsible for the luminosity of both classes, then SN Ib must eject about four times less ^{56}Ni than SN Ia. The light curve shapes are similar near peak light although SN Ib decay more slowly at later times (Panagia et al. 1986). Simple arguments based on light diffusion near maximum light suggest that if the rise time, photospheric velocities and opacities are similar for SN Ia and SN Ib, then SN Ib eject approximately the same mass, and kinetic energy. Wheeler and Levreault argued that if the kinetic energy is the same and the thermonuclear input as measured by the amount of ^{56}Ni ejected is substantially less, the mechanism of explosion must not be purely thermonuclear, as hypothesized for SN Ia, but something different, e.g. core collapse. As we shall see, there are now conflicting arguments concerning the mass ejected in SN Ib; the above argument may be in error, for instance by ascribing similar opacities to SN Ia and b. The energetic argument may thus not follow directly, even though the presumption of core collapse may prove true.

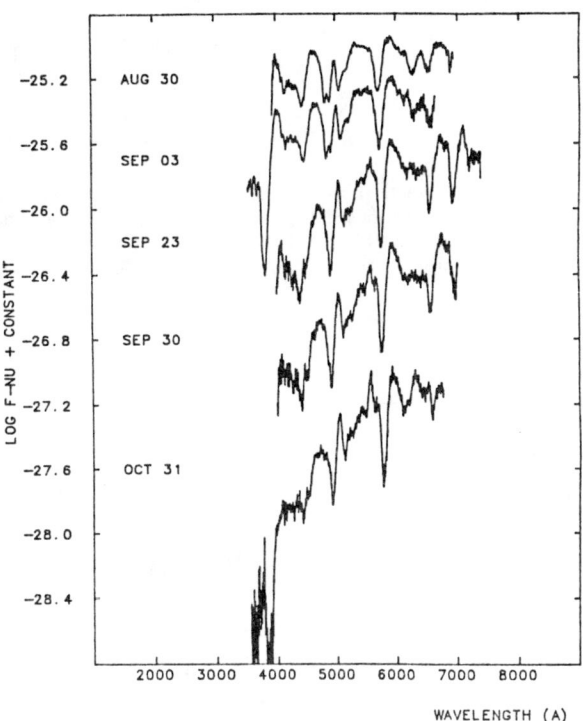

Figure 3 - Spectra of the Type Ib supernova SN 1984l are shown from near maximum light until about two months later. Note the homogeneity of the spectra and the distinct absorption at about 5800 Å.

Figure 3 shows a time series of spectra of SN 1984l, some obtained by B. Margon and R. Downes with the KPNO 2.1 m and others with the 2.7 m at McDonald Observatory (Harkness et al. 1986). These spectra cover the period from near maximum light until about two months later. Note that the spectra change very little over the epoch shown here. Figure 4 shows the spectra of SN 1981b (SN Ia) and SN 1984l (SN Ib) at maximum light to contrast the differences of the spectra. SN 1984l does not have the 6150 Å feature that typifies SN Ia. It does, however, have a distinct feature with a minimum at 5800 Å. Figure 4 also shows the Margon and Downes spectrum obtained on 23 September 1984, approximately three weeks after maximum light. The minimum at

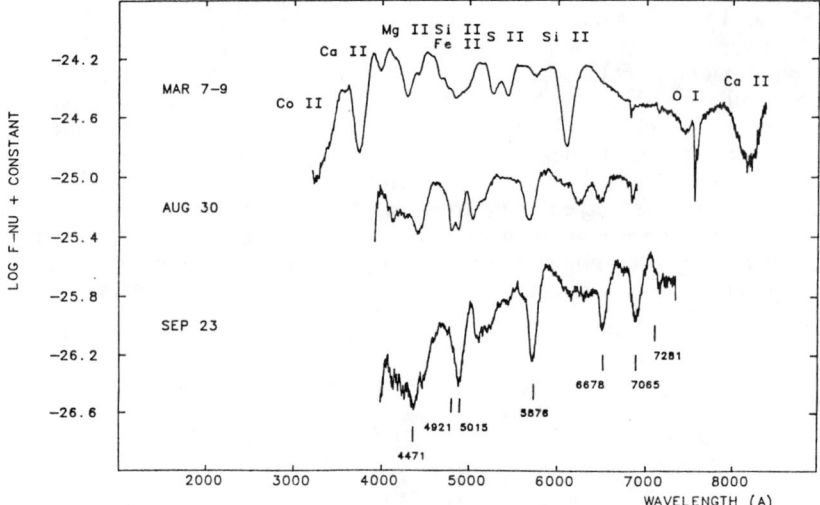

Figure 4 - The optical spectrum of the Type Ia supernova SN 1981b near maximum light (top) is compared with that of the Type Ib supernova SN 1984l also near maximum light (middle). The bottom spectrum of SN 1984l about 3 weeks after maximum shows particularly distinct absorption features. Also shown for comparison are the expected locations of He I lines blue shifted by 7500 km/s.

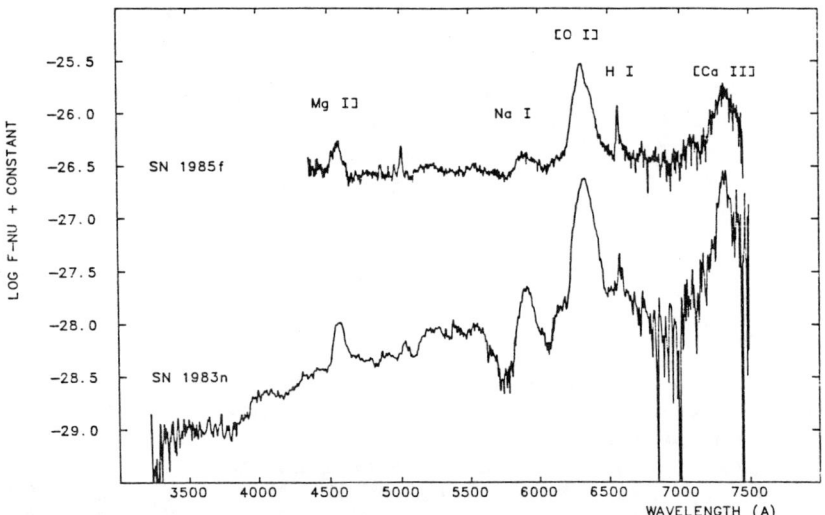

Figure 5 - The spectra of the Type Ib supernova SN 1983n obtained approximately 8 months after optical maximum (bottom) shows distinct emission lines of O, Na, Mg, and Ca, in distinct contrast with the spectrum for the first two months (see figures 3 and 4). The very similar spectrum of SN 1985f (top) suggests that event was also a member of the class SN Ib.

6300 Å has disappeared, but all other minima have grown more distinct. Also shown is a series of fiducial lines marking the expected blue shifted minima of lines of He I moving at 7500 km/s. The agreement between the location and spacing of these predicted He features, particularly the 5876, 6678, and 7065 Å lines, with the observed spectrum is suggestive evidence for the first identification of He in a supernova.

Figure 5 shows a spectrum of the SN Ib SN 1983n approximately 8 months after maximum light (Gaskell et al. 1986). The spectrum at this stage is very different from that given at earlier phases in Figure 3 for an event, SN 1984l, that had quite similar spectra to SN 1983n at that phase. The late-time spectra are also very different from those of SN Ia and SN II at similar epochs. The spectrum is dominated by emission lines, particularly [O I] $\lambda\lambda$6300, 6364. Figure 5 also shows a spectrum of SN 1985f (Gaskell et al.). Whereas this event was unprecedented at discovery (Filippenko and Sargent 1985), the similarity of the two spectra in Figure 5 strongly argues that SN 1985f is also a member of the class SN Ib discovered ~250 days after maximum light (Gaskell et al. 1986). Begelman and Sarazin (1986) analyze the emission lines of SN 1985f and conclude that $\gtrsim 5$ $M_\odot$ has been ejected, but that amount of mass may confict with the constraint of the rather narrow width of the peak of the light curve.

The helium rich ejecta evident in the spectra for the first two months and the oxygen-rich spectra at eight months strongly suggest that SN Ib supernovae have their origin in a star very reminiscent of a Wolf-Rayet star. Thus, although the spectra of SN Ia and SN Ib display some qualitative similarity, a closer examination leads to a very different physical model. We regard this as relevant to the question of uniqueness of supernova models in matching the detailed observations, though precisely how is unclear.

IV. TYPE Ic SUPERNOVAE

SN 1983v in NGC 1365 was observed by R. Cannon on 30 November at the AAT and by E. Barker and A. Cochran on 8 December 1983 at McDonald Observatory. The spectrum does not display the 6150 Å absorption characteristic of SN Ia. Branch et al. (1986) argue that the spectrum of SN 1983v resembles the spectra of the SN Ib supernovae SN 1983n and SN 1984l except that it is Doppler blueshifted by an extra ~10,000 km/s. We note that the spectra of SN 1983v also do not display a feature corresponding to the strong 5800 Å absorption characteristic of the spectra of SN Ib. We suggest that SN 1983v represents yet a third distinct class of hydrogen deficient supernova, Type Ic, and speculate that the spectrum consists essentially entirely of Fe moving at ~20,000 km/s. We are modeling such a configuration and if this hypothesis proves reasonable it may be evidence for a supersonic detonation event. SN 1983v was at about 13th magnitude at discovery. If this was maximum light it was comparable to an average SN Ib or SN II and dimmer than a SN Ia making it difficult to reconcile with the complete detonation of a white dwarf. Any photometric record of this supernova would be of great interest. Once again SN 1983v will provide a test of our ability to begin with a spectrum and determine a unique model which can account for the observed properties.

V. SN Ia AND H$_0$

If one accepts the ability of the carbon detonation model to reproduce the optical spectrum at maximum light as evidence of the veracity of the model, then one must also accept the constraints the model places on the luminosity of SN Ia and hence on the distance scale.

The model predicts that the amount of ^{56}Ni produced and hence the luminosity due to radioactive decay is directly related to the strength of the thermonuclear explosion. Those quantities are a function of the rate of propagation of the deflagration in the model, the physics of which is ill-determined. Fortunately for the present argument, we can regard the amount of ^{56}Ni produced as the parameter to be contrained and do not need to explicitly address the details of how it is manufactured.

Sutherland and Wheeler (1984) argued that if the explosion is too weak the kinematics of the expansion as reflected in the photospheric velocity and the width of the light curve will be too meagre. They argued that $\gtrsim 0.8$ M$_\odot$ of carbon and oxygen must be consumed in the explosion to reproduce the kinematics deduced for SN Ia. Arnett, Branch and Wheeler (1985) argued that $\gtrsim 1/2$ of the burned fuel would be converted to ^{56}Ni and hence that $\gtrsim 0.4$ M$_\odot$ of ^{56}Ni must be ejected in the carbon deflagration model of SN Ia to be consistent with the observations. This amount of ^{56}Ni produces a blue magnitude at maximum light of $M_B = -19.1$. Adopting the calibration $M_B = -18.4 + 5 \log H_0/(100 \text{ km/s / Mpc})$ then gives $H_0 < 71$ km/s/Mpc. A model which gives a good representation of the photospheric velocity at maximum light and of the light curve is one with ~ 0.6 M$_\odot$ of ^{56}Ni which gives $M_B = -19.6$ and $H_0 = 58$ km/s/Mpc.

Arnett, Branch and Wheeler (1985) give an upper limit to the amount of ^{56}Ni of 1.4 M$_\odot$, as if an entire white dwarf were converted to ^{56}Ni. This amount of ^{56}Ni would correspond to $M_B = -20.4$ and $H_0 = 38$ km/s/Mpc. This lower limit to H_0 is not, in fact, realistic in the face of the observational constraints. The observation of intermediate mass elements implies that the entire star can not be converted to nuclear statistical equilibrium and ^{56}Ni. Detailed models need to be constructed to get a realistic lower limit to the absolute abundances of the intermediate mass elements. A crude guess is that at least 0.2 M$_\odot$ of intermediate mass elements must be ejected implying that the upper limit on the amount of ^{56}Ni ejected in the context of the carbon deflagration model is 1.2 M$_\odot$ and hence $H_0 > 41$ km/s/Mpc. More detailed calculations may well show that the amount of intermediate mass elements must be $\gtrsim 0.4$ M$_\odot$ in which case the maximum amount of ^{56}Ni is ~ 1.0 M$_\odot$ corresponding to $H_0 \sim 45$ km/s/Mpc. We note that a Baade-Wesselink analysis of SN 1981b by Branch et al. (1983) gave $M_B = -20.7 \pm 0.6$, which exceeds the realistic limit expected from the carbon deflagration model. Clearly the constraints imposed by more detailed models will be very useful in setting a more rigorous lower limit on H_0 in the context of the carbon deflagration model.

VI. CONCLUSIONS

The ability to calculate self-consistent supernova atmospheres provides a crucial new tool to relate dynamical models and observations of supernovae.

The success of the carbon deflagration explosion in matching the optical spectra at maximum light suggests that it should be taken seriously as a basis for interpreting SN Ia. If so, the model sets an upper limit of 71 km/s/Mpc on H_0 and an absolute lower limit of 38 km/s/Mpc. A more realistic lower limit is in the range 41-45 km/s/Mpc allowing for the fact that an appreciable amount of intermediate mass elements must be ejected. The calculations which best fit the light curves and spectra give $H_0 \sim 60$ km/s/Mpc.

The link between the spectral fit of the carbon deflagration model and constraints on H_0 is still rather indirect. The model would be on firmer ground if there were a satisfactory explanation for the ultraviolet deficit. Even then, one would prefer direct evidence for radioactive decay. The fact that radioactive decay calculations also fit the late time spectrum and that there is evidence for a decaying feature of Co is supportive (Woosley, Axelrod, and Weaver 1984), but does not necessarily single out the carbon deflagration model and place absolute constraints on the amount of ^{56}Ni and hence the luminosity. Confirmation of the Co line at 3300 Å would give welcome support to the whole picture, a goal that will be greatly aided by the Hubble Space Telescope. Even better would be a direct detection of the gamma rays expected from the radioactive decay of nickel and cobalt, for which the Gamma Ray Observatory may provide some constraint.

Is it possible that the model is wrong, despite its remarkable ability to reproduce the optical spectrum? Could there be alternative structures with a similar composition of intermediate mass elements, but $\lesssim 0.2$ $M_\odot$ of ^{56}Ni ejected? Is it possible that some very different strongly non-LTE configuration could reproduce the present LTE spectrum? It is very difficult to prove the uniqueness of an astrophysical model but clearly it would be worthwhile to expend some effort in constructing ad hoc alternative models to explore the sensitivity of the agreement to the assumed model.

The development of a more detailed understanding of supernovae of Type Ib and perhaps Type Ic says something of our ability to trace a spectrum to a physical model, and hence about uniqueness. As the physical origin of these events becomes clarified there may be useful feedback on the relation of SN Ia to the distance scale. For instance, models of SN Ib are crudely consistent with the ejection of ~0.2 $M_\odot$ of ^{56}Ni if their absolute magnitude is ~ -18.5, which is consistant with H_0 ~50 km/s/Mpc. If H_0 ~100 km/s/Mpc then the amount of ^{56}Ni would have to be $\lesssim .05$ $M_\odot$, which would have an impact on the model.

We are grateful for useful conversations with Mark Aaronson, David Branch, Stirling Colgate, Gerard de Vaucouleurs, Willy Fowler, Alan Sandage and Gustav Tammann which have helped us to put this work in context. This research was supported in part by NSF Grant AST 8413301.

REFERENCES

Arnett, W. D., Branch, D., and Wheeler, J. C. 1985, *Nature*, **314**, 337.
Begelman, M. C. and Sarazin, C. L. 1986, *Ap. J. (Letters)*, in press.
Bertola, F. 1964, *Annals d' Astrophysique*, **27**, 319.
Bertola, F., Mammano, A., and Perinotto, M. 1965, *Contributions of the Asiago Observatory*, #174.
Branch, D. 1986, *Ap. J. (Letters)*, in press.
Branch, D., Cannon, R., Axon, D., McDowell, J., and Godwin, P. 1986, in preparation.
Branch, D., Doggett, J. B., Nomoto, K. and Thielemann, F.K. 1985, *Ap. J.*, **294**, 619.
Branch, D., Lacy, C. H., McCall, M. L., Sutherland, P. G., Uomoto, A., Wheeler, J. C., and Wills, B. J. 1983, *Ap. J.*, **270**, 123.
Elias, J. H., Matthews, K., Neugebauer, G. and Persson, S. E. 1985, *Ap. J.*, **296**, 379.
Filippenko, A. V. and Sargent, W. L. W. 1985, *Nature*, **316**, 407.
Gaskell, C. M., Cappellaro, E., Dinerstein, H., Garnett, D., Harkness, R., and Wheeler, J. C. 1986, *Ap. J. (Letters)*, submitted.
Graham, J. R., Meikle, W. P. S., Allan, D. A., Longmore, A. J., and Williams, P. M. 1986, *M.N.R.A.S.*, in press.
Harkness, R. P. 1985, in *Supernovae as Distance Indicators*, ed. N. Bartel (Berlin: Springer-Verlag), p. 183.
Harkness, R. P. 1986, in *Radiation Hydrodynamics*, ed. D. Mihalas (Dordrecht: Reidel), in press.
Harkness, R. P., Wheeler, J. C., Margon, B., Downes, R., Kirshner, R. P., Uomoto, A., Dinerstein, H., and Garnett, D. 1986, in preparation.
Nomoto, K., Thielemann, F.-K., and Yokoi, K. 1984, *Ap. J.*, **286**, 644.
Panagia, N., et al. 1986, preprint.
Sutherland, P. G. and Wheeler, J. C. 1984, *Ap. J.*, **280**, 282.
Uomoto, A. and Kirshner, R. P. 1985, *Astron. Ap.*, **149**, L7.
Wheeler, J. C. and Levreault, R. 1985, *Ap. J. (Letters)*, **294**, L17.
Wooseley, S. E., Axelrod, T. S., and Weaver, T. A. 1984, in *Stellar Nucleosynthesis*, eds. C. Chiosi and A. Renzini (Dordrecht: Reidel), p. 263.

CALIBRATION OF THE INFRARED TULLY-FISHER RELATION AND THE GLOBAL VALUE OF THE HUBBLE CONSTANT

M. Aaronson
Steward Observatory
University of Arizona
Tucson, Arizona 85721
U.S.A.

ABSTRACT. Among existing secondary distance indicators, the IR Tully-Fisher relation is probably the most reliable available for measuring the far flung Hubble flow. Absolute calibration of the method, however, remains highly uncertain owing to the poor state of nearby galaxy distances. A coordinated ground-based and space attack on this problem, primarily involving Cepheids, holds out great promise for a measurement of the Hubble constant good to ten percent.

1. INTRODUCTION: WHY TULLY-FISHER?

The classical approach of measuring the expansion rate via the study of resolvable stellar content and subsequent calibration of distance indicators based on global galaxian properties remains today still the most viable one at our disposal. The well-known failure of this avenue to lead to no better than a factor of two in the value for H_0 is in the mind of this author more a reflection of the inadequacy of available tools, rather than an indictment of the classical philosophy itself. With the development of new secondary indicators (particularly the Tully-Fisher method discussed in this article), the now widespread use of CCD's (and concomitant software algorithms for performing accurate crowded field photometry), and the imminent launch of the Hubble Space Telescope (HST), it does not seem too optimistic to predict that a believable value for H_0 -- i.e., one good to 10%, say -- will soon be within our grasp.

A specific path for achieving this happy goal must still be set forth. It is argued here that the primary indicator of choice is Cepheids, while the premier secondary indicator is the infrared Tully-Fisher relation (or IRTF). I believe it is fair to say that there is general, though by no means universal, consensus in this regard. While one can and should build as many checks and balances as is feasible into the construction of any distance scale edifice, the weight of present evidence seems to make a compelling case for the emphasis to be placed on the Cepheid-IRTF route.

With regard to other primary indicators, the only viable alternatives are RR Lyraes and novae. The former are simply too faint, even from space, for any effective secondary calibration to be made. The serendipitous nature of the latter makes an attack from space impractical. Novae, however, are bright, and their detection from the ground has been claimed as far away as M87 (Pritchet and van den Bergh 1985). These authors are undertaking a full-scale ground-based assult on the Virgo cluster with novae using the superior seeing capabilities of the CFHT. Even so, the observations are extremely difficult, and one worries about the differences, if any, between novae in early-type galaxies and the calibrating novae in M31 and the Milky Way, but the results will be eagerly awaited.

There appear to be a somewhat larger number of secondary indicators to choose from. As one possible means of ranking these, Aaronson and Mould (1986) list five criteria: a physical basis, measurables that are quantitatively and not subjectively determined, observables requring a minimal of corrections, applicably over a wide range of distance, and small scatter. Traditional methods such as H I region diameters and luminosity classes do not, to say the least, stack up very well under such scrutiny. The IRTF method, though, meets all of the desired qualities.

The leading alternatives to IRTF are supergiants and type I supernovae as standard candles. The principal problem with both is that the true intrinsic scatter remains unclear (as well as perhaps the physical basis involved). For the supergiants, the number of Cepheid calibrating distances is simply too small and unreliable. As a consequence, the form of the red supergiant luminosity function is hotly contested (cf. Humphreys 1983; Sandage 1983). This question bears directly on whether a true physical cut-off exists in the brightest red supergiants, or whether one is seeing a mere statistical effect, as with the blue ones. Using supergiants also presents some rather tricky practical problems, as they must somehow be distinguished from foreground objects. Variability is one means to do so, but this in itself means that a well-sampled light curve should be obtained to properly define a fiducial magnitude (and also, if desired, to eliminate "violent variables" as in Sandage 1984). Although a large number of type I supernovae are known, present-day tests of their reliability as standard candles rest upon a small subset with "well-observed" light curves confined to early-type galaxies, in order to circumvent problems with internal reddening and with a new class of peculiar type I objects that has been identified as occurring in just spirals, at least so far (e.g., Uomoto and Kirshner 1985). On the other hand, the absolute calibration of type I supernovae is most readily accomplished with later-type systems; clearly, considerable caution must be exercised in this whole area.

In contrast, observations of hundreds of galaxies in the Local Supercluster and especially in a dozen galaxy clusters, where distance effects are minimized (e.g., Aaronson and Mould 1983; Aaronson et al. 1986) have convincingly established an observed scatter in the IRTF method of $\sigma \sim 0.4 - 0.5$ mag. We are very far from any kind of similar rigorous testing of supergiants or supernovae. While the IRTF method is not perfect (no distance indicator is!), the only substantial doubts that have been raised relate to possible dependence of the relation on morphological type or location. The significance of type effects in the Tully-Fisher diagram remains controversial, but it does seem clear that if the types are confined to the range Sab-Sdm, any segregation in the infrared is minimized (see Figure 2 of Aaronson and Mould 1986).

Similarly, the available evidence suggests that environmental factors have little influence on the IRTF relation. This point is perhaps best illustrated with Figure 1, which shows the velocity-distance relation for 11 nearby galaxy clusters from Aaronson et al. (1986) derived using IRTF moduli. (As discussed by these authors, the velocity of the nearest cluster, Virgo, has been adjusted for an infall motion of $\sim$ 300 km s^{-1}, while the other velocities have been corrected for anisotropy of the 3 K background. Aaronson et al. 1986 have shown that the latter can be accounted for by a combination

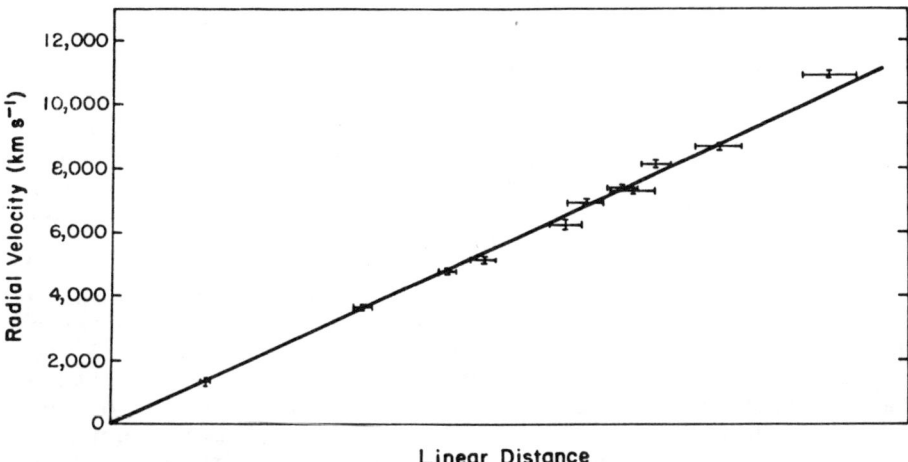

Figure 1. The velocity-distance relation for eleven galaxy clusters, using moduli determined from the IR/H I relation. The velocity of Virgo (the lowest redshift cluster) has been corrected for infall, and the remaining velocities have been corrected for the dipole anisotropy of the microwave background.

of infall and a comparably-sized bulk Supercluster motion). The clusters in Figure 1 range from very spiral poor objects like Coma, to very spiral rich ones such as Pisces. The formal scatter in Hubble ratio for the 11 cluster sample is only $\pm$ 1 km s^{-1} Mpc^{-1}, leaving little room for any environmental effects of note. There is also little evidence of Malmquist bias in Figure 1; cluster work has in fact been emphasized with the IRTF in order to avoid this thorny problem.

There have been claims in the literature of differences between the optical and infrared results obtained with the Tully-Fisher method, or of weaknesses in the technique due to the fact that different authors can apply it and come up with very different values for H_0. However, as discussed by Aaronson et al. (1986), these claims are in all cases merely a reflection of the inappropriate use of face-on galaxies and/or the adoption of differing distances to the calibrating systems. For example, by adopting de Vaucouleurs' nearby distance scale, the results in Figure 1 lead to $H_0 \sim 100$ km s^{-1} Mpc^{-1}, while the older Sandage-Tammann moduli produce $H_0 \sim 70$ km s^{-1} Mpc^{-1}. This value can be pushed lower still if the larger moduli to M33 and M81 advocated by Sandage and Tammann (1984) are employed. On the other hand, the most recent CCD and IR observations suggest an expansion rate value near the upper end of the quoted range (see Aaronson et al. 1986). The author believes that this conflict can be resolved by a concentrated attack using both ground-based and space means. In the next section, we discuss the propects for obtaining a definitive zero-point for IRTF via Cepheids, while in the section to follow, we consider the calibration problem presented by the Cepheids themselves.

2. CALIBRATING IR TULLY-FISHER

The one-sigma scatter in the IRTF suggests that of order 20 galaxies are needed to define a zero point good to ~ 0.1 mag. Ideally, one would like to sample evenly over morphological type (or velocity width, which is roughly the same thing); inclination constraints must also be met. About 40 potential calibrators within the reach of HST have been listed by Aaronson and Mould (1986) by selecting a sample of objects with existing IR and H I data and having types in the range Sab-Sdm, inclinations from 50 - 80°, corrected velocity widths between 200 - 600 km s^{-1}, and either galactocentric redshifts < 800 km s^{-1} or flow-model adjusted relative Virgo distances $d/d_V < 0.8$. The sample could be increased considerably by even a small relaxation of the distance constraints, but the point is clear -- many suitable calibrators are out there. In fact, the orbital characteristics of HST itself may be the most limiting factor involved.

Cepheids observations in a number of these systems can be pursued from the ground, most notably in the Local, Sculptor, and M81 groups, and several teams are at present engaged in such work with modern digital techniques. The author and W. Freedman, J. Graham, B. Madore, and J. Mould have recently begun a CCD search for Cepheids at CTIO and Las Campanas in the Sculptor objects NGC 247, 253, and 7793 -- all quite suitable IRTF calibrators. BVRI observations are being obtained, with 10 min exposures on the CTIO 4-m telescope of sufficient length to reach an adequate magnitude depth. A photograph of the two search areas in NGC 247 is illustrated in Figure 2, along with a CCD exposure at V for one of these fields. The ease with which the galaxy resolves should be noted. NGC 7793 is also resolved (though perhaps not quite as well as NGC 247) and we anticipate that a few more years of work will lead to good Cepheid moduli to these objects. The resolution of

NGC 253, though not negligible, is rather poorer, and it will be of interest to see if we can pull some Cepheids out of the muck.

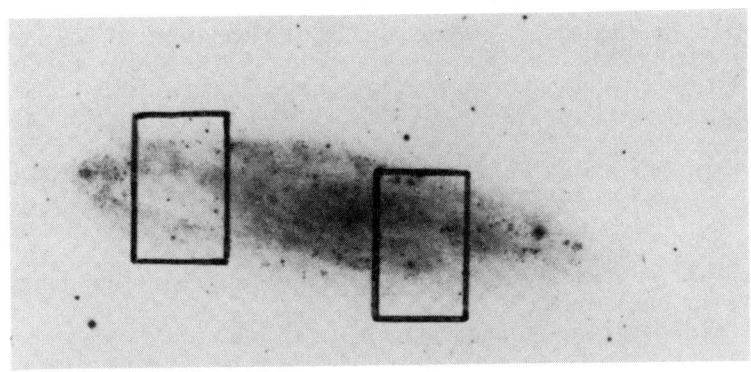

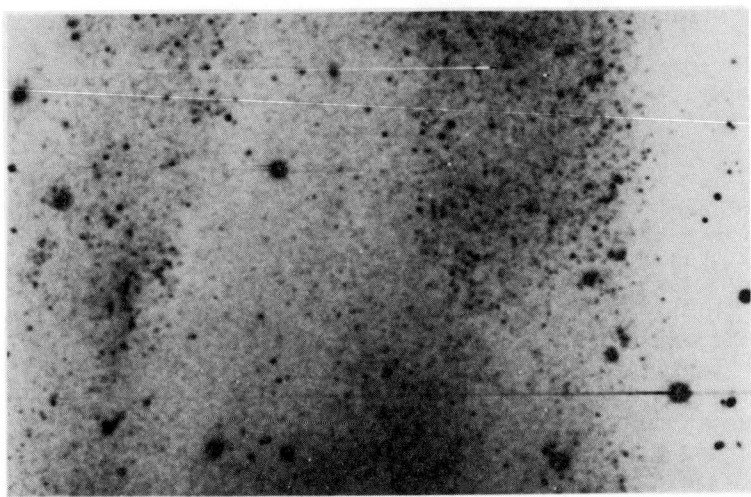

Figure 2. The upper panel shows the two areas in NGC 247 being surveyed for Cepheids. The lower panel is a ten minute CCD exposure at V taken in moderate (1.3") seeing. Note the good resolution into stars.

A dramatic illustration of the large ground-based potential provided by CCD's comes from the work of Cook, Aaronson, and Illingworth (1986), who report the first definite detection of Cepheids in M101. Their survey reaches ~ 2 mag deeper than earlier photographic studies, all of which failed to find Cepheids. We note in passing that because of its face-on nature, the modern-day importance of M101 lies not with IRTF, but in the proper calibration of brightest red supergiant luminosity on parent galaxy absolute magnitude. The M101 group does contain three edge-on systems which have been used as IRTF calibrators, but these appear to show considerable scatter. This might be a reflection of increased dispersion in the relation at low velocity widths; only NGC 5585 in fact has a corrected width above our nominal cut-off of 200 km s^{-1}. On the other hand, depth effects may be relevant, as they are now recognized to be in both the Sculptor and M81 groups.

It is also important to stress the need for multi-color Cepheid observations in order to contain possible internal absorption effects. In particular, Freedman (1985) has used BVRI data for a small sample of M33 Cepheids to demonstrate a typical internal absorption of $A_V \sim 0.5$ mag. This is obviously a major effect which can no longer be neglected, as in past work.

We can expect ground-based efforts to at best yield good moduli to only about one-third of the number of desired IRTF calibrators. For the remainder, we must call upon the resources of HST. Aside from the obvious resolution gain, HST also offers a considerable advantage in search efficiency, for the spacing of epochs can be much more ideally chosen than from on the ground. For instance, over the cycle that may be available for viewing a galaxy in roughly the same orientation, twelve suitably placed epochs ought to yield very good phase diagrams and mean magnitudes for variables in the 10 - 50 day period span.

It is of interest to consider whether a value for H_0 might be obtained directly from the Cepheids, without recourse to IRTF and the less than ideal constraints this puts on the kind of spiral to observe. For instance, Cepheids could be observed in several Virgo cluster systems. It remains to be seen, however, if Virgo is within reach of HST. Furthermore, depth effects could be substantial, and to check this one really needs to (unwisely) monopolize the effort on Virgo by trying to observe a large number of spirals. Finally, it should be remembered that the uncertainty in the Virgo redshift introduces a non-negligible error into the problem. For these reasons, I do not believe Virgo alone is the way to go.

Another alternative, discussed by Kraan-Korteweg (1986), is to observe galaxies in so-called "null" regions where the effects of Virgocentric infall are canceled out, so that the observed redshift can be used directly to calculate the expansion rate, but there are dangers with this approach as well. For example, the location of the null regions are dependent on the assumption of a spherically symmetric flow model, which is known to be a poor description of the Local Supercluster. Also, streaming notions may play an important role in the velocity field, so that for instance several galaxies at a redshift of 800 km s^{-1} having a peculiar velocity of 200 km s^{-1} would lead to a value for H_0 in error by 25%. For similar reasons, the complementary proposal of Kraan-Kortewig's involving measure of the infall velocity by observing systems in critical "infall" regions is also open to question. The present author believes the infall problem is one that can only be sensibly tested from the ground, with a sample of at least several hundred galaxies.

3. CALIBRATING THE CEPHEIDS

The aforementioned M101 results of Cook, Aaronson, and Illingworth (1986) provide a good illustration of the problem here. De Vaucouleurs (1978) reports an M101 modulus of 28.5 mag, while the Sandage and Tammann (1974) result adjusted to the same Hyades distance is 29.6 mag -- a full magnitude different! (Note that the lower value implies a brightest red supergiant luminosity that is independent of parent galaxy magnitude, while the higher one leads to a steep dependence. As discussed by Humphreys et al. 1986, the higher value, if correct, may undercut the physical basis on which red supergiants are believed to be standard candles. It would also further hinder use of these objects as distance indicators, since a priori knowledge of the galaxy luminosity would then be required.) Now Cook, Aaronson, and Illingworth report a relative LMC-M101 modulus of 10.8 mag. Unfortunately, the LMC modulus itself remains uncertain to half a magnitude -- direct main sequence fitting gives 18.2 mag (Schommer, Olszewski, and Aaronson 1984; see also the talk by R. Schommer elsewhere in this volume), while the latest Cepheid results calibrated by main sequence fitting within the Galaxy gives 18.7 mag (Caldwell and Coulson 1985). Furthermore, it is not out of the question that the M101 Cepheids are internally reddened by as much as 0.2 mag (multi-color data are still needed). Thus, we are left with an M101 modulus somewhere between 28.8 - 29.5, a range almost as big as the difference in the de Vaucouleurs and Sandage-Tammann values. The quoted M101 modulus is at present based on just two Cepheids, but it is amusing that this is not the major source of error in the results!

This example should serve to illustrate that proper calibration of the Cepheid PL zero-point remains a critical issue. The traditional ground-based approach via galactic main-sequence fitting is beset with problems (e.g. Schmidt 1984), and it remains unclear to what extent these can be overcome by the newer work in this area being pursued. A more promising approach might be a full scale attack with the Baade-Wesselink method using modern photometry, velocity data, and stellar atmosphere codes.

Considerable opportunities are also again offered by HST. GTO time is already assigned to get astrometry to the Hyades and solve this problem once and for all. A calibration of the Cepheids could then come from direct main sequence fitting to the LMC and judiciously chosen intermediate age clusters in M31 and M33. LMC clusters with a range of abundances and ages could be observed to test the understanding of our stellar evolution codes with regard to these two parameters. Abundance affects on the Cepheids remain a source of concern. A handle on this may come from observations of Cepheids radially placed in M31 that are now being undertaken by the author, W. Freedman, and B. Madore.

A population II check with HST is also feasible. One starts with astrometry to nearby stars to define the subdwarf main sequence (also a GTO program), and then uses galactic globular main sequence fitting to derive the mean horizontal branch luminosity as a function of abundance. A direct check on such a calibration would be provided by observations of the horizontal branch position in M31 clusters of varying line strength. Assuming things are copacetic, horizontal branch distances to the LMC, M31, and M33 along with the Pop I results would then give a total of six separate estimates of the Cepheid PL zero point, which we can all pray are in reasonable agreement, defining it to < 0.1 mag.

The program outlined here and in the previous section is probably 400 - 500 hours of HST time (though somewhat less in GO time, as a good part of the work will be done by GTO's). The investment seems well worth it, as the outcome should be a believable Hubble constant, and we can all then get on with our lives.

It is a pleasure to thank W. Freedman, J. Gunn, B. Madore, and J. Mould for helpful discussion. Preparation of this article was partially supported with NSF grant AST 83-16629.

REFERENCES

Aaronson, M., Bothun, G., Mould, J., Huchra, J., Schommer, R. A., and Cornell, M. E. 1986, Ap. J., March 15, in press.
Aaronson, M., and Mould, J. 1983, Ap. J., **265**, 1.
───────. 1986, Ap. J., April 1, in press.
Caldwell, J. A. R., and Coulson, I. M. 1985, preprint.
Cook, K. H., Aaronson, M., and Illingworth, G. 1986, Ap. J. (Letters), February 15, in press.
de Vaucouleurs, G. 1978, Ap. J., **224**, 710.
Freedman, W. L. 1985, in IAU Colloquium No. 82, Cepheids: Observation and Theory, ed. B. F. Madore (Cambridge: Cambridge University Press), p. 225.
Humphreys, R. M. 1983, Ap. J., **269**, 335.
Humphreys, R. M., Aaronson, M., Lebofsky, M., McAlary, C. W., Strom, S. E., and Capps, R. W. 1985, A. J., submitted.
Kraan-Korteweg, R. C. 1985, preprint.
Pritchet, C., and van den Bergh, S. 1985, Ap. J. (Letters), **288**, L41.
Sandage, A. 1983, A. J., **88**, 1569.
───────. 1984, A. J., **89**, 630.
Sandage, A., and Tammann, G. A. 1974, Ap. J., **194**, 223.
───────. 1984, NATURE, **307**, 326.
Schmidt, E. G. 1984, Ap. J., **285**, 801.
Schommer, R. A., Olszewski, E. W., and Aaronson, M. 1984, Ap. J. (Letters), **285**, L53.
Uomoto, A., and Kirshner, R. P. 1985, Astr. Ap., **149**, L7.

DISTANCES FROM HI OBSERVATIONS OF NEARBY GROUPS AND CLUSTERS

W. K. Huchtmeier
Max-Planck-Institut für Radioastronomie
Auf dem Hügel 69
5300 Bonn 1, F.R.G.

ABSTRACT. Galaxies in the Local Group and some nearby groups have been used to calibrate the Tully-Fisher (1977) relation (TF) between blue luminosity (M_B) and the 21 cm line width ($\Delta\nu$). Using these best known extragalactic distances it is hoped to get an efficient tool for deriving distances to galaxies whose magnitudes and line widths have been measured. In addition the galaxy's inclination (i) is needed for correcting both quantities, the magnitude for internal extinction and the line width for projection effects. The line width has to be corrected for finite instrumental resolution and gaussian velocity dispersion. The corresponding sample and the reduction techniques have been discribed in detail by Richter and Huchtmeier (1984). In addition to that sample of 66 galaxies we used more member galaxies of those seven groups and added three more groups to a total of 124 galaxies. Including another 50 field galaxies from the nearby sample (i.e. galaxies with $v_o < 500$ km/s, assuming $H_o = 50$ km s^{-1} Mpc^{-1}) did not change the original relation

$$M_{B_T}^{o,i} = -7.12 \log \Delta\nu_{o,i} - 1.73$$

The assumption of a universal TF seems to be justified by the Virgo cluster sample of 144 galaxies (of types later than S0). A least squares fit yielded a TF:

$$M_{B_T}^{o,i} = -6,95 \log \Delta\nu_{o,i} - 2,34$$

with no significant differences for subsamples with inclinations smaller than 45° or greater. Imposing the slope of the calibration sample to the Virgo sample yields the assumed distance to about 10 per cent.

Some different samples of galaxies have been chosen: About 30 galaxies each of the Hydra I cluster and the Coma cluster and 30 isolated galaxies from the sample of Vettolani et al. 1985.

These data were compatible with a value of H_0 of 50 to 60 km s^{-1} Mpc^{-1}. A sample of ~40 Sa galaxies (Huchtmeier 1982 and new observations) has been used to check for a possible type-dependence of the TF. (Only few galaxies of this type are included in the nearby sample and the Virgo sample)

Using the corrections for internal extinction as given by Sandage and Tammann 1981 the Sa sample does not differ significantly (twice the rms noise) from the TF as defined above. Changing this correction considerably could modify the situation.

REFERENCES

Richter, O.-G., Huchtmeier, W. K.: 1984, Astron. Astrophys. 132, 253
Sandage, A., Tammann, G. A.: 1981, A Revised Shapley-Ames Catalogue of Bright Galaxies, Carnegie Institution of Washington, Publ. 635 (RSA)
Tully, R. B., Tisher, J. R.: 1977, Astron. Astrophys. 54, 661
Vettolani, G., de Souza, R., and Chincarini, G.: 1985, Astron. Astrophys. in press

21cm LINE WIDTHS AND DISTANCES OF SPIRAL GALAXIES

R.C. Kraan-Korteweg, L. Cameron, and G.A. Tammann*
Astronomisches Institut der Universitaet Basel

* Associate European Southern Observatory

ABSTRACT. The Tully-Fisher relation is analysed for the sample of field spirals with known IR magnitudes m_H. They exhibit a pronounced distance effect in the M(IR) vs. log Δv_{21} relation <u>and</u> in the logΔv_{21} vs. M(IR) relation. This means that in addition to the Malmquist effect another bias must be present. A discussion of possible bias sources reveals that, in addition to type differences, the line width Δv_{21} itself has probably influenced the sample. Indeed, a nearly complete sample of Sbc, Sc galaxies does not show any distance bias in the (Malmquist-free) log v_{21} vs. M(B) relation. This demonstrates that even the latter relation can be applied only to well defined samples of type-restricted galaxies, if unbiased distances are aimed at.

1. INTRODUCTION

The correlation of blue (B) and infrared H (IR) absolute magnitudes with the inclination-corrected 21cm line width Δv_{21} of spiral galaxies (Sa to Sm) has been widely used in the past to determine quite discrepant distances to the Virgo cluster and to galaxies within the Virgo complex. To investigate the origin of these discrepancies, the M(IR) versus logΔv_{21} relation of a sample of field galaxies with known IR magnitudes m_H is analysed; as expected from the internal dispersion of the relation the sample is dominated by the Malmquist effect (Section 2). However, even in the Malmquist-free logΔv_{21} versus M(IR) relation of the same galaxy sample a systematic distance effect persists; the origin of this additional bias is discussed in Section 3. A newly defined, nearly complete sample of Sbc, Sc shows no bias beyond the Malmquist effect (Section 4). Conclusions are presented in Section 5.

2. THE CORRELATION OF M(IR) VERSUS $\log\Delta v_{21}$

The sample of 230 spirals (Sa to Sm) with $M_B < 13\overset{m}{.}4$ and inclination $i > 45°$ is considered for which apparent IR magnitudes in the H band and 21cm line widths are available from Aaronson et al. (1982). Relative distances of these galaxies are taken from Kraan-Korteweg (1985), which are calculated from their observed recession velocities for three selfconsistent Virgocentric infall models, characterized by the local infall velocity $v_{VC} = 0$, 220 and 440 km s^{-1}, respectively. 19 galaxies are members of the Virgo cluster. The influence of small random velocities ($\sigma_v < 70$ km s^{-1}) of nearby field spirals should be small and unsystematic. For the determinations of the absolute magnitudes the Hubble constant was arbitrarily set at $H_o = 55$ km s^{-1} Mpc^{-1}. Depending on the true value of H_o, all absolute magnitudes M(IR) may require a correction by a <u>constant</u> amount.

The correlation of M(IR) versus $\log\Delta v_{21}$ for the case $v_{VC} = 220$ km s^{-1} is shown in Fig.1 for four different distance intervals. The distances x are expressed in units of the Virgo cluster distance $x = 1$. For a correlation of the form

$$M(IR) = a \log\Delta v_{21} + b \qquad (1)$$

the nearest interval yields $a = -10.33$, $b = 4.34$. The three remaining distance intervals show progressively smaller values of b (= ΔM), which implies that the mean luminosity of a galaxy increases with increasing distance at a given line width.

The progression in Fig.1 is <u>not</u> due to the specific infall model. Trial runs with widely different values of v_{VC} i.e. $v_{VC} = 0$ and 440 km s^{-1}, yield essentially the same result; the progression is slightly less pronounced for the case $v_{VC} = 0$ and is somewhat enhanced for $v_{VC} = 440$ km s^{-1}.

The most obvious interpretation of the illustrated progression of luminosity with distance is the Malmquist bias. The next Section shows, however, that other effects must also contribute to the progression.

Whatever the explanation of this systematic trend is, the present sample of spirals with known IR magnitudes m_H and line widths <u>cannot</u> be used for the determination of distances via eq.(1). A direct application would lead for field galaxies to progressively lower luminosities - hence smaller distances and larger values of H_o - with increasing distance.

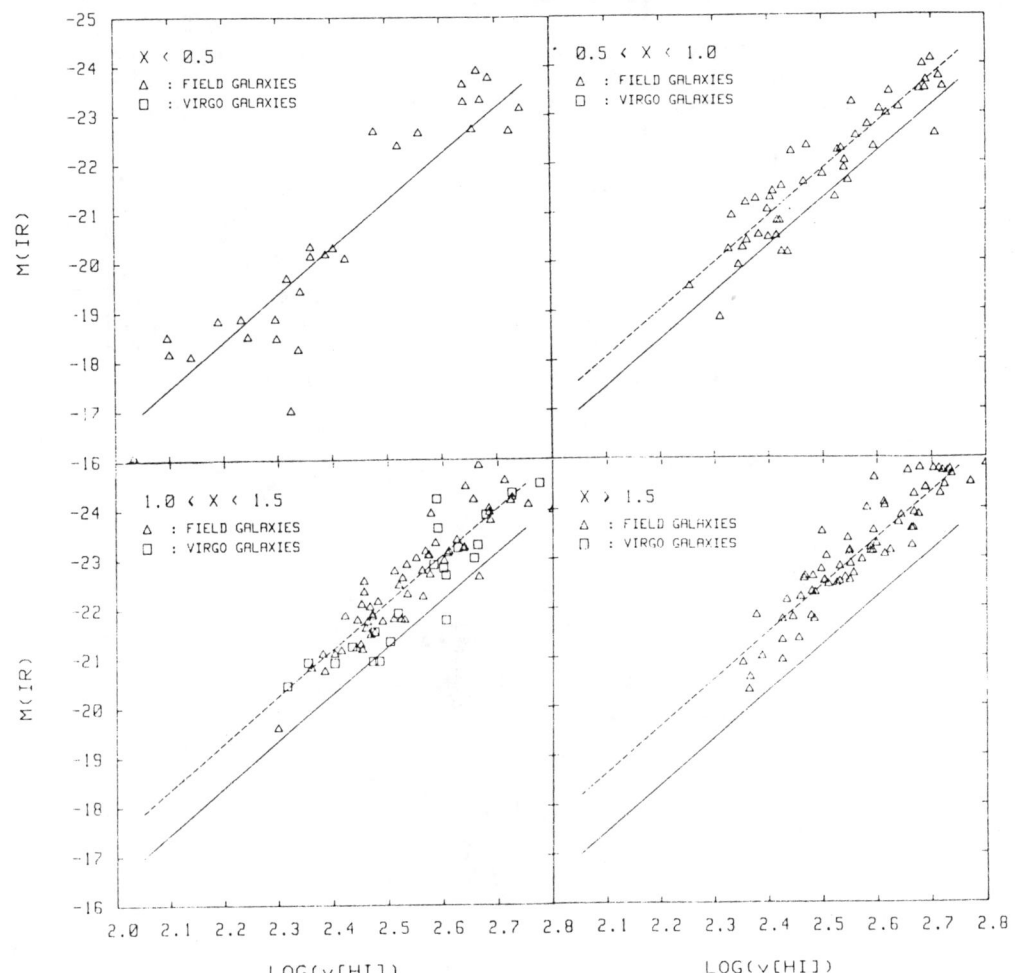

Fig.1. The correlation M(IR) versus $\log \Delta v_{21}$ for 211 field spirals and 19 Virgo cluster spirals in four different distance intervals. x is the distance in units of the Virgo cluster distance. The full drawn line in each panel is the best fit in the distance interval $x < 0.5$. The dashed lines are the respective best fits in the three more distant bins.

3. THE CORRELATION OF $\log v_{21}$ VERSUS $M(IR)$

The regression of $\log\Delta v_{21}$ on M is independent of the Malmquist bias because the expected mean value of Δv_{21} of a galaxy with absolute magnitude M is <u>independent</u> of the applied selection criterium, as long as this selection is independent of the value of Δv_{21}.

Contrary to this expectation, however, the sample of IR spirals shows a systematic distance effect also in the $\log\Delta v_{21}$ versus M diagram. The deviations $\Delta\log\Delta v_{21}$ from a fiducial line of the form

$$\log\Delta v_{21} = cM + d \qquad (2)$$

are shown in Fig.2 for the 137 Sbc, Sc galaxies of the IR sample. It becomes obvious that distant spirals of a fixed absolute magnitude have smaller 21cm line widths than nearby ones. The effect is quite insensitive to the adopted slope c (here c = -0.078).

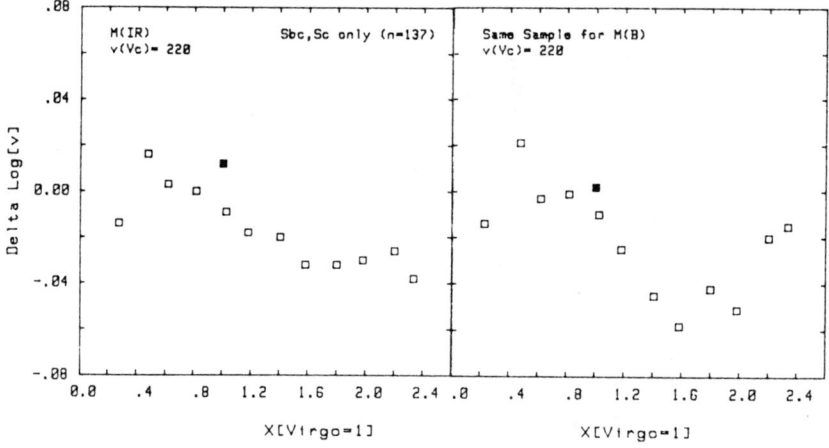

Fig.2. The mean shift in line width, $\Delta\log\Delta v_{21}$, at fixed absolute magnitude as a function of the distance x for the Sbc, Sc spirals of the IR sample. (a) For infrared magnitudes; (b) for blue magnitudes.

In Fig.2 the IR sample was restricted to Sbc, Sc galaxies in order to minimize possible differences between galaxian types. Type differences exist at the level of $\Delta M(IR)$ $0^m\!.5$ for constant line width and of $\Delta\log\Delta v_{21}$ ~0.05 for constant luminosity, as shown in Fig.3 (cf. for blue light Roberts, 1978; Rubin et al., 1982; Giraud, 1985).

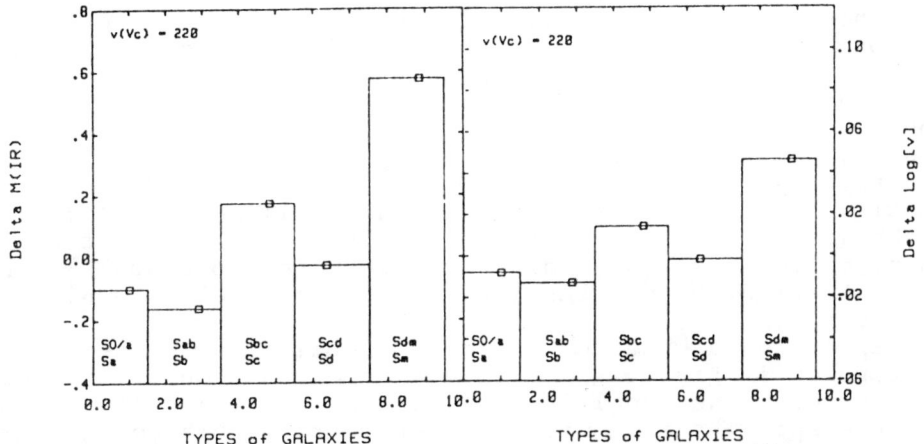

Fig.3. (a) The mean deviation $\Delta M(IR)$ from a fiducial line in the M(IR) versus $\log\Delta v_{21}$ diagram according to galaxian type for the IR sample of spirals. (b) The mean deviation $\Delta\log\Delta v_{21}$ from a fiducial line in the $\log\Delta v_{21}$ versus M(IR) diagram as a function of the galaxian type.

Possible explanations for the unexpected behaviour of the Sbc, Sc sample galaxies in Fig.2 are, e.g.: (1) The selfconsistent Virgocentric infall model with v_{vc} = 220 km s^{-1} is incorrect. - (2) The linear fiducial line, from which the values $\Delta\log\Delta v_{21}$ are read, is incorrect and should be replaced by a non-linear relation. - (3) The physical properties of spiral galaxies may change with the Virgocentric distance r_v. - (4) The apparent IR magnitudes m_H may contain observational errors which increase systematically with distance. - (5) The quadropolar tidal velocity field, imposed on the Virgo complex by more distant mass concentrations (Lilje et al., 1985), may falsify the calculated magnitudes M(IR) and increasingly so with increasing distance. - (6) The sample of spirals with IR magnitudes may be biased in the sense, that it was not selected independently of the 21cm line width. - (7) Other possibilities, like a distance-dependent error of the adopted inclinations, seem to be too far fetched to be considered further.
The possibilities (1) to (5) are considered in turn.

(1) Other infall parameters (v_{VC} = 0 or 440 km s^{-1}) lead to effects quite similar to the one shown in Fig.2a. This proves that the result is insensitive to the adopted infall model. (2) Attempts to detect a significant deviation from linearity of the logΔv_{21}-M(IR) relation for the complete sample or for any subset have remained unsuccessful. (3) The effect of Fig.2a becomes <u>less</u> clear when the distances x from the observer are substituted by Virgocentric distances r_V; this makes it unlikely that the effect is due to an environmental effect. (4) If the apparent IR magnitudes m_H were afflicted by distance-dependent observational errors, the logΔv_{21}-M_B relation of the same galaxies in <u>blue</u> light should be free of this effect; the corresponding plot (Fig.2b), however, still shows the progression, although slightly less pronounced. (5) The effect of a quadropolar velocity field should average out for a all-sky sample; i.e. it should increase the scatter, but it should not introduce a systematic effect with distance. Moreover, one would expect to find the progression with distance also in Fig.4 (see below).

Excluding at the moment the possibility that the progression in Fig.2 is due to the superposition of several small effects, we are left with possibility (6) only, viz. that the spiral sample with IR photometry is biased against broad line widths at large distances. This possibility is checked in the next Section.

4. AN UNBIASED SAMPLE OF SPIRALS

A new sample of spirals was defined by taking all Sbc, Sc galaxies in the RSA with m_B < 12^m.0 and i ⩾ 45°. For 136 out of 147 such galaxies mean 21cm line widths and inclinations were taken from Huchtmeier et al. (1983); i.e. the sample is complete to within 93 %; it is therefore - if at all - only weakly influenced by the size of Δv_{21}. Apparent blue magnitudes $m_B^{o,i}$ were taken from the RSA, and they were transformed into absolute magnitudes M_B analoguously to Section 2 (H_o = 55, v_{VC} = 220 km s^{-1}). This sample defines the coefficients in eq.(2) to be c = -0.082, d = 0.88. Fig.4 illustrates that the constant term d (ordinate) is <u>independent</u> of distance. The Virgo cluster square (black) is based on only 6 galaxies and may not be representative, because most bright Sbc, Sc cluster members happen to have i < 45°.

5. CONCLUSIONS

The present investigation has considered only the 230 spiral galaxies with m_B < 13^m.4 and with known IR magnitudes. This restriction has no influence on the following conclusions. The distance-dependent bias described in Section 2 and 3 persists, if all 300 galaxies, including those with $m_B > 13\overset{m}{.}4$, with known IR magnitudes are used.

The specific conclusions are as follows:

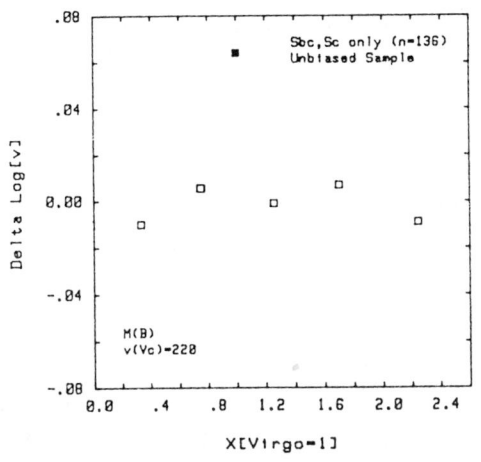

Fig.4. The mean shift $\Delta\log\Delta v_{21}$ at constant luminosity M_B as a function of distance x for a nearly complete sample of Sbc, Sc galaxies. No systematic distance effect exists.

(1) The mean absolute M(IR) magnitude of a sample galaxy with fixed line width Δv_{21} becomes brighter with increasing distance by as much as $\sim 1^m$. Equation (1) can therefore not be used in a simple way to determine distances.

(2) The mean line width of a sample galaxy with fixed absolute magnitude M(IR) decreases with increasing distance. Because the $\log\Delta v_{21} = f(M)$ relation (eq.2) is free of the Malmquist bias, additional bias sources must be at work. The most probable additional bias is that the sample is deficient of galaxies with broad line widths, which indeed yield lower signal-to-noise ratios.

(3) Late-type spirals of a given line width seem to be brighter on average than early-type spirals.

(4) A nearly complete RSA sample of Sbc, Sc galaxies with $m_B < 12\overset{m}{.}4$ is, of course, also affected by the Malmquist effect, but its $\log\Delta v_{21} = f(M_B)$ relation (in blue light) is free of a distance-dependent bias. It is best represented by eq.(2) if c = -0.082 and d = 0.88 (assuming $H_o = 55$). The two local calibrators, M33 and NGC 2403, require - in fortuitously close agreement - $d = 0.89_5$ (using the same distances as Aaronson et al., 1985), which corresponds to a best estimate of $H_o = 60$. This way of deriving H_o is attractive because it bypasses entirely the Virgo cluster, and hence is less sensitive to the adopted value of v_{vc} and does not

depend on the asumption that the intrinsic properties of cluster members and of (calibrating) galaxies in groups and in the field are identical.
(5) Because of the intrinsic dispersion of the Tully-Fisher relation ($\sigma_M \sim 0\overset{m}{.}5$ at constant line width) its application to apparent-magnitude-limited samples leads necessarily to a rubber band distance scale, which is compressed at large distances. The inverse $\log\Delta v_{21}$ versus M relation is in principle free of this problem, but also here a careful sample selection is of great importance.

One of the authors (G.A.T.) is indebted to Dr P.L.Schechter and A.Yahil for illuminating discussions. The authors gratefully acknowledge financial support from the Swiss National Science Foundation. They thank Mrs. M.Saladin for her editorial work on this manuscript.

REFERENCES

Aaronson, M., Bothun, G., Mould, J., Huchra, J., Schommer, R.A., and Cornell, M.E. 1985, Reprints Steward Obs. No.601.
Aaronson, M., Huchra, J., Mould, J.R., Tully, R.B., Fisher, J.R., van Woerden, H., Gross, W.M., Chamaraux, P., Mebold, U., Siegman, B., Berriman, G., and Persson, S.E. 1982, Ap.J.Suppl.$\underline{50}$, 241.
Giraud, E. 1982, Astron.Astrophys.$\underline{153}$, 125.
Huchtmeier, W.K., Richter, O.-G., Bohnenstengel, H.-D., and Hauschildt, M. 1983, ESO Sci. Preprint No.250.
Kraan-Korteweg, R.C. 1985, Astronomisches Institut der Universitaet Basel, Preprint Series No.18.
Lilje, P.B., Yahil, A., and Jones, B.J.T. 1985, Nordita Preprint No.34.
Roberts, M.S. 1978, A.J.$\underline{83}$, 1026.
Rubin, V.C., Ford, W.K., Thonnard, N., and Burstein, D. 1982, Ap.J.$\underline{261}$, 439.

MALMQUIST BIAS AND VIRGOCENTRIC MODEL IN TULLY-FISHER RELATION

L. Bottinelli, P. Fouqué, L. Gouguenheim, G. Paturel,
P. Teerikorpi
Observatoire de Paris, section de Meudon
92195 Meudon Cedex
France

ABSTRACT. The influence of the virgocentric model in the study of Malmquist bias in TF relation is investigated from field galaxies and Virgo cluster.

The amount of bias expected at a given distance when applying the TF relation to a magnitude limited sample of galaxies has been established by Teerikorpi (1984); in a previous work (Bottinelli et al., 1986, hereafter BGPT), this bias has been clearly identified from a large sample of galaxies by using kinematic distances d_v derived from a virgocentric infall model. From 41 galaxies situated within their own unbiased distance range we found the mean logarithmic value, in de Vaucouleurs scale, $H_o=72\pm3$. This result was obtained from a mean velocity of Virgo cluster (VC) with respect to the Local group (LG) $V_o=980$ and an infall velocity of LG towards VC $v=350$ km s^{-1}, but has been shown to be rather insensitive to the values of these parameters. It is our purpose here to study in more details the influence of the virgocentric model on the determination of H_o.

1. FIELD GALAXIES

It must be emphasized that the virgocentric model is used essentially for identifying the galaxies in the unbiased distance range, but has no critical influence on the determination of the value of H_o.
  In order to illustrate this point, we use here the sample of 300 galaxies, studied by de Vaucouleurs et al. (1981), who have determined the solar motion relative to the frame of reference defined by galaxies in 3 distance intervals. From their parameters of the Sun motion, we have determined the corrected velocity V'_c of each galaxy. The corresponding Hubble ratio $H=V'_c/d_{TF}$ is plotted in fig.1 against d_v, deduced from the virgocentric model, expressed in terms of the VC distance. The galaxies are binned in 3 luminosity groups represented by 3 different symbols. It can be seen that, as expected, at a given d_v, the less luminous galaxies are affected by a larger bias. This result is similar to that obtained by BGPT for sosies galaxies. It explain also the reason why similar H values are

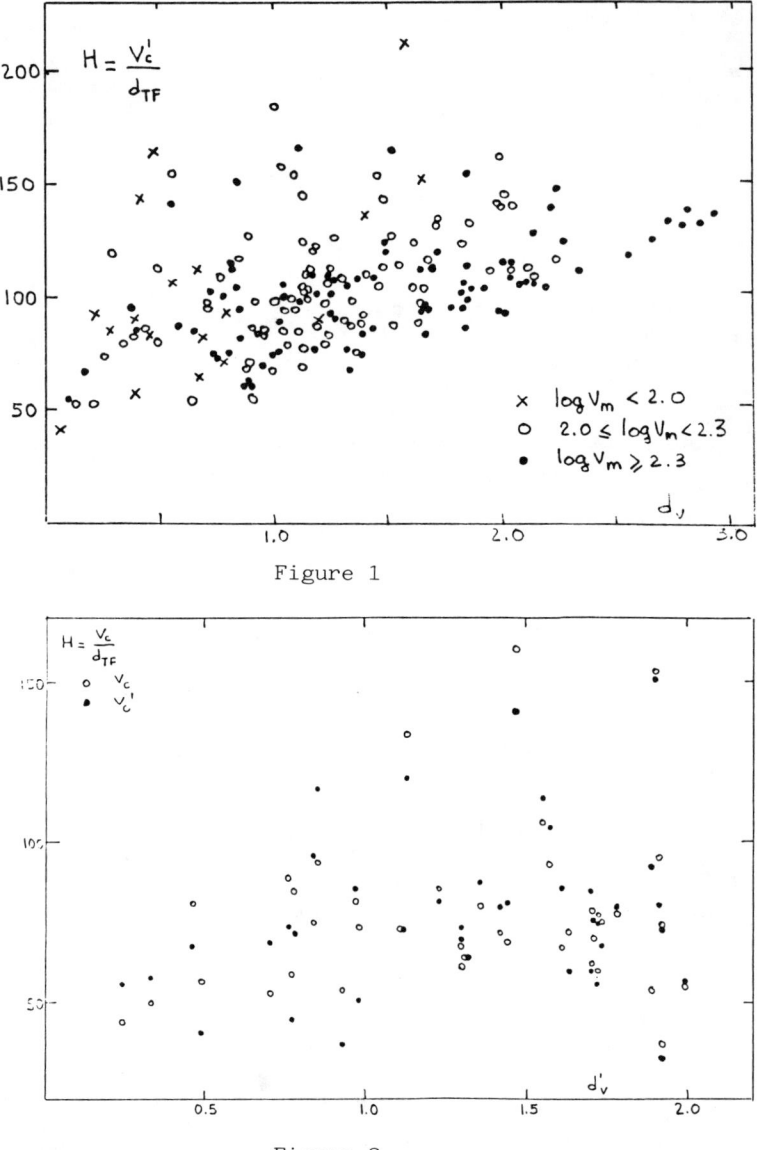

Figure 1

Figure 2

obtained in each of the 3 (biased) distance ranges: the bias is expected to increase with the distance but, because it increases also with smaller luminosities, the two effects play on an opposite side because the percentage of low luminosity galaxies is larger at small distances.

The H values in the unbiased distance range determined either here, using de Vaucouleurs et al. model, or in BGPT in which we used the corrected velocity V_c computed from the virgocentric model are compared in fig.2. They do not differ significantly. We obtain here a mean logarithmic value $H_o=73\pm4$ instead of 72 ± 3 km s^{-1} Mpc^{-1}.

Whether the virgocentric model gives a plausible description of the velocity field cannot be checked with strong accuracy because we cannot disentangle at large distances the effects due to the bias and to the velocity field variation. It can be seen in fig.3 that the mean observed value of H in a given direction in the supergalactic plane does not differ by an amount larger than 10% from the theoretical value expected from the virgocentric model.

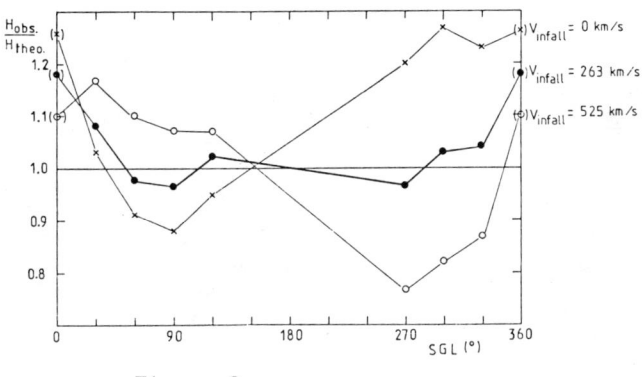

Figure 3

2. VIRGO CLUSTER

In the velocity field model, the VC has a central role. As complementary to our previous work, we have investigated the VC distance from TF relation (Teerikorpi et al., 1986). We used the same <u>direct</u> regression as previously in the distance determination and we <u>investigated</u> the bias due to the incompleteness towards fainter galaxies.

Considering the simple case where exists a sharp cut-off magnitude m_1 down to which the sample is complete and an exponential luminosity function, we obtained the following expression of the difference ΔM between the average value $<M>$ observed at a given log V_m and the local unbiased value M: $\Delta M=-(2\sigma/\sqrt{2\pi})\exp(-A^2)/(1+\text{erf}(A))$, where $A=(M_1-M)/\sqrt{2}\sigma$ and σ is the dispersion of the absolute magnitude at an observed value of log V_m. In a typical situation we expect that $\Delta M\simeq0$ to a range of large log V_m corresponding to luminous galaxies. This implies an unbiased plateau in a H versus log V_m diagram. We have selected 50 galaxies in the VC area (de Vaucouleurs and Pence, 1979; Kraan-Korteveg, 1982) for which the 21-cm line width are collected mainly from our catalogue (Bottinelli et al., 1984, 1985a) and the VC survey of Helou et al. (1984). Fig.4 shows the plot of $H=1330/d_{TF}$ vs. log V_m together with the bias curve obtained for $m_1=13$, $\sigma=0.5$ and $H_o=76$.

The plateau is identified for log $V_m \gtrsim 2.18$ and leads to a logaritmic mean value $H=77\pm6$ (n=23) or $H=82\pm6$ (n=16) when eliminating 7 galaxies which are given as background or foreground by different authors (Huchra,

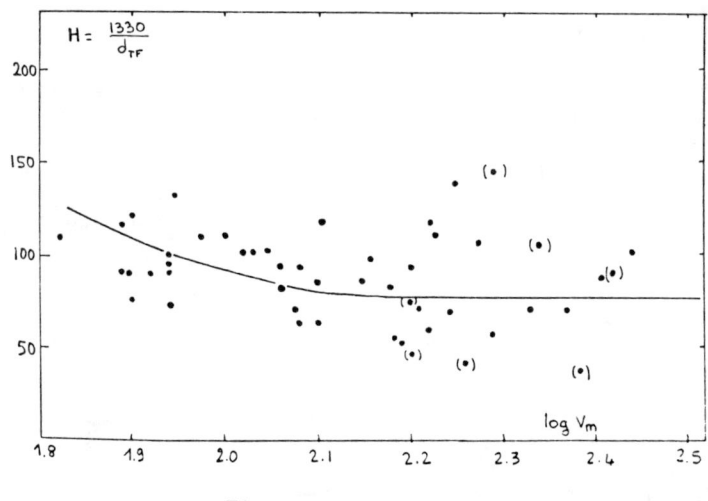

Figure 4

1985; Sulentic, 1977; Tully and Fisher, 1977). These values are larger than the 72± 3 obtained for field galaxies. They imply an infall velocity of the LG smaller than 350. Adopting the value of 220 (Tammann and Sandage, 1985) leads to H_0= 74 instead of 82.

REFERENCES

Bottinelli,L., Gouguenheim,L., Paturel,G., Vaucouleurs,G.de 1984 Astron. Astrophys. Suppl. Ser. 56,381
Bottinelli,L., Gouguenheim,L., Paturel,G., Vaucouleurs,G.de 1985a Astron. Astrophys. Suppl. Ser. 59,43
Bottinelli,L., Gouguenheim,L., Paturel,G., Vaucouleurs,G.de 1985b Astrophys.J. Suppl. Ser. 59,293
Bottinelli,L., Gouguenheim,L., Paturel,G., Teerikorpi,P. 1986 Astron. Astrophys. in press
De Vaucouleurs,G., Pence,W.D. 1979, Astrophys.J. Suppl. Ser. 40,425
De Vaucouleurs,G., Peters,W.L., Bottinelli,L., Gouguenheim,L., Paturel,G., 1981 Astrophys.J. 248,408
Helou,G., Hoffman,G.L., Salpeter,E.E. 1984 Astrophys.J. Suppl. Ser.55,433
Huchra,J. 1985 in ESO Workshop of the Virgo Cluster of Galaxies p.181
Kraan-Korteweg,R.C. 1982 Astron. Astrophys. Suppl. Ser. 47,505
Sulentic,J. 1977 Astrophys.J. Lett. 211,L59
Tammann,G.A., Sandage,A. 1985 Astrophys.J. 294,81
Teerikorpi,P. 1984 Astron. Astrophys. 141,407
Teerikorpi,P., Bottinelli,L., Fouqué,P., Gouguenheim,L., Paturel,G. 1986 in preparation
Tully,R.B., Fisher,J.R., 1977 Astron. Astrophys. 54,661

THE LUMINOSITY INDEX AS A DISTANCE INDICATOR
AND THE STRUCTURE OF THE VIRGO CLUSTER

Gérard de Vaucouleurs
Department of Astronomy
University of Texas
Austin, TX 78712

ABSTRACT. The large scatter ($\sigma \approx 1.0$ mag) in the magnitude - luminosity index correlation (RSA scale) presented by Sandage, Binggeli and Tammann in support of their claim that the luminosity index is unsuitable as a distance indicator is shown to be the result of a variety of systematic and accidental errors and to confusion between the Virgo cluster proper (the S-cloud) and adjacent, but unrelated clouds at different distances. Two tests are presented which confirm the small scatter ($\sigma \approx 0.4$ mag) of the corrected magnitude-luminosity index correlation (RC2 system) in the Virgo S cloud and in the Hercules supercluster.

I. INTRODUCTION. In support of their repeated claims that the "short" distance scale leading to $H_0 \approx 100$ is the result of Malmquist bias, Sandage, Tammann and collaborators have recently published a table and a plot of what they represent to be "de Vaucouleurs' luminosity index", Λ, versus apparent total magnitude, B_T, of spiral galaxies supposedly in "the Virgo cluster" (Sandage, Binggeli and Tammann 1985 = SBT) (Figure 1). If the large scatter, $\sigma = 1.04$ mag at Λ = const., in this graph were indeed applicable to the author's luminosity index it would make it unsuitable as a distance indicator because of the large Malmquist bias (proportional to σ^2 for a gaussian luminosity function). A detailed analysis of the SBT data shows that they are affected by large systematic and accidental errors, inconsistencies, misclassifications and, especially, unwarranted mixing of different galaxy clouds at different distances and in different directions which completely negate their conclusions, or at least make them entirely inapplicable to the author's <u>corrected</u> luminosity index, Λ_c. New tests confirm the small value, $\sigma \approx 0.4$, of the absolute magnitudes derived from it and justify its use as one of the many distance indicators leading to the short distance scale.

II. STRUCTURE OF THE VIRGO CLUSTER. It is obvious and essential that if <u>apparent</u> magnitudes are to be used to test a correlation between a luminosity indicator and <u>absolute</u> magnitudes, the test objects should all be very nearly at the same distance from us, i.e. in a cluster. It should be also obvious that the "cluster" must be a bona fide concentration of galaxies in a small volume of space, not a spurious, accidental projection effect of galaxies at different distances along the line of sight. One such volume-limited sample is the S-cloud of the Virgo cluster (dV. 1961, 1973, 1982) which is concentric with the more concentrated E cloud. There has been (and still is) much confusion between this cluster and various galaxy groupings in different directions and at different distances in the general Virgo area marking the central regions of the Local Supercluster (dV. 1956, 1958). In particular, members of the M group (Ftaclas et al. 1984) and of the W cloud (dV. 1961) which have velocities in excess of 2,000 km s^{-1} and are at about twice the distance of the S cloud, should not be mixed with Virgo cluster members as was done by SBT. Then, objects that overlap with the VC in velocity range, but are in different directions, such as the X cloud (dV. 1961), should not be included either. This cloud, making up the northern part of the so-called "southern extension" and included by SBT in their survey area, is not even a real "cloud" in space, but a projection effect of galaxies scattered at different distances in the supergalactic plane (similar to the Ursa Major cloud north of Virgo). It is even doubtful that the S' cloud, making up the southern half (south of +11° declination) of the traditional Virgo cluster, is at the same distance as the S cloud (dV. 1961), because it has a distinctly higher mean velocity and its member galaxies are ≈ 0.4 mag fainter than those in the S cloud at the same Λ. These different components of the Virgo region are identified in Figure 2 within the area covered by the SBT survey (the S cloud extends slightly beyond the survey area as indicated by three additional bright galaxies; see dV. 1982). Restricting the SBT data to the bona fide S cloud members (Figure 3) immediately reduces the scatter to ≈ 0.6 mag and the amplitude of any Malmquist bias to one third of their claim. This, however, is still a gross exaggeration, because of the large systematic and accidental errors in their data.

III. IDENTIFICATION AND CLASSIFICATION ERRORS. Detailed cross-checking of the data in Table III, Paper IV of SBT with their general catalogue in Table II, Paper II and comparisons with other sources of types and luminosity classes discloses much confusion: erroneous identifications (7 or 8 objects), inconsistencies between paper II and paper IV in classifications (5 objects), magnitudes (6 objects), luminosity index (2 objects, one entered twice in Table III, paper IV with different values), unwarranted L classes assigned to Sa or near edge on galaxies (2 objects), possible T misclassifications (7 objects), possible L misclassifications (10 objects). It is probably significant that the majority of the most discrepant points in Figure 11, Paper IV (Figure 1 here) arise from these errors, including the most discrepant point (Λ = 0.9, B_T = 15.7) which is undocumented.

IV. SYSTEMATIC ERRORS IN THE SBT DATA. The dashed line in Figure 1 represents, according to SBT, the author's adopted relation for the Virgo cluster, which has a slope - 3.0, and differs from their (solid line) slope of -4.86. This difference is fictitious, because the SBT Λ index, simply defined as 0.1 times the sum of the T and L values on the RSA system uncorrected for inclination, differs greatly and systematically from the author's index on the RC2 system corrected for inclination effects. Apart from the neglect of the inclination corrections, the difference arises from a 13 % scale difference between the Sandage and van den Bergh luminosity classifications and from systematic errors in the RSA Hubble types, particularly for galaxies of type Sc and later (too many late types classified as Sc). Here it is not merely a question of different systems, leading to scale differences, but of truly incorrect classifications as can be objectively demonstrated by comparing the RC2 and RSA types to color index and hydrogen index (i.e. HI - luminosity ratios). The net result is that the Λ index (RSA system) has a scale error of 20 % (all-sky average) to 60 % (Virgo area) relative to the Λ_c index (RC2 system). The dashed line in Figure 1 ignores this fact; replotting it on consistent scales eliminates the difference. Apart from such systematic effects the excessive scatter of the SBT graph is caused by larger accidental errors in their magnitudes (eyeball estimates at $B_T > 14$), in their morphological types (but curiously not in their luminosity classes, as demonstrated by separate studies of the T and L classes in RSA and other catalogues), and by their neglect of the important inclination corrections to magnitudes and luminosity classes (cf. RC2 and dV. 1979). (Their comment that inclination corrections are unimportant applies only to the scatter in their Figure 1, dominated by distance mixing; it is demonstrably invalid when classification errors are the dominant sources of scatter).

V. THE TRUE DISPERSION OF THE MAGNITUDE - LUMINOSITY INDEX RELATION. The true dispersion of the correlation between absolute magnitude and corrected luminosity index, Λ_c, in the RC2 system is illustrated in Figure 4 for two strictly volume-limited samples, the Virgo S cloud (dV. 1982) and the Hercules Supercluster (Buta and Corwin 1986). Both use precise photoelectric magnitudes, corrected for inclination as in RC2, types on the RC2 revised Hubble system, and luminosity classes in the RC2 system (after van den Bergh for Vir S), also corrected for inclination. In the unbiased parts of the diagrams ($\Lambda_c < 1.0$) the standard deviations from the mean relations having the standard slope -3.0 are 0.43 mag (Vir S) and 0.37 mag (Her). Note that the magnitude cutoffs, near 12.5 for Vir S and 16.0 for Her, do not cause the small scatter. That is, the lower envelopes of the scatter diagrams are well above the magnitude cutoffs at all $\Lambda_c < 1.0$. The correlations are even tighter for the nearby calibrating galaxies which are not subject to magnitude limitations at all $\Lambda_c < 1.5$ as originally indicated (dV. 1979). If $\sigma \approx 0.4$ mag the amplitude of any Malmquist bias should be less than 15 % of that claimed by SBT and it will be further reduced to a negligible amount by matching the unbiased parts of the correlations in different clusters as discussed elsewhere (Buta and dV. 1983, Buta and Corwin 1986).

REFERENCES

Buta, R. and Corwin, H.G. 1986, Ap.J., submitted.
Buta, R. and de Vaucouleurs, G. 1983, Ap.J. 266, 1.
de Vaucouleurs, G. 1956, Vistas in Astr., 2, 1584; 1958, A.J., 63, 253;
    1961, Ap.J. Suppl., 6, 213; 1973, Astr. Ap., 28, 109;
    1979, Ap.J., 227, 380; 1982, Ap.J., 253, 520.
Ftaclas, C., Fanelli, M.N. and Struble, M.F. 1984, Ap.J., 282, 19.
Sandage, A., Bingelli, B. and Tammann, G.A. 1985, A.J., 90, 395.

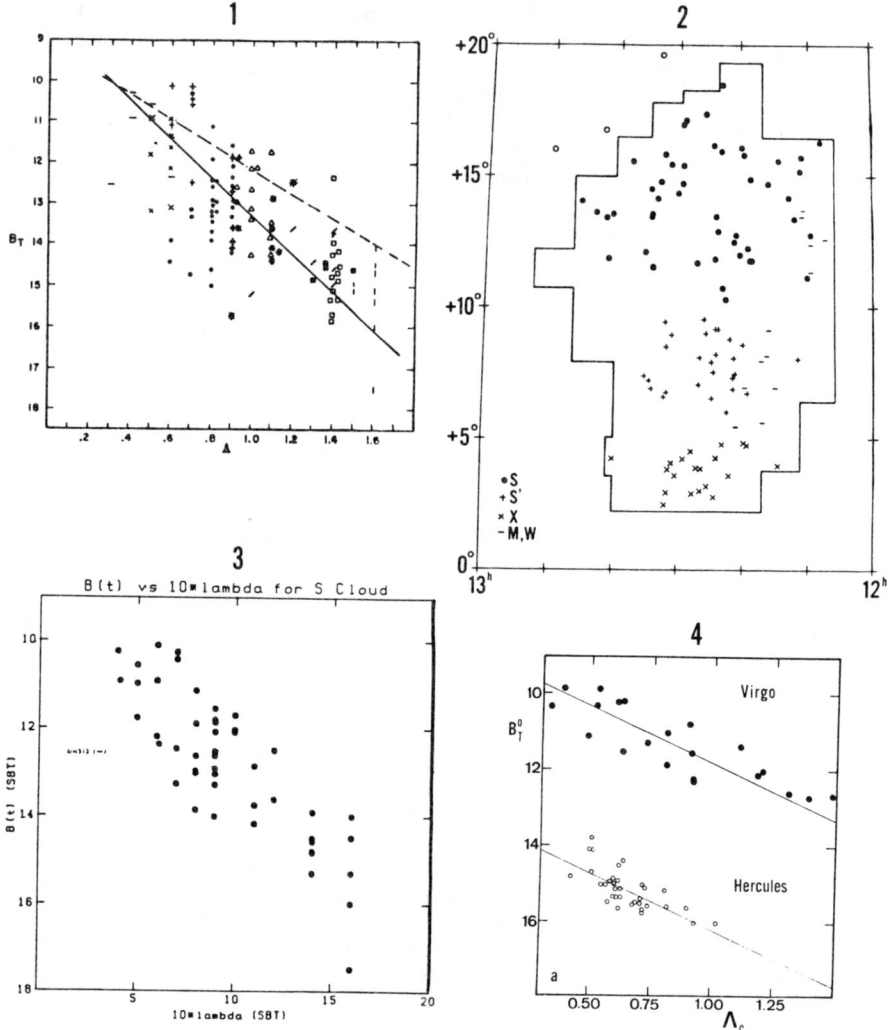

INFRARED COLOUR-LUMINOSITY RELATIONS FOR FIELD GALAXIES

R.S. Ellis, B. Mobasher, Durham University, England.
R.M. Sharples, Anglo-Australian Observatory, Australia

ABSTRACT

Using a new deep field sample we find a tight infrared J-K relation for early type galaxies. The slope agrees with that seen in rich clusters and should be useful as a stable distance indicator provided precision photometry is possible. In Bj-H we find that the scatter and type-dependence of the zero point for spirals can be reduced by incorporating a surface brightness correction. These developments suggest that colour-luminosity studies will continue to be a valuable check on velocities fields derived from the Tully-Fisher method.

1. INTRODUCTION

Despite the success of the IRTF relation as a distance indicator for statistical samples of intermediate depth, at least one independent method is desirable to check velocity fields derived from it. The colour-luminosity (c-L) relation has the advantage that it employs face-on spirals with little or no internal reddening. The scatter on the relation is no worse than that for the TF method, but recent work (Bothun et al. 1985a) suggests its zero points vary from cluster to cluster. Here we present some new results supporting the continued use of c-L relations.

2. NEW DATA

The Durham/AAT galaxy redshift sample (hereafter AARS, Peterson et al. (1986)) is a magnitude limited sample in 5 Schmidt fields totalling 329 galaxies with photometry and redshifts to a limit of Bj $\sim$ 17. JHK photometry of a complete subsample of AARS galaxies in 3 fields has been achieved using the UK Infrared Telescope and used to construct optical-to-infrared colours for 195 galaxies of known luminosity and morphology.

The advantages of a distant sample are: firstly, only small corrections are needed to maintain colours at a large fixed effective aperture of log (A/D0) = 0. Secondly, luminosities are unaffected by corrections for local motion (the survey depth is ∿ 20,000 km/sec). Thirdly, all Hubble types are represented with a field distribution different from that seen in the TF samples. By 'field' we mean a sample covering a volume representative of the Universe on large scales - groups and clusters are contained with adequate representation.

All colours presented have been reduced to those applicable at log (A/D0) = 0 and zero redshift. Infrared K-corrections were found directly from the data with the aid of nearby and more distant photometric catalogues. The reduction is described in more detail by Mobasher et al. (1986).

3. DISCUSSION

Two points emerge from the sample. Firstly, we find a J-K c-L relation for E/S0s whose scatter is limited largely by observational uncertainties. We explore its potential briefly using rich cluster data. Secondly, we review the possible type-dependence of the B-H relation for spirals and discuss a method for reducing this via infrared surface brightness correlations.

A. J-K relations

A colour such as J-K is largely unaffected by residual star formation, internal reddening, population differences and colour gradients - problems plaguing the optical indicators. Although we determine significant J-K c-L relations for E/S0s and spirals (c.f. Bothun et al., 1984 for spirals) the slope for E/S0s is much weaker. However, the Bj-K/J-K colour-colour vector follows closely that expected for a single age population with metallicity as the driving factor. The weak E/S0 relation is offset largely by its small scatter - barely exceeding the observational errors.

In Figure 1 we combine rest-frame J-K data for 40 AARS E/S0 galaxies with those for 76 E/S0s from published work (Frogel et al. (1978), Persson et al. (1979)). Galaxies in rich clusters (including Virgo)

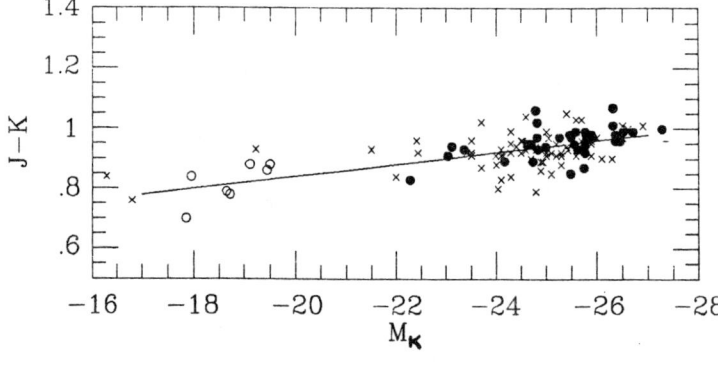

Fig. 1. J-K c-L diagram for AARS E/S0s(●), published E/S0s (x) and Virgo dwarf Es(o) with the AARS regression (Ho = 50 throughout).

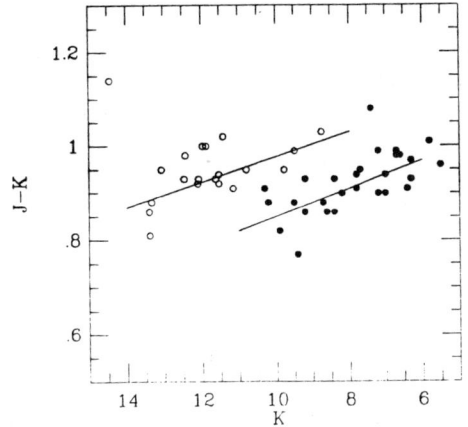

Fig. 2. J-K vs K for Virgo (●) and Coma (o) cluster E/S0s with independently determined regressions.

have been removed from the latter sample. The c-L slopes are summarised in Table 1; the small scatter is apparent.

TABLE 1 Field and Cluster (J-K) - M_K relations

Sample	$d(J-K) / dM_K$	rms	r	N
AARS E/S0	0.023 ± .007	0.04	0.50	40
F→P E/S0	0.014 ± .006	0.05	0.43	76
AARS spirals	0.038 ± .006	0.08	0.59	95
Virgo E/S0	0.030 ± .007	0.05	0.66	29
Coma E/S0	0.027 ± .009	0.04	0.61	19

The Virgo dwarf colours (Bothun et al. 1985b) are also in reasonable agreement with extrapolation of the AARS slope despite being anomalously blue for their luminosity in Bj-K. Even if dwarf Es have a distinct population the effect is minimal in the infrared.

Figure 2 shows the J-K colours for the rich cluster samples of Aaronson et al. (1981). They claimed that Virgo E/S0s contain a cool stellar component not present in Coma. The J-K slopes are similar to that in the AARS (Table 1). By fitting to the cluster data we derive a Virgo-Coma relative distance modulus of 3.81 ± 0.23 in excellent agreement with independent estimates (e.g. Dressler 1984). Either the anomalous component in Virgo has produced no effect in the infrared or the claim is not significant. Further accurate photometry based on reliable diameters is needed to resolve this question.

Thus the J-K relation is a good indicator (by virtue of its immunity to environmental variations) if high precision is possible e.g. using imaging detectors. The AARS photometry achieved a precision of 0.04 mag in J-K. With slightly better results, cluster moduli should be determined to 0.10-0.15 mag.

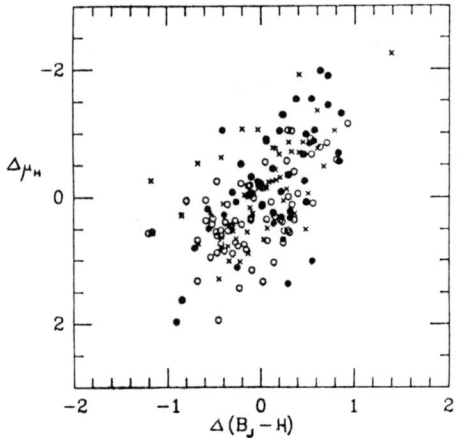

Fig. 3. Residuals in Bj-H and μ_H derived from luminosity-dependent regressions for the combined AARS and published spiral samples. Sa-Sab (●), Sb-Sbc (x), Sc-Scd (o), Sd-Sdm (*)

B. Bj-H relations

To investigate environmental and type-dependences of the Bj-H relation we combine the 95 AARS spirals (mostly early types) with 87 (mostly later types) from the catalogues of Fisher and Tully (1981) and Aaronson et al. (1982). Table 2 shows the results. Early type spirals are systematically redder than the later types presumably because of bulge contamination. Neither sample has measured bulge/disk ratios; instead we introduce a mean infrared surface brightness parameter μ_H defined within the aperture DO. A luminosity correlation is also found for μ_H (c.f. van den Bergh 1981). However, the residuals about the μ_H and Bj-H luminosity regressions are correlated (Figure 3). This effect is not induced systematically from photometric or diameter uncertainties; e.g. a change of 0.1 in log(A/DO) produces a change of only 0.04 in Δ(Bj-H) and 0.1 in $\Delta\mu_H$. Correction for this second parameter reduces considerably the rms scatter and type-dependences in the Bj-H relation (Table 2).

TABLE 2 (Bj-H) - M_H relations

Class	(Bj - H)*	d(Bj-H)/dM_H	rms	N
Sa-Sab	2.88 (2.66)	0.36 (0.36)	0.42 (0.28)	31
Sb-Sbc	2.76 (2.65)	0.20 (0.26)	0.43 (0.35)	57
Sc-Scd	2.57 (2.64)	0.31 (0.35)	0.42 (0.34)	78
Sd-Sdm	2.59 92.70)	0.31 (0.28)	0.36 (0.43)	16

* at M_H = -23.0 (Ho = 50). Entries in parenthesis are corrected for μ_H correlation.

The correlation of residuals is also found in the spiral cluster samples of Bothun et al. (1985c) used by Aaronson et al. (1985) in their recent solution for local motion. The dispersion of the mean colour zero point of 7 clusters reduces to 0.09 mag when the correction is made, implying the distance modulus of a single cluster can be determined to better than 0.25 mag. This still falls short of that for the TF solution (better than 0.1 if the infall model is correct). The important point, however, is that the improved c-L relation can provide an independent check on the derived model.

REFERENCES

Aaronson, M., Persson, S.E., Frogel, J., 1981, Ap. J., 245, 18.
Aaronson, M., Huchra, J., Mould, J., Schechter, P., and Tully, R.B., 1982, Ap. J., 258, 64.
Aaronson, M., Bothun, G.D., Mould, J., Huchra, J., Schommer, R., and Cornell, M.E., 1985, preprint.
Bothun, G.D., Romanishin, W., Strom, S.E., and Strom, K.M., 1984, Astron. J., 89, 1300.
Bothun, G.D., Mould, J., Schommer, R., and Aaronson, M., 1985a, Ap. J., 291, 586.
Bothun, G.D., Mould, J., Wirth, A., and Caldwell, N., 1985b, Astron. J., 90, 697.
Bothun, G.D., Aaronson, M., Schommer, R., Huchra, J., Mould, J., and Sullivan, W., 1985c, Ap. J. Suppl., 57, 423.
Dressler, A., 1984, Ap. J., 281, 512.
Fisher, J.R., and Tully, R.B., 1981, Ap. J. Suppl., 47, 139.
Frogel, J., Persson, S.E., Aaronson, M., and Matthews, K., 1978, Ap. J., 220, 75.
Mobasher, B., Ellis, R.S., and Sharples, R.M., 1986, MNRAS (submitted).
Persson, S.E., Frogel, J., Aaronson, M., 1979, Ap. J. Suppl., 39, 61.
Peterson, B.A., Ellis, R.S., Efstathiou, G., Shanks, T., Bean, A.J., Fong, R., and Zou, Z-L, 1986, MNRAS (in press).
van den Bergh, S., 1981, Ap. J. (Lett)., 248, L9.

MINIMIZING THE SCATTER IN THE TULLY-FISHER RELATION

G. D. Bothun
Caltech

The observed scatter in the H-band Tully-Fisher relation is 0.45 - 0.50 mag (Aaronson and Mould 1983, Aaronson et al. 1986). In order to yield more reliable information concerning the local form of the velocity-distance relation, it is desirable to obtain relative distances to galaxies that are accurate to 10%. The present scatter in the Tully-Fisher relation thus needs to be reduced by a factor of 2. In this contribution, we report on the sources of scatter in the Tully-Fisher relation and elaborate on a possible scheme which yields a scatter of only 0.25 mag.

Scatter is due to observational uncertainties in combination with the intrinsic width of the Tully-Fisher relation. The principle sources of observational scatter and their magnitudes are listed below:

Observational Source of Scatter	1-sigma value
H-band Photometric Errors	0.05 mag
Typical Line Width Errors	0.18 mag
Errors in log $A/D(0) = -0.5$	0.10 mag
Inclination errors	<0.05 mag for $i > 60$

Thus it would appear that observational errors, in most cases, could only contribute 0.25 mag to the observed scatter. Are we to conclude that the intrinsic scatter is thus 0.35 - 0.4 mag? Figure 1 summarizes this present situation by showing the combined H-band TF relation for all galaxies with inclination > 60 degrees that are in the cluster sample of Aaronson et al. (1986). The scatter in Figure 1 is 0.43 mag.

There are, of course, other sources of scatter that need to be considered. For instance, many of the clusters which Aaronson et al. (1986) select are mainly loose aggregates of galaxies and thus depth effects can introduce scatter since the galaxies aren't all at the same distance. If the typical line-of-sight depth of these clusters is 10 Mpc then this will introduce 0.1-0.2 mag of scatter over the redshift range encompassed by these clusters (note that this effect is much more pronounced in the Virgo supercluster - see paper by Shaya in these proceedings). A second problem, peculiar to the H-band Tully-Fisher relation, is related to metallicity variations at fixed line width.

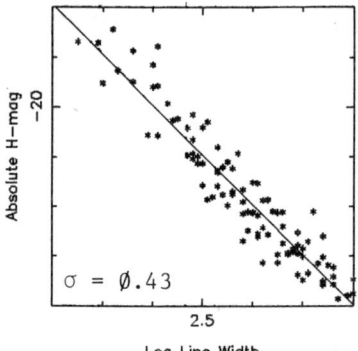

Figure 1: H-band Tully-Fisher relation for sample of Aaronson et al. cluster spirals with inclination > 60 degrees.

Briefly, Bothun et al. (1984) have established a near-IR color magnitude relation for spirals which they interpret as evidence for the existence of a disk metallicity - disk luminosity relation. In fact, the existence of this relation demands curvature in the H-band Tully-Fisher relation which Aaronson et al. (1986) now explicitly take into account. The IR light from stellar populations (e.g. giants) is very sensitive to metallicity since it determines the effective temperature of the giant branch. Thus, variations in [Fe/H] at fixed line width will produce variations in the M-Giant/K-Giant ratio which in turn produce variations in the H-band M/L ratio. Differences in the amount of AGB light will produce a similar effect. Hence, probable variations in the amount of cool stellar component in galaxies of similar line width is another important source of scatter. This problem is less worrisome at wavelengths shorter than H (e.g. I).

But the largest source of scatter involves the choice of aperture. In the case of the H-band work of Aaronson et al. (1986), the standard IR aperture is defined in terms of blue isophotes. This is a particularly cumbersome method of standardizing the aperture. Moreover, this aperture doesn't correspond to anything physical within the galaxy. In an ideal world, one would like to use an aperture that corresponds to the radius at which the rotation curve of the galaxy flattens out. Lacking that, we have searched for an alternative by using I-band CCD data which allows us the opportunity to choose our aperture a posteriori.

In what follows we hope to make it quite clear that the major source of the scatter in the Tully-Fisher relations is due to variations in the form of the light profile of disk galaxies that have the same line width. These variations will produce scatter in the magnitude-line width relation no matter which aperture is used to measure the light. However, by using the information contained in the light distribution it will be possible to minimize this scatter through a judicious choice of aperture as a function of line width and/or surface brightness. In more physical terms, it is possible that these differences in intensity distributions are related to differences in the form of the rotation curve. For instance, low surface brightness galaxies may have a turn-over radius which is a larger fraction of the isophotal radius than high surface brightness galaxies (e.g. Rubin et al. 1985).

The basic problem is illustrated in Figure 2 which presents surface brightness profiles in the Gunn I-band of disk galaxies which have identical line width. In Figure 2 we see that large line width galaxies have surface brightness profiles which diverge beyond a surface brightness level of ~ I = 20.5 mag. sq. arcsec. The situation for low line width galaxies is even more severe since the dynamic range in disk surface brightnesses is quite large. By examining the nature of disk galaxy luminosity profiles as a function of line width, we have devised a scheme to scale the aperture with the surface brightness of the galaxy. In particular, bigger apertures are used for galaxies of progressively lower surface brightness.

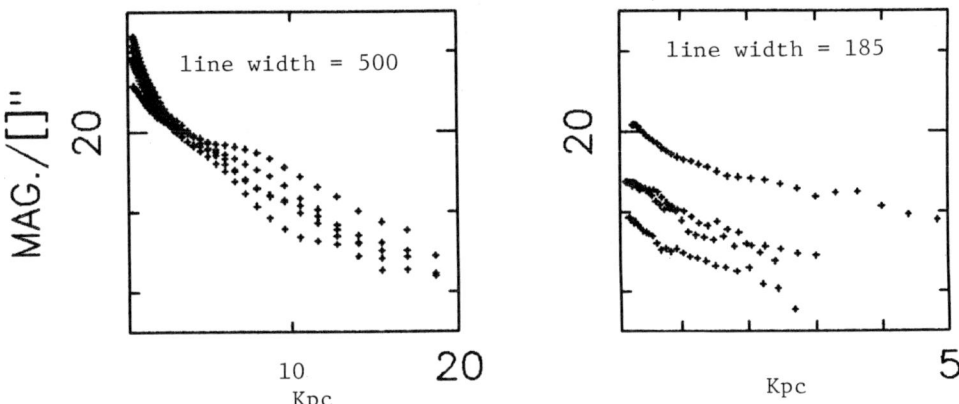

Figure 2: Gunn I-band surface brightness profiles of galaxies that have identical line widths.

The results of applying this scheme are shown in Figure 3 where we see that it is possible to produce a magnitude-line width relation in which the scatter is 0.2 - 0.25 mag. The samples illustrated is a sample of spirals in the Pisces cluster. For comparison, in Figure 3 we show the corresponding H-band Tully-Fisher relation for the identical sample. After going through the exercise of minimizing the scatter in the Tully-Fisher relation it becomes clear that the use of total magnitudes will increase the scatter. This can be seen in Figure 4, where we show that the scatter increases significantly when total magnitudes are used. This again is a reflection of the divergent nature of intensity profiles for galaxies of the same line width. In this case the sample is a collection of pure-disk field galaxies located at the South Supergalactic Pole. Distances were assigned by using a uniform Hubble flow. For the Pisces sample (Figure 3) the total I mag TF relation has a scatter of 0.38 mag.

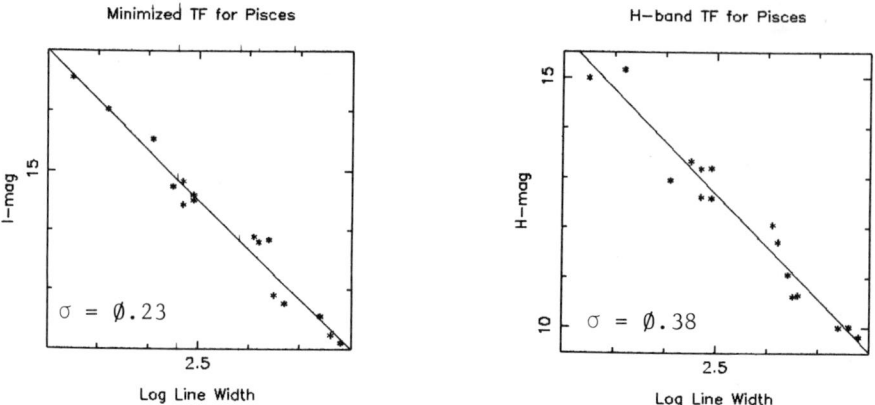

Figure 3: Comparison of minimized I-band CCD TF relation with H-band TF relation for the same sample.

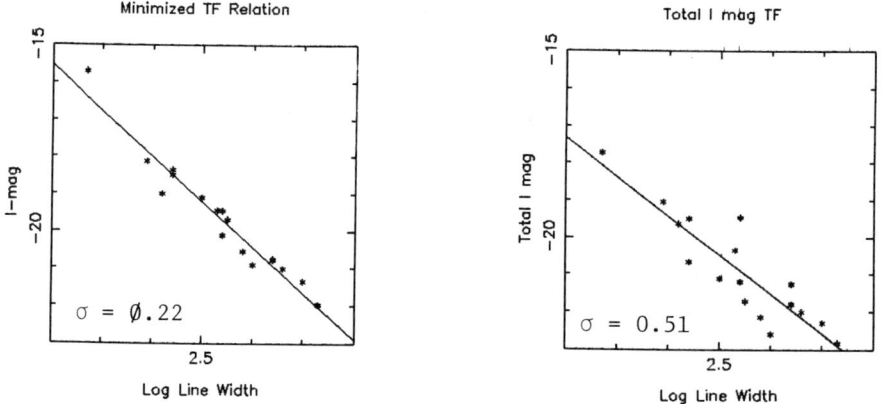

Figure 4: Comparison of minimized I-band CCD TF relation with total mag. TF relation for the same sample.

We have tried other means of selecting the aperture including a) using an aperture in which the surface brightness is a constant, b) using an aperture in which the bulge-to-disk ratio is contant, c) using an aperture that encloses a standard number (1 or 2) of disk scale lengths and d) using a metric aperture. In all cases, except the latter, the scatter was not improved. We do note with interest that, in addition to the minimization scheme successfully employed in Figure 3 and 4, simply using a metric aperture for the Pisces sample of 5 kpc produced a TF relation with a scatter of ∅.22 mag!

The essence of the problem is best illustrated in Figure 5 in which we plot the relation between line width and central surface brightness of the sample of field spirals. Although there is a correlation, there is very large scatter. In essence, this plot vividly demonstrates the wide range in disk surface brightness which is encountered at all line widths. Some of this range may be related to the episodic nature of star formation in disk galaxies. That is, disks will fade between bursts of star formation. If it were possible to find a disk color term that was proportional to the amount of fading, then one could use that as a second parameter in the Tully-Fisher relation. To date, the residuals from the magnitude line-width relation are not well correlated with color since broad band colors are also sensitive to mean disk age and metallicity.

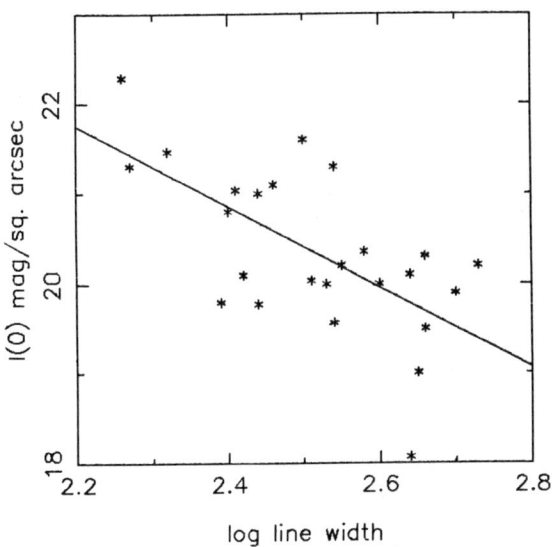

Figure 5: Relationship between line width and central surface brightness for a sample of pure disk galaxies.

In sum, it seems clear that there exists no unique luminosity profile for disk galaxies of a given line width. However, it is possible to use the luminosity profile of the galaxy itself to help in choosing the aperture through which the magnitude is determined. In order to produce a Tully-Fisher relation that has the least amount of scatter, one will have to resort to the method of "sosie" which has been advocated by de Vaucouleurs and collaborators. That is, only galaxies which have similar luminosity profiles as a function of line width should be used. This may make the Tully-Fisher relation difficult to reliably calibrate within an accuracy of ∅.2 mag. For instance, M33, is a substantially higher surface brightness galaxy for its line width than the bulk of the H I selected low surface brightness, low luminosity galaxies in the cluster sample of Aaronson et al. (1986).

Thus, to obtain a more reliable calibration will simply require a larger set of nearby calibrating galaxies. In that way, it may be possible to construct fiducial surface brightness profiles for disk galaxies of given line widths and then to select both calibrating galaxies and distant cluster galaxies which best match that fiducial profile.

I would like to thank M. Aaronson, J. Mould, J. Huchra and R. Schommer for continuing to making it a pleasure to work on the Tully-Fisher relation. I further thank B. Tully for his inspirational organization of this meeting.

REFERENCES:
Aaronson, M., and Mould, J., 1983, Ap.J. 265,1.
Aaronson, M., Bothun, G., Mould, J., Huchra, J., Schommer, R., and Cornell, M., 1986, preprint.
Bothun, G., Romanishin, W., Strom, S., and Strom, K., 1984, A.J. 90, 1300.
Rubin, V., Burstein, D., Ford, W., and Thonnard, N., 1985, Ap.J. 289,81.

I VERSUS B-I COLOR-MAGNITUDE RELATION FOR GALAXIES AND I BAND DISTANCES

Michael J. Pierce
University of Hawaii
Institute for Astronomy
2680 Woodlawn Drive
Honolulu, Hawaii 96822 USA

INTRODUCTION

The great success of the Hubble-Sandage system of galaxy classification indicates that galaxies as a whole are remarkably similar. As a result, the global properties of galaxies have proven to be extremely useful as diagnostic tools for examining their stellar and gaseous content. This fortunate occurrence has allowed the formulation of the Tully-Fisher (T-F) relationships for estimating the distances to galaxies (Tully and Fisher 1977; Aaronson et al. 1979). The difference in the slopes of the blue and infrared Tully-Fisher relations implies that correlations must exist between color and 21-cm line width as well as color and absolute magnitude (Tully et al. 1982).

This contribution relates the application of CCD photometry in the B and I passbands to the above relations. The data set constitutes 62 galaxies found within the Virgo 6° core with $V_o <$ 3000 km sec and $M_B <$ -19; they should all be at the same distance. This data is a subset of a volume-limited sample of 220 galaxies within the Local Supercluster that is currently being examined.

OBSERVATIONS/CALIBRATIONS

Images of the program galaxies were acquired at the Cassegrain focus of the 0.61-m University of Hawaii Planetary Patrol Telescope using the Galileo/Institute for Astronomy CCD camera (Hlivak et al. 1982). The combination of a small telescope and focal reducer provides a field of view of 13 arcmin on a side with a scale of 1.6 arcsec per pixel. Single point residuals from calibration stars were typically 0.02 mag. (1 σ).

Apparent magnitudes for the galaxies were determined through aperture photometry following the interactive removal of interfering stars. The aperture size was not strictly controlled but was about 1.5 times the μ_B = 26 mag isophotal diameter. The same size aperture was used for both the B and I images. The magnitudes were then corrected for internal and Galactic extinction following Tully and Fouqué (1985).

The extinction in the I passband is assumed to be 35% that at B, according to a Whitford reddening law. The 21-cm line widths (corrected for inclination) were obtained from the Atlas of Nearby Galaxies (Tully and Fisher 1986).

RESULTS

Figure 1 illustrates the Tully-Fisher diagram for the I passband. The correlation is clear with a scatter (1 σ) of about 0.4 mag. Similar scatter was obtained for the B passband. Note that similar scatter is reported in other passbands (this conference). As the photometric errors should only be about 0.04 mag, and there is no evidence for improper extinction corrections the scatter must be intrinsic.

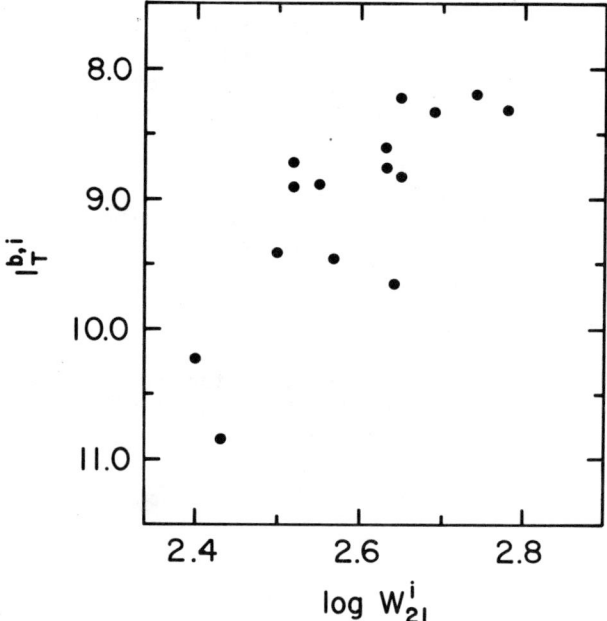

Figure 1

Figure 2 illustrates the color-magnitude diagram for the entire sample. The distinction between the gas-rich and gas-poor branches is clear. This distinction was first found by Tully et al. (1982), who noted that no galaxies were found to lie between the two branches and discussed the possibility of the rapid evolution of spirals from the gas-rich branch to the gas-poor branch, becoming S0 galaxies. Of the

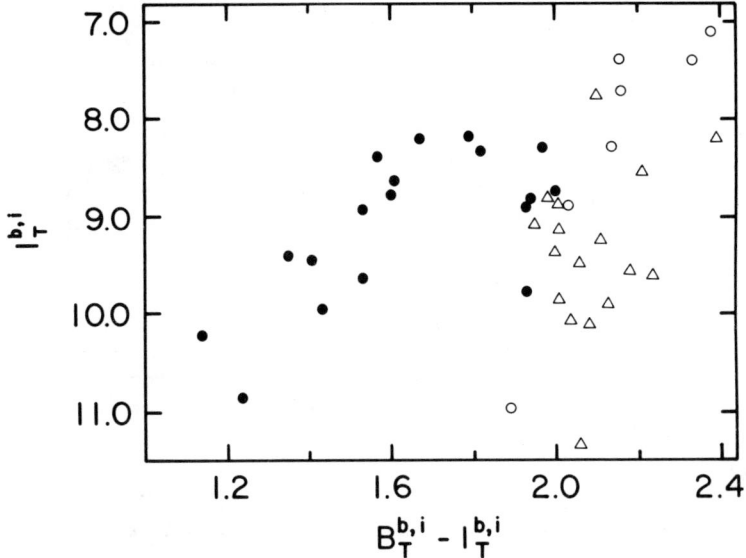

Figure 2. The Virgo sample. ●'s indicate spirals, △'s are S0s, and the o's are ellipticals.

five deviant spirals in Figure 2, two (NGC 4579 and 4450) are classified as anemic spirals by van den Bergh 1976), and all are H I deficient (Giovanelli and Haynes 1983). I will extend the term "anemic" to include all these galaxies. These "anemics" may be the missing transition objects. Note that the scatter along the gas-rich branch is again about 0.4 mag, making this relation appear as good as the T-F relations.

A clue to the nature of the scatter is found in Figure 3. Here the B-I color is plotted against the 21-cm line width. If we neglect the field galaxies for the moment, the scatter is only about 0.06 mag, comparable to the photometric errors and a factor of 7 better than the T-F relations. For the field sample the scatter is about 0.1 mag, still a factor of 4 better than the T-F relations. Clearly the scatter in the T-F relations must be in the magnitudes and not in the line widths. This scatter may be due to variations in M/L for a given mass range (line width). Evidently, the color of a galaxy is more closely correlated with mass (21-cm line width) than is luminosity. Perhaps the mass of a galaxy specifies the subsequent star formation rate as a function of time and hence its present color. There appears to be a group of deviant galaxies in Figure 3 that are about 0.3 mag too red for a given mass, consistent with lower star formation rates. The three in Virgo are what I referred to in Figure 2 as "anemic." Although they are deficient in H I, they do not appear to be deficient in molecular material (Kenney and Young 1986).

The smaller scatter for the Virgo galaxies is a mystery. However,

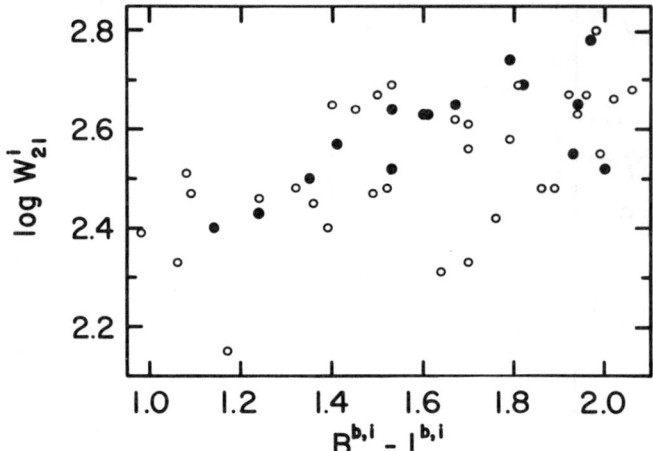

Figure 3. The ●'s are the galaxies within Virgo, while the o's are a field sample.

we might speculate that the Virgo spirals constitute a distinct family of galaxies such that their recent star formation rate is related to their mass in the same manner (whatever that may be). This concept is consistent with the suggestion by Tully and Shaya (1984) that the Virgo spirals have all just recently fallen in. If the scatter in Figure 3 represents the superposition of different families of galaxies, each of which has scatter comparable to that for Virgo, then the color-magnitude diagram (Figure 2) might be strongly influenced by a particular group origin or history. These arguments (admittedly thin) would point to environmental variations in the T-F relations. This possibility is currently being investigated. Although the I band T-F is at least as good as that of the H band, particularly since the magnitudes are much easier to get, it seems premature to derive distances with it until the intrinsic scatter is better understood. Tully and I plan to extend the Virgo sample to fainter limits and investigate whether structural parameters such as the bulge-to-disk ratio can improve the scatter in the correlations.

REFERENCES

Aaronson, M., Huchra, J. P., and Mould, J. R.: 1979, Astrophys. J. 229, 1.
Giovanelli, R. and Haynes, M. P.: 1983, Astron. J. 88, 881.
Hlivak, R. J., Pilcher, C. B., Howell, R. R., Colucci, A. J. and Henry, J. P.: 1982, Proc. Soc. Photo-Opt. Instru. Eng. 331, 96.
Kenney, J. D. and Young, Y. S.: 1986, preprint.
Tully, R. B. and Fisher, J. R.: 1977, Astron. Astrophys. 54, 661.

Tully, R. B. and Fisher, J. R.: 1986, <u>Atlas of Nearby Galaxies</u>, in preparation.
Tully, R. B. and Fouqué, P.: 1985, <u>Astrophys. J. Suppl.</u> **58**, 67.
Tully, R. B., Mould, J. R. and Aaronson, M.: 1982, <u>Astrophys. J.</u> **257**, 527.
Tully, R. B. and Shaya, E. J.: 1984, <u>Astrophys. J.</u> **281**, 31.
van den Bergh, S.: 1976, <u>Astrophys. J.</u> **206**, 883.

PHOTOMETRY OF GALAXIES AND THE LOCAL PECULIAR MOTION

Natarajan Visvanathan
Mount Stromlo and Siding Spring Observatories
The Australian National University
Woden P.O. ACT 2606 Australia

ABSTRACT. Independent distances to 21 clusters and groups extending up to a redshift of 11,000 Km/sec are presented. Analysis of these data with their redshifts reveals the presence of a component of the motion of the Local Group towards Virgo with an amplitude of 300 Km/sec. The nearby field galaxies' redshift/distance data are consistent with a Local Group motion towards Virgo of 260 Km/sec. The farther cluster data give some support for the presence of a motion of approximately 300 Km/sec towards the Centaurus-Hydra complex. A distance modulus of 31.34 $\pm$ 0.15 to Virgo derived previously has been used to obtain true distances to all of the clusters. The redshifts of the clusters corrected for our motion towards Virgo are combined with their distances to determine a value for the global Hubble Constant of 70 $\pm$ 5 Km/sec/Mpc.

1. INTRODUCTION

Attempts to determine a global value for the Hubble Constant have been met with two major difficulties. 1. The presence of peculiar motions in the local region which affects the cosmological redshifts of the galaxies in the level of hundreds of Km/sec (Silk 1974; Peebles 1976; Wilkinson 1983). 2. The distances to nearby galaxies are not known well enough (de Vaucouleurs 1978) leading to systematic errors in the distances to Virgo and other clusters. As the redshifts for many clusters and groups are known well enough, to solve the above two problems we need accurate independent distances to these clusters and groups.

Independent distances to spiral are determined through Cepheid luminosity-period relation (Visvanathan 1985) and luminosity-HI width relation (Visvanathan 1983). While luminosity-colour (Visvanathan and Sandage 1977) and luminosity-velocity dispersion (Dressler 1984) relations are used for ellipticals. Thus the determination of accurate luminosity plays an essential role in the determination of distances. Normally the luminosities are measured in the B and V region and corrected for dust and metallicity variations. However, luminosities measured in near-infrared wavebands are less affected by these factors, leading to more reliable distances (Aaronson, Huchra and Mould 1979; Visvanathan 1981). To help disentangle the peculiar motion from the observed redshift of the galaxies, we need distances with an error of the order of 0.1 mag. Even though the error in distance determination for a single galaxy by the above methods is about 0.3-0.4 mag, by observing many galaxies in a cluster the error could be reduced to the desired level.

2. PHOTOMETRY AND DISTANCES

We have chosen to do photometry of the galaxies in the 1.05 micron near infrared waveband (Iv) with a InGaAsP tube because of the advantages mentioned in the previous section. Also it enables us to use conventional photometric techniques, multiple and large apertures as well as different wavebands sequentially.

Multiple aperture photometry in Iv for 266 spirals and V-Iv colours for 60 ellipticals in the local region and farther clusters have been obtained.

The individual aperture magnitudes of spirals have been corrected for: 1. full galaxy size 2. internal absorption in the galaxy, 3. galactic extinction and 4. redshift (Visvanathan 1983). These are then averaged to represent the corrected luminosity of the galaxy. Distances to these galaxies have been derived through luminosity-HI width relation. The HI width data at the 20% level have been taken from the available literature and corrected for inclination and the relativistic Doppler effect (1+Z). Only galaxies that are inclined more than 45 deg were selected for observation. For each cluster and group an average of the distances of its members is taken and these are given with respect to the Virgo cluster. Those clusters for which distances are computed are given in Table 1. Distances are given under the column $\Delta(m-M)_{HI}$.

The V-Iv colour of ellipticals observed with different apertures has been corrected for redshift as well as galactic extinction and averaged for each galaxy. The V magnitudes on the other hand were additionally corrected for galaxy size to derive a total magnitude for the galaxy. The corrected V-Iv colour and the V total magnitude are used to form a colour magnitude relation for each cluster separately. For those clusters with V-Iv data the distances were derived by comparing the CM relation of the cluster with that of Virgo and are given in Table 1 as $\Delta(m-M)_{Iv}$.

In Table 1 we have also included the clusters for which distances $(\Delta(m-M)_u$ were computed through the V-(u-V) relation (Visvanathan 1981a).

Overall in Table 1 we have 21 clusters for which independent distances are available. For Fornax and Centaurus clusters we have distances from all the three methods and it is encouraging to see they are in good agreement. In the case of Peg 1 cluster we have two consistent distances. For analysis we have used the average of these values and will be referring to this as "photometric distance": $\Delta(m-M)_p$ (given in column 5 of Table 1).

Table 1
Data on 21 Groups and Clusters

Group Cluster	Δ(m-M)$_{HI}$	Δ(m-M)$_{IV}$	Δ(m-M)$_u$	Δ(m-M)$_p$	v_0 Km/sec	Δ(m-M)$_R$	Δ(m-M)$_p$-Δ(m-M)$_R$	v_{pec} Km/sec	Cos θ Deg	Cos A Deg	(m-M)°	v_{cos} Km/sec	H km/sec/Mpc
Virgo	+0.0	+0.0	+0.0	+0.0 ± 0.1	1019	0.0	+0.0	+295	+1.00		31.34	1319	71.2
Fornax	-0.16	-0.07	+0.12	-0.04 ± 0.1	1428	0.73	-0.77	-138	-0.68		31.30	1224	67.3
Grus	-0.00			-0.00 ± 0.1	1609	0.99	-0.99	-295	-0.82		31.34	1363	73.5
Eridanus	-0.11			-0.11 ± 0.2	1613	1.00	-1.11	-364	-0.76		31.23	1385	78.6
G 5846			+0.91	+0.91 ± 0.1	1808	1.25	-0.34	+180	+0.81		32.25	2051	72.8
G 7144			+0.55	+0.55 ± 0.2	1919	1.37	-0.82	-226	-0.63		31.89	1730	72.5
Cetus II			+0.63	+0.63 ± 0.2	1935	1.39	-0.76	-179	-0.97		31.97	1644	66.3
G 1209			+1.11	+1.11 ± 0.2	2217	1.69	-0.58	- 26	-0.83		32.45	1968	63.7
G 5077			+1.58	+1.58 ± 0.2	2515	1.96	-0.38	+205	+0.87		32.92	2776	72.3
I 4797			+1.54	+1.54 ± 0.2	2653	2.08	-0.54	+ 18	-0.19		32.88	2596	68.9
G7196			+1.48	+1.48 ± 0.2	2844	2.23	-0.75	-246	-0.63		32.82	2655	72.5
A 1060		+2.18		+2.18 ± 0.1	3259	2.52	-0.34	+327	+0.66	+0.97	33.52	3457	68.3
Centaurus	+2.24	+2.32	+2.40	+2.32 ± 0.1	3407	2.62	-0.30	+418	+0.60	+0.88	33.66	3587	66.5
N 6769 gr	+2.59			+2.59 ± 0.2	3953	2.94	-0.35	+378	-0.29	+0.32	33.91	3866	63.9
PEG I	+2.06	+2.08		+2.07 ± 0.1	3992	2.96	-0.89	-583	-0.89	-0.81	33.41	3725	77.5
Sersic 129-1		+2.70		+2.70 ± 0.1	4230	3.09	-0.39	+326	-0.22	+0.39	34.04	4164	64.8
Cancer	+2.72			+2.72 ± 0.2	4556	3.25	-0.53	+ 42	+0.51	+0.35	34.06	4709	72.6
G 507			+2.94	+2.94 ± 0.2	5128	3.51	-0.57	- 39	-0.69	-0.82	34.28	4921	68.6
ZW 74-23	+3.27			+3.27 ± 0.1	5893	3.81	-0.54	+307	+0.93	+0.35	34.61	6083	72.8
Coma			+3.71	+3.71 ± 0.1	6947	4.17	-0.46	+307	+0.96	+0.19	35.05	7235	70.7
Hercules	+4.89			+4.72 ± 0.1	11000	5.14	-0.42	+550	+0.77	-0.10	36.06	11231	68.9
Super cluster													

3. RESULTS

3.1. Infall Velocity of the Local Group

In Table 1 along with the distances the systemic velocity of each group and cluster taken from the literature is also shown. This value is already corrected for the standard solar motion relative to the Local Group of 300 sin l cos b Km/sec. The error in v_o is about 5%. From the redshift data, we have calculated a new "redshift distance" for each cluster $\Delta(m-M)_R = 5 \log(v_o/v_o \text{ Virgo})$ and these are shown in the sixth column.

It is obvious that the photometric distance should equal the redshift distance if the clusters including Virgo have no other motions except that due to uniform cosmological expansion. Yet as can be seen in the eighth column of Table 1, the differences are all negative, with an average of -0.55 mag indicating that the redshift distances are all larger. As the redshifts of the clusters were normalised to that of Virgo, this anomaly is due to the fact that the observed redshift of Virgo is lower by a factor of 1.29 compared to its cosmological counterpart. The simplest interpretation of this is that our Local Group has a peculiar motion towards Virgo of 295 Km/sec, thus providing the lower value of the observed redshift of Virgo.

The consequence of our motion towards Virgo is that clusters surrounding Virgo should reflect our peculiar motion. The peculiar motion component should vary as a function of its polar angle (θ) with Virgo. This effect can be detected more easily with nearby clusters, where it makes up a substantial portion of its redshift. Using our newly derived redshift of the Virgo cluster (1314 Km/sec) and the photometric distance of the clusters, the cosmological redshift for each cluster is derived. The difference between this redshift and the observed one, representing the peculiar motion component (v_{pec}), is calculated and plotted as a function of the polar angle $\cos\theta$ in Figure 1.

The straight line in the figure represents a least squares fit to the points with a slope of 353 ± 35 Km/sec. This demonstrates that our peculiar motion towards Virgo is the reason for the lower observed redshift of Virgo. Further confirmation of our motion towards Virgo comes from the observed redshift of Grus (1609 Km/sec) which is at the same distance but the opposite direction as Virgo. This is much higher than that of Virgo (1019 Km/sec) and the average of these two values (1314 Km/sec) agrees well with the cosmological redshift of Virgo.

The field spirals' data within a redshift range of 2000 Km/sec confirm our motion towards Virgo with a value of 260 Km/sec. The average value for the Local Group motion comes out as 300 ± 38 Km/sec.

3.2. Search for a Local Group Motion Towards Centaurus-Hydra Complex.

It is well known that the anisotropy of the microwave background indicates that our Local Group has a peculiar motion towards a direction 44 deg from Virgo with an amplitude of about 600 Km/sec

(Wilkinson 1983). As we have just established that our Local Group has a motion towards Virgo of 300 Km/sec, it is reasonable to expect a further peculiar motion of the Local Supercluster towards the Centaurus-Hydra complex of about 400 Km/sec (Tammann and Sandage 1984). Then the vectoral addition of these two components will explain the microwave background results.

In an attempt to identify the latter component in our data, we have plotted in Figure 2 the peculiar motion component in each cluster (v_{pec}) against cos A, where A is the polar angle of the cluster with the Centaurus-Hydra complex. We have chosen to plot only the farther

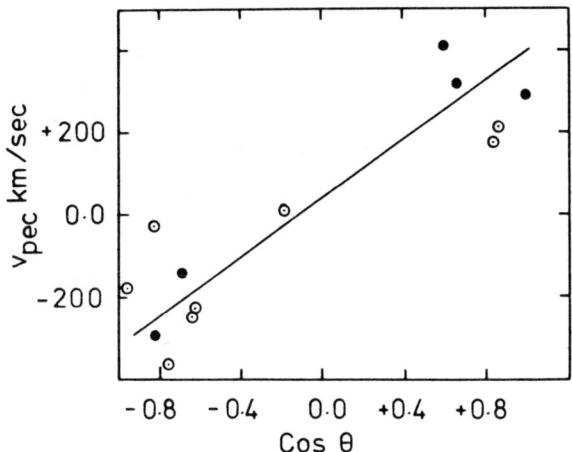

Fig. 1   The peculiar motion of each nearby cluster plotted against Cos θ, θ being the polar angle of the cluster from Virgo.

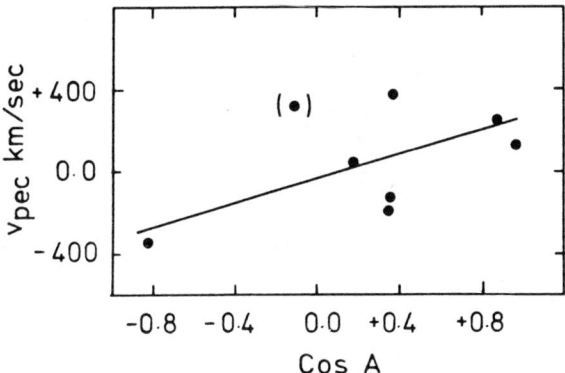

Fig. 2   The peculiar velocity of the farther clusters, after correction for our motion of 300 Km/sec towards Virgo plotted against Cos A, A being the polar angle of the cluster from the Centaurus-Hydra Complex.

clusters with redshifts greater than 3000 Km/sec. The clusters N6769 and G507 are excluded from the plot because of high errors in their distances.

A straight line fitted to the points in Figure 2, indicates a peculiar motion towards Centaurus-Hydra complex of about 300 ± 130 Km/sec. Thus, this component and the Virgo infall component of the Local Group when added vectorially, show agreement with the velocity and direction inferred from the microwave background dipole anisotropy. Solution of our data by the standard least-squares method gives a value of 430 Km/sec for the Local Group motion, in the direction $l = 286$ and $b = 32$, confirming our previous conclusion.

4. CONCLUSIONS

Our distances to 21 clusters when combined with their published redshifts indicate a Local Group motion towards Virgo of 300 ± 38 Km/sec. This value is in general agreement with that obtained by de Vaucouleurs et al. (1981: 230 ± 50), Tammann and Sandage (1984: 200 ± 50), and Aaronson et al. (1982: 331 ± 41). Our search for an infall component of the Local Group towards Centaurus-Hydra complex gives some support to the existence of such a motion with 300 Km/sec.

True distances to all the clusters were calculated using our derived distance to Virgo of 31.34 (Visvanathan 1986). The cosmological redshifts were derived by correcting the observed redshifts for our peculiar motion towards Virgo of 300 Km/sec. The final column of Table 1 shows the Hubble constant (H) derived for each cluster. The average value for the global Hubble constant is 70 ± 5 Km/sec/Mpc.

5. REFERENCES

Aaronson, M., Huchra, J., and Mould, J. 1979, Astrophys. J., **229**, 1.
Aaronson, M., Huchra, J., Mould, J., Schechter, P. L., and Tully, R. B. 1982, Astrophys. J., **258**, 64.
de Vaucouleurs, G. 1978, Astrophys. J., **223**, 730.
de Vaucouleurs, G., Peters, W. L., Bottinelli, L., Gouguenheim, L., and Paturel, G. 1981, Astrophys. J., **248**, 408.
Dressler, A. 1984, Astrophys. J., **281**, 512.
Peebles, P. J. E. 1976, Astrophys. J., **205**, 318.
Silk, J. 1974, Astrophys. J., **193**, 525.
Tammann, G. A., and Sandage, A. 1984, in the preprint of the Astronomical Institute of the University of Basel, No. 6.
Visvanathan, N. 1981, Journal Astrophys. Astr., **2**, 67.
. 1981a, Proc. Astr. Soc. Aus., **4**, 172.
. 1983, Astrophys. J., **275**, 430.
. 1985, Astrophys. J., **288**, 182.
. 1986, Proc. Astr. Soc. Aus., 6, (in press).
Visvanathan, N., and Sandage, A. 1977, Astrophys. J., **216**, 214.
Wilkinson, D. T. 1983, in IAU Symposium No. 104, Early Evolution of the Universe and Its Present Structure, Eds. G. O. Abell and G. Cincarini (Dordrecht: D. Reidel).

THE MOTION OF THE SUN AND THE LOCAL GROUP AND THE VELOCITY DISPERSION OF "FIELD" GALAXIES

Otto-G. Richter[1]
Space Telescope Science Institute
Homewood Campus
3700 San Martin Drive
Baltimore, MD 21218, U.S.A.

[1] Affiliated to: Astrophysics Division,
Space Science Department of E.S.A.

ABSTRACT. Distances to a complete sample of nearby galaxies with radial velocities less than 500 km/s computed via the blue Tully-Fisher relation are used to redetermine the vector of the solar motion, the streaming motion of the Local Group of galaxies and the velocity dispersion of field galaxies. The Hubble expansion is found only outside the Local Group. An upper limit of 120 km/s for the velocity dispersion of "field" galaxies within 10 Mpc is derived.

1. INTRODUCTION

In the absence of independent distance determinations the distances to external galaxies are computed from their radial velocity - suitably corrected to a supposedly cosmic reference frame - assuming a certain value for the Hubble constant. Currently Hubble constants of either 50 or 100 km/s/Mpc are adopted. Also, there are two concurrent procedures for the correction for solar motion. The correction found by Yahil et al. (1977, YTS) differs by up to 87 km/s from that with the standard values (recommended by the I.A.U. since 1976; e.g. deVaucouleurs et al., 1976, RC2). This difference is important for studies of the kinematical and dynamical state of the nearer part of the local supercluster. In particular, estimates for the so-called infall velocity of the Local Group toward the Virgo cluster and the value of a possible peculiar velocity of the Local Group may be affected by different correction procedures for radial velocities.

Here preliminary results of a new reduction of all available data for galaxies in the catalogue of nearby galaxies presented by Kraan-Korteweg and Tammann (1979, hereafter KKT) are presented. The Tully-Fisher relation (henceforth TF) given by Richter and Huchtmeier (1984, RH) is used to derive distances to all spiral and most irregular galaxies in the KKT sample. A full account of this work will be presented elsewhere (Richter et al., 1986).

2. THE DATASET

The galaxies constituting the KKT sample have been observed in the 21cm line of neutral hydrogen (Huchtmeier and Richter, 1986). The optical information, magnitudes in particular were taken from KKT. Sandage and Tammann (1981, RSA), or deVaucouleurs et al. (1981). In some cases values have been averaged if they all were uncertain. Diameters - used to determine the inclinations and the internal absorption corrections - were taken from various sources and transformed into the RC2 system. The RSA convention for the corrections for galactic and internal absorption (cf. RH) was adopted. The detailed data used below is described extensively in Richter et al. (1986).

In Fig. 1 the revised blue Tully-Fisher relation for the galaxies in the Local Group is shown. It is evident that the zero-point b for the blue TF - given by RH

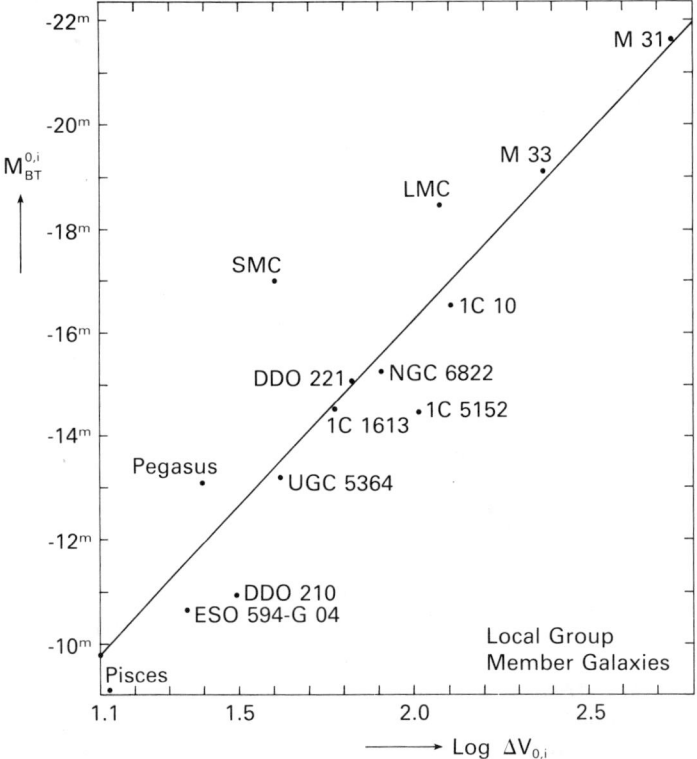

Fig. 1 The blue Tully-Fisher relation for Local Group galaxies. The SMC and LMC data are still quite uncertain.

as -2.12 - needs small adjustment in order to better fit the data. Keeping the slope a fixed at -7.17 (RH) one finds $b = -1.88 \pm 0.38$.

For the purpose of this study the KKT galaxies belonging to the Virgo cluster had been excluded.

3. THE REDUCTION AND FIRST RESULTS

The first step in the whole reduction procedure was to redetermine the solar motion with respect to the Local Group. As the basic sample all galaxies which are certain, probable or possible members of the Local Group (YTS, KKT, see also Richter, 1984) were taken.

At first, a dipole velocity field on top of a linear expansion field was fitted to the heliocentric velocities and (previously accepted) distances of the 14 spiral and irregular galaxies in the Local Group. The solution from this very first run did not indicate any significant expansion or contraction for this special sample. It was not dependant on the inclusion of certain galaxies. Omitting e.g. the two galaxies with the largest residuals, viz. M33 and the Pegasus dwarf, from that sample, produced no different solutions. Using the distances determined from the TF for the spiral and irregular galaxies gives solutions that, again, show no significant expansion or contraction term for the Local Group sample. Therefore, new solutions without an expansion term, based on a pure dipole field, appeared adequate. Because they are then independant of any adopted distance, one can add also the elliptical galaxies with known radial velocities to produce a larger sample and an improved fit. The dubious object UGCA 435 (= Pal 3, a globular cluster and a HVC in superposition ?) was excluded yielding a resulting solar motion of 302 ± 17 km/s toward $l^{II} = 98°.0 \pm 4°.8$ and $b^{II} = -2°.9 \pm 3°.3$.

It was checked whether the low accuracy of some of the radial velocities for the elliptical galaxies influenced the results. NGC 147 has currently the most uncertain radial velocity. Its removal lead again to a better fit. Therefore, the final sample consists of 24 galaxies in the Local Group of galaxies. Also this sample does not show expansion or contraction. The final resulting parameters are $v_\odot = 312 \pm 16$ km/s, $l^{II} = 99°.3 \pm 4°.2$, and $b^{II} = -3°.3 \pm 2°.9$. This solution is stable against the exclusion of individual galaxies. Changes in the length and direction of the solar velocity vector generally remain within less than 0.5σ of combined errors. It should be noted in particular, that the exclusion of all galaxies which are satellites of our Galaxy does not change the result. Also the exclusion of the companions of M 31 does not alter the situation. This fact could be used to argue for small masses of the two dominant galaxies in the Local Group.

In the second reduction step all heliocentric radial velocities were corrected according to the preferred solution via the formula
$$\Delta v = -50.1 \cdot \cos l^{II} \cdot \cos b^{II} + 307.1 \cdot \sin l^{II} \cdot \cos b^{II} - 17.7 \cdot \sin b^{II}.$$
Now all galaxies used in the previous section to generate this solution were excluded from the sample and a linear flow model was fitted to the outer KKT sample. (A linear model is sufficient because the sample galaxies have distances less than 1/2 the Virgo cluster, i.e. the suspected apex, distance.) Because one now needs independent distance determinations via the TF we have to exclude all KKT galaxies without observed HI profiles (mostly ellipticals and S0's). A somewhat generalized

form of Schechter's original (1980) formula was used by letting
$$v_{Apex} = H_0 \cdot D_{Apex} - w_{Flow} .$$
This modification allows to determine the location of the apex of the flow rather than having to assume those three parameters. The parameter a was taken from RH as -7.17. and b was set to -1.88 (see above).

Before the model could be applied to the data an assumption had to be made for i_{min}, the minimum inclination angle used to calculate corrected linewidths. Using $i_{min} = 35°$ for irregular galaxies, $i_{min} = 25°$ for magellanic spirals, and $i_{min} = 15°$ for earlier types produced an acceptable fit. The high value for irregulars affected only a few galaxies and it may indicate a rather broad range of intrinsic flattenings for this morphological type.

In performing the fit one can either assume a Hubble constant H_0 and let the zero-point b of the TF vary, or one fixes b and leaves H_0 as a free parameter. Both are scaling parameters and can not be determined simultaneously. Because we think b is better determined at the moment we prefer the second alternative. Assuming $\gamma = 2$ (e.g. YTS) one gets the following result. The Local Group moves with a velocity of $w_{Flow} = 356.3 \pm 57.0$ km/s toward an apex at $l^{II}_{Apex} = 270°.05 \pm 7°.81$, $b^{II}_{Apex} = 62°.55 \pm 3°.50$, and $D_{Apex} = 20.91 \pm 3.74$ Mpc. H_0 is then found to be 44.5 ± 8.2 km/s/Mpc. The quoted errors include the uncertainty in the zero-point b. The residual r.m.s. velocity for our total sample is 131 km/s. This velocity dispersion is larger than expected as is the infall velocity w. In addition, it is somewhat uncomforting to find that all residuals greater than twice the r.m.s. are positive, i.e. the predicted radial velocity is too small.

In order to improve the results a second fit was performed with both a dipole and a flow component in order to test the existence of a peculiar velocity of the entire Local Group. In this second fit one finds $w_{flow} = 382.2 \pm 58.3$ km/s toward an apex at $l^{II}_{Apex} = 273°.48 \pm 6°.64$, $b^{II}_{Apex} = 57°.35 \pm 3°.78$, and $D_{Apex} = 20.84 \pm 3.80$ Mpc, basically identical to the previous solution. Here now $H_0 = 42.1 \pm 7.9$ km/s/Mpc. The peculiar velocity of the Local Group is found as $v_{pec} = 87.4 \pm 16.8$ km/s toward $l^{II} = 313°.15 \pm 55°.03$ and $b^{II} = -62°.08 \pm 13°.88$. Again, all quoted errors include the uncertainty in the zero-point b. Now the residual r.m.s. velocity is reduced to 119 km/s.

4. DISCUSSION

The motion of the sun determined in section 3.1 is based on 24 galaxies with highly accurate radial velocity information. Because no expansion term is present for the Local Group the model uses only observed parameters and it is unlikely that this result can be improved much further beyond the current accuracy. The maximum difference of the adopted correction formula against the standard (RC2) correction is $+54$ km/s toward $l^{II} = 352°$ and $b^{II} = 19°$.

Incidentally, the apex of the flow determined in the previous section lies on the great circle connecting the centre of the Virgo cluster (M 87) and the centre of the Hydra I cluster (NGC 3311); roughly 18° from Virgo and about 31° from Hydra I. Note, that Hydra I itself lies not far from the apex of the microwave background radiation (cf. Lubin and Villela, this volume). Altogether, this is taken as evidence that within our sample of nearby galaxies differential effects due to bulk motion toward the Hydra/Centaurus supercluster are felt. Although in reasonable agreement with the result by Aaronson et al. (1982), the flow velocity found here is substantially higher than reported by Sandage and Tammann (1984). The result from this distance limited sample is not regarded as being good enough to make any further comment on the discrepancies found in the past.

However, significance is assigned to the formal residual velocity dispersion of about 120 km/s found for the model which includes a peculiar motion of the Local Group. In calculating the predicted radial velocity one makes use of several parameters that are themselves uncertain by a certain amount, in particular inclinations and corrected magnitudes. Therefore, the formal velocity dispersion found here represents an upper limit to the true (one-dimensional) velocity dispersion within volumes of space up to 20 Mpc in diameter.

The velocity dispersion of 48 km/s found inside the Local Group is certainly a lower limit to the cosmic velocity dispersion. Because it is anomalously small as compared to other nearby groups which we see from the distance rather than from the inside, the true velocity dispersion of "field" galaxies is estimated as 90 ± 20 km/s. This would be compatible with a peculiar velocity of the entire Local Group of the same order.

REFERENCES

Aaronson,M., Mould,J., Huchra,J., Schechter,P.L. and Tully,R.B.: 1982, *Astrophys. J.*, **258**, 64

deVaucouleurs,G., deVaucouleurs,A. and Buta,R.: 1981, *Astron. J.*, **86**, 1429

deVaucouleurs,G., deVaucouleurs,A. and Corwin,R.: 1976, *A Second Reference Catalogue of Bright Galaxies*, Univ. of Texas Press (RC2)

Huchtmeier,W.K. and Richter,O.-G.: 1986, *Astron.Astrophys.Suppl.Ser.* in press

Kraan-Korteweg,R.C. and Tammann,G.A.: 1979, *Astron. Nachr.*, **300**, 181 (KKT)

Richter,O,-G.: 1984, *The ESO Messenger No.*, **35**, 17

Richter,O.-G. and Huchtmeier,W.K.: 1984, *Astron. Astrophys.*, **132**, 253 (RH)

Richter,O.-G., Tammann,G.A. and Huchtmeier,W.K.: 1986, *Astron.Astrophys.* submitted

Sandage,A. and Tammann,G.A.: 1981, *A Revised Shapley-Ames Catalog of Bright Galaxies*, Carnegie Institution of Washington, Publication 635 (RSA)

Schechter,P.L.: 1980, *Astron. J.*, **85**, 801

Yahil,A., Tammann,G.A. and Sandage,A.: 1977, *Astrophys. J.*, **217**, 903 (YTS)

THE MOTION OF THE LOCAL GROUP RELATIVE TO THE NEAREST CLUSTERS OF GALAXIES

Jeremy Mould
Palomar Observatory
California Institute of Technology
Pasadena, CA 91125
U.S.A.

ABSTRACT. The motion of the Local Group relative to an ensemble of galaxy clusters between 4000 and 10000 km/s is within 250 km/s of that inferred from the microwave dipole anisotropy. The one-dimensional peculiar velocities of the component clusters are less than 350 km/s with 90% confidence.

1. SOLUTION FOR THE MOTION OF THE OBSERVER

One of the most basic experiments one can perform in astronomical kinematics is to take a set of isotropically distributed reference objects located at r_i, assume (as the simplest case) they are in random motion, and solve for the motion of the observer v_{obs} by fitting:

$$v_i = -v_{obs} \cdot \hat{r}_i \qquad (1)$$

where v_i are the set of observed radial velocities. For example one can measure the solar motion with respect to a local standard of rest.

One can do a similar experiment to determine the motion of the Local Group relative to a distant reference frame by adding an expansion term:

$$v_i = H_o r_i - v_{obs} \cdot \hat{r}_i \qquad (2)$$

The most recent published example is by de Vaucouleurs and Peters (1983). Our own work in this area is about to appear (Aaronson et al. 1986, hereafter ABMHSC).

2. VELOCITIES AND DISTANCES OF THE NEAREST CLUSTERS

Exactly what one learns from this kind of analysis depends on how the experiment is set up. We decided to abide by the following rules:
  1) get outside the Local Supercluster to establish the reference frame,

2) use a homogeneous set of distance indicators,
3) employ clusters of galaxies to construct the reference frame.

The first rule obviates the need to model the Virgocentric flow, a problem addressed in a previous paper (Aaronson et al. 1983, AHMST). The second rule sacrifices possible additional accuracy to avoid biases due to the distribution of alternative distance indicators. The third rule allows distances of very high quality ($0.4/\sqrt{n_i}$ mag, where n_i is the number of galaxies in the cluster) and avoids the need to deal with the Malmquist bias. The Malmquist bias is defined strictly here as the apparent brightening of a spatially distributed population due to the larger volume contributing to the statistically brighter sample. A cluster of 3 degree radius has a depth in the line of sight of 0.1 mag, much less than the scatter in the Tully Fisher relation of 0.4 mag.

Data for 140 galaxies in 10 clusters within 11000 km/s have been published by Bothun et al. (1985). Some modifications to the isophotal diameters used in the measurement of infrared magnitudes are discussed by Cornell et al. (1986). From this database ABMHSC have determined distances to the ten clusters by means of the infrared Tully Fisher relation. Mean velocities, V_i (good to 1 or 2%), were accumulated from all available sources. Ten data points were used to determine the four unknowns in equation (2). The first "unknown" (H_0) is just a reflection of the absolute calibration of the Tully Fisher relation adopted by ABMHSC. If one were to arbitrarily double the ten cluster distances, one would simply "derive" a value of H_0 half as large, but not change v_{obs} at all. In other words, the Tully Fisher relation is being used to determine that Pisces is 0.64 times as far away as Abell 400 to 7% accuracy, for example. The distances derived were tested for the possible selection effects discussed at this meeting. In our sample faint, low ΔV galaxies do not appear closer due to magnitude cutoff effects. In a 21 cm flux limited sample it is more relevant to ask if low H I flux galaxies appear to show systematic distance discrepancies. None was observed. We did notice a tendency for low surface brightness galaxies to appear systematically more distant than the cluster mean. This is a 3σ effect in the present data. Correcting it would make H_0 larger, but not significantly affect $\underline{v}_{obs}$.

3. COMPARISON WITH THE MICROWAVE ANISOTROPY

Our solution is $\underline{v}_{obs}$ = 780 ± 190 km/s toward l = 255° ± 17 and b = 18° ± 13. This vector is not significantly different from the velocity inferred from the microwave dipole anisotropy $\underline{v}_\mu$ = 600 km/s toward l = 268° and b = 27°. It is significantly different from the motion of the Local Group within the Local Supercluster measured by AHMST: $\underline{v}_{in}$ = 300 ± 40 km/s toward Virgo. The difference must correspond to a motion of the whole Local Supercluster with respect to the ensemble of clusters beyond it. The uncertainty in $\underline{v}_{in}$ is actually larger than quoted because of systematic uncertainties, such as the size of the Local Group peculiar velocities, the possibility of differential rotation in the Supercluster and the effect of tides induced by external mass concentrations (Lilje et al. 1986). In subtracting $\underline{v}_{in}$ from $\underline{v}_\mu$ to obtain the best estimate of the bulk motion of the Local Supercluster, $\underset{\sim}{v}_{LSC}$, these effects have been included as uncertainties: $\underset{\sim}{v}_{LSC}$ = 420 ± 80 km/s toward l = 277° ± 11, b = -3° ± 6.

As has been noted in a number of recent papers, this difference vector lies in the general direction of the Hydra Centaurus supercluster. The centroid of the component groups and clusters listed by Hopp et al. (1985) is at 3540 km/s redshift in the direction l = 290°, b = 25°.

To induce this motion a point mass at Hydra Centaurus would therefore require almost ten times the mass of Virgo. Figures 1 and 2 use diameter limited catalogs by Nilson (1973) and Lauberts et al. (1982) affording the best comparison of Virgo and Hydra Centaurus without redshifts. The difference between v_{LSC} at b = -3° and Hydra Centaurus at b = 25° would seem to call for additional mass behind the zone of avoidance. This is not apparent in the IRAS pictures presented at this meeting by Meiksin and Yahil and collaborators, but the IRAS counts may be diluted by the long tail on the luminosity function.

What is really needed to evaluate Hydra Centaurus is a southern equivalent of the CfA survey. For the time being we note that the mass of the Virgo supercluster inside the Local Group circle is ~0.7 x 10^{15} h_{100}^{-1} $M_\odot$ (from Tammann and Sandage 1985), while the virial mass contained in just the component clusters of Hydra Centaurus can be estimated at ~1.4 x 10^{15} h_{100}^{-1} $M_\odot$ from the data of Hopp et al. There seems to be a shortfall, but until more redshifts are available, the book should not be closed on the subject yet.

4. PECULIAR VELOCITIES ON LARGE SCALES

Peculiar velocities are a useful constraint on the spectrum of irregularities from which the structure in the Universe was formed. The present data offers information on peculiar velocities of volumes with radius somewhere between their current 3 h_{100}^{-1} Mpc size and their nearest neighbor separation of 16 h_{100}^{-1} Mpc.

ABMHSC made two points about the velocity residuals for the ten clusters. First, that they must be less than 500 km/s; second, that equation (2) fit better than it had a right to, even in the absence of peculiar velocities (i.e., χ^2 per degree of freedom was less than one). Since there have been claims for very large peculiar velocities at this meeting, let us look more closely at the current data.

The velocity residuals from equation (2) are very low: $\langle (v_i - V_i)^2 \rangle^{1/2}$ = 200 km/s. However, if we take the view that our result for $\underset{\sim}{v}_{obs}$ is not significantly different from $\underset{\sim}{v}_\mu$, then we cannot really distinguish that very quiet result from the picture drawn in Figure 3, where the residuals are 370 km/s rms with respect to $\underset{\sim}{v}_\mu$. (Note the systematic trend in Figure 3: this is what we are calling insignificant.)

The expected scatter is composed of σ_v the cluster redshift uncertainty, σ_R the cluster centroid uncertainty, and v_p, the peculiar velocity. From Table 6 of ABMHSC $\langle \sigma_v^2 + H_0^2 \sigma_R^2 \rangle^{1/2}$ = 450 km/s. Because of subclustering, however, σ_v and σ_R are statistical overestimates of the

Figure 1. The distribution of UGC galaxies larger than 3' diameter. The Virgo cluster is shown. The supergalactic plane proceeds northwards through Ursa Major. The scale on this figure is a third that of Figure 2.

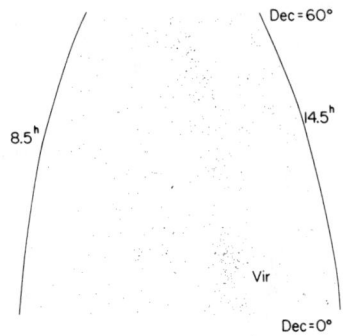

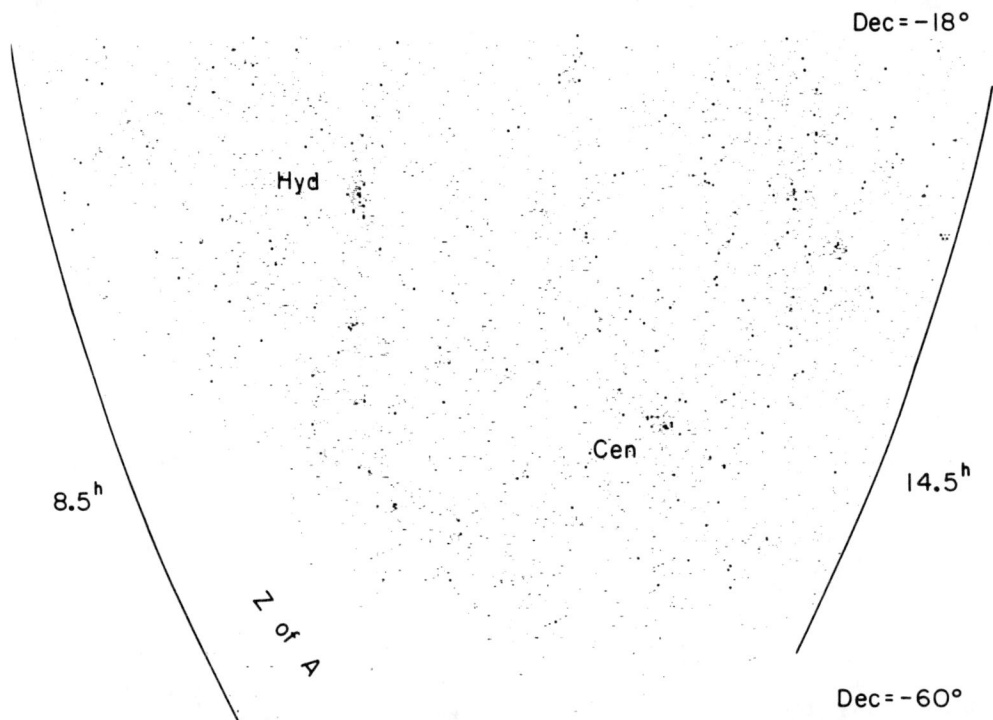

Figure 2. The distribution of ESO galaxies larger than 1'. The Hydra and Centaurus clusters are shown. The zone of avoidance is due to the southern Milky Way.

true uncertainties. Subclustering was demonstrated by ABMHSC for two clusters, but it must affect all clusters to a greater or lesser extent. If we replace the observed scatter in the Tully Fisher relation for each cluster by the smallest value of the true scatter we can imagine (0.3 mag), then $\langle H_o \sigma_R \rangle > 300$ km/s for the present sample. Neglecting $\sigma_v^2 \ll H_o^2 \sigma_R^2$ and converting to 90% confidence statistics, we can difference these new estimates in quadrature, obtaining $\langle v_p^2 \rangle^{1/2} <$ 350 km/s. This result is consistent with the peculiar velocity of the Local Supercluster, which is $420/\sqrt{3} = 240$ km/s, when scaled to one dimension.

Finally, since we claim nondetection of $|\underset{\sim}{v}_{obs} - \underset{\sim}{v}\mu| = 250$ km/s, we can put an upper limit on motion of the whole 100 h_{100}^{-1} Mpc radius volume under investigation of 150 km/s (1-D). Note that this volume (as illustrated in Figure 3) is a large contiguous volume, although it may appear contorted in a Galactic projection. It is simply that third of the sky which passes within 20° either side of the zenith here in Hawaii (or Puerto Rico). But we do have reduced sensitivity to its motion perpendicular to the plane of Figure 3. This needs to be remedied.

5. ACKNOWLEDGEMENTS

The work reported here is a joint project with M. Aaronson, G. Bothun, J. Huchra, R. Schommer and M. Cornell. NSF support is gratefully acknowledged. It is a pleasure to thank the Institute for Astronomy at the University of Hawaii for its hospitality, which allowed this report to be written, and Louise Good and Dottie Rosinsky for typing the manuscript.

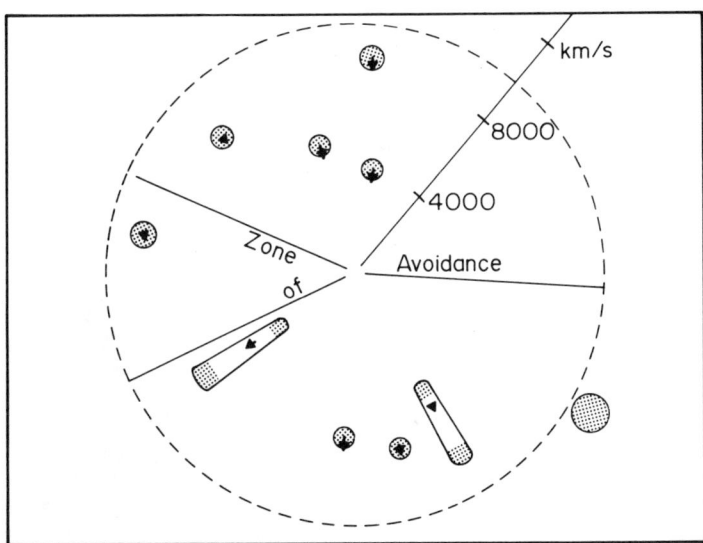

Figure 3. The distribution of clusters in the Arecibo plane. Velocity residuals in the microwave frame are shown by triangles or arrows. Two clusters are shown as spatially resolved.

REFERENCES

Aaronson, M., Bothun, G., Mould, J., Huchra, J., Schommer, R. and Cornell, M. 1986, Ap.J. (in press).

Aaronson, M., Huchra, J., Mould, J., Schechter, P. and Tully, R.B. 1982, Ap.J. 258, 64.

Bothun, G., Aaronson, M., Schommer, R., Mould, J., Huchra, J. and Sullivan, W. 1985, Ap.J. Suppl 57, 443.

Cornell, M., Aaronson, M., Bothun, G. and Mould, J. 1985, in preparation.

de Vaucouleurs, G. and Peters, W. 1984, Ap.J. 287, 1.

Hopp, U. and Materne, J. 1985, Astr. and Astrophys. Suppl. 61, 93.

Lamberts, A., Holmberg, E., Schuster, H. and West, R. 1981, Astr. and Astrophys. Suppl. 46, 311.

Lilje, P., Yahil, A. and Jones, B. 1986, preprint.

Nilson, P. 1973, Upsalla General Catalogue of Galaxies, Uppsala Astr. Obs. Ann., 6.

Tammann, G. and Sandage, A. 1985, Ap.J. 294, 81.

DISTANCES AND MOTIONS OF 2000 GALAXIES IN THE PISCES-PERSEUS SUPERCLUSTER

M.P. Haynes and R. Giovanelli
National Astronomy and Ionosphere Center
Cornell University
Ithaca, NY 14853

ABSTRACT. Twenty-one centimeter HI line observations have been made of over 2000 spiral galaxies in the region of the Pisces-Perseus supercluster, bringing to 3000 the number of known redshifts in that direction. Preliminary analysis of the redshift distribution shows a wealth of three-dimensional structure, including the prominent high density supercluster ridge, numerous lesser connective enhancements and also voids.

I. INTRODUCTION

The determination of the detailed large-scale structure in the local universe can be achieved via major observational efforts to obtain redshifts of a large number of galaxies. The connectivity of the major galaxy groupings in the region extending from the Pegasus to the Perseus (A426) cluster and including as well A194, A262, A347, A400 and Pisces, was apparent to early observers. The sky distribution of galaxies contained in the Catalog of Galaxies and Clusters of Galaxies (Zwicky et al. 1961-8; CGCG) vividly illustrates the enhancements in surface density seen in the Pisces-Perseus region. Indeed, the identification of enhancements over the average density in the local galaxian distribution is better obtained from the CGCG than from deeper surveys because the depth of the CGCG best magnifies enhancements corresponding to a magnitude of about +15 and such features, if they exist, will not be so diluted by the "noise" contributed by the background (Giovanelli et al. 1986). At the same time, only the availability of redshift information can prove that the surface density enhancements are not Martian canals.

The true complexity of the structure in the Pisces-Perseus region was suggested by Einasto et al. (1980) and witnessed by Gregory et al. (1981) as a preponderance of redshifts in the range around 5000 km/s, even with their limited sample. Because of its low galactic latitude, this region was not been included in the Center for Astrophysics survey (Huchra et al. 1983). Indeed, the effects of galactic extinction are severe in the northern and easternmost portions of the survey region.

The Pisces-Perseus supercluster is an especially attractive candidate for a major redshift survey via the 21 cm line of neutral hydrogen because it lies well within the redshift range (z < .04) accessible to modern hydrogen line spectrometers and because most of the region lies within the declination range viewable with the Arecibo 305 m telescope. At such a distance, spirals of even modest optical luminosity are easily detected at Arecibo in the 21 cm line; a significant number of 21 cm spectra have also be obtained with the N.R.A.O. 91 m telescope at Green Bank in the region north of +38 degrees declination. The 21 cm data contribute not only accurate redshifts, but also measurements of relative HI content and profile width, the latter to be used as a secondary distance indicator via the Tully-Fisher relation. The restriction of the survey to 21 cm line observations necessarily introduces a bias in that the observed galaxies all belong to morphological classes later than S0a. The HI detection rate for UGC galaxies in the major survey region ($+3^\circ <$ Dec. $< +35^\circ$) is better than eighty percent, and a good estimate of the selection effects can be made (Giovanelli and Haynes 1985). The morphological segregation seen in regions of different galaxian density ensures that the spiral component is in fact the best tracer of the lower density, large-scale structure (Giovanelli et al. 1986).

II. THE REDSHIFT DISTRIBUTION

At the time of this conference, our survey was nearing completion; this report is to be considered preliminary and deals only with the redshift distribution. Three figures are included in this presentation; all are cone diagrams of the redshift distribution of galaxies with known recessional velocity V_o less than +12000 km/s found in specific declination zones but covering, as the angular coordinate, the right ascension range from 22 to 4 hours. The velocity has been corrected for solar motion with respect to the Local Group assuming a velocity of 300 km/s towards $l = 180^\circ$, $b = 0^\circ$. Galaxies with redshifts available from other sources are also included.

Figure 1 shows the observed redshift distribution of all 2556 galaxies with known redshifts in the entire search area, covering all declinations from 0 to +50 degrees. Perhaps three features are most striking in Figure 1. First, the Perseus cluster shows itself evident as the "finger of God" at R.A. = 03h15m; it is even more evident in Figure 2. Second, the Pisces-Perseus supercluster is responsible for the large concentration of galaxies in the redshift range from +4500 to +6000 km/s. Third, a conspicuous absence of galaxies -- a void -- is present at a redshift of about +3200 km/s and covering at least 15 degrees of right ascension. Note that this void extends at least fifty degrees in declination and some 1500 km/s in velocity space. It should be emphasized that there are no experimental reasons why galaxies of average luminosity should not have been found by this survey in that empty volume, if such truly exist there. In the region identified as the void, that is, in which we find no galaxies, we would have expected, given the magnitude distribution of our sample, to have found 20 objects with redshifts in the range 2500 to 4000 km/s. While the precise outline of the void is quite uncertain, it appears to be both large and elongated. It is not a single, spherically-symmetry empty volume.

Although Figure 1 is useful in providing an overall impression of

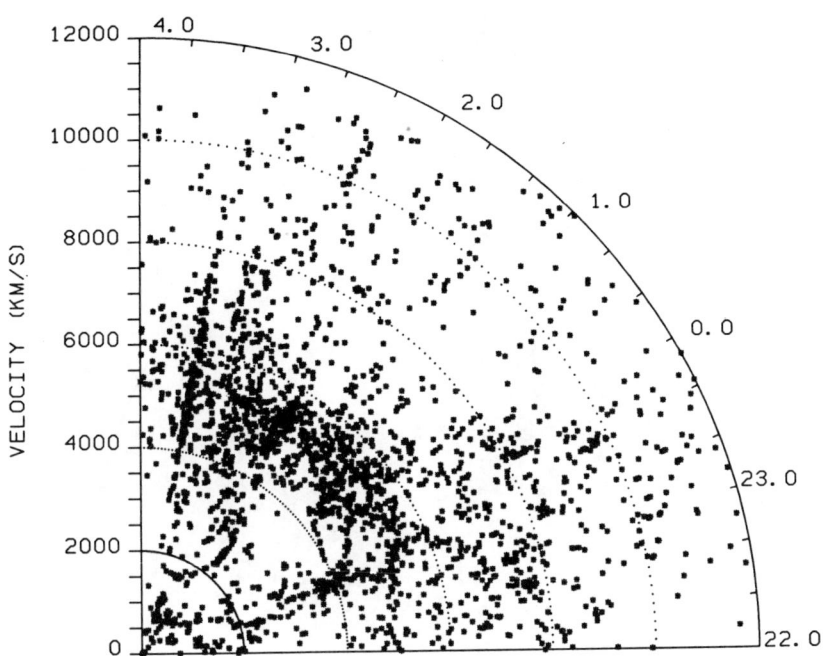

Figure 1. Cone diagram for all galaxies with known redshift in the range $0 < V_o < +12000$ km/s in the declination range from 0 to +50 degrees. The angular coordinate is right ascension in hours.

connectiveness and structure, the information it contains is, of course, blurred by the large area of sky covered. When combined also with the sky distribution of CGCG galaxies, cone diagrams of smaller regions allow the identification of separate connective enhancements -- "filaments". The velocity dispersion characteristic of individual enhancements is generally much lower than that seen when galaxies over a wide area are combined. When adequate account of the apparent orientation to the line of sight is taken for the filament extending from the Pegasus cluster to the concentration in Pisces around NGC 383, the observed velocity dispersion, over a length of 17/h Mpc, is only 128 km/s. The velocity gradient along the filament is about 1000 km/s.

Figures 2 and 3 show cone diagrams, identical in form to Figure 1, but separately for the declination zones +30 to +50 degrees (Fig. 2) and 0 to +30 degrees (Fig. 3). This separation has been chosen to emphasize the difference between the region dominated by the main ridge of the supercluster, north of 30 degrees, and the lower density extension to the south. The canonical Pisces-Perseus supercluster, connecting Perseus, A262, and Pisces, lies in the northern portion of our survey region, in

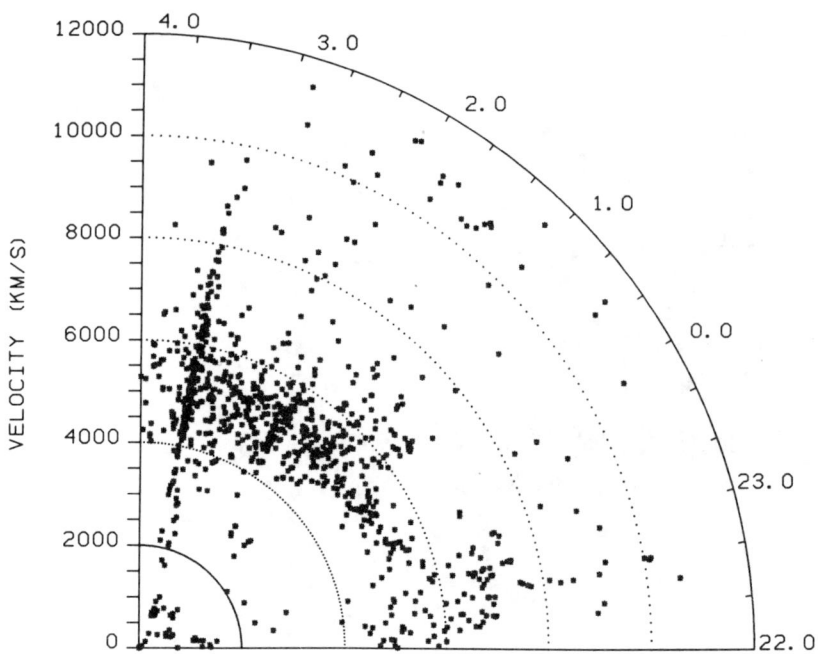

Figure 2. Cone diagram showing the redshift distribution of all galaxies included in Figure 1 but which have declinations between +30 and +50 degrees.

the declination range from +30 to +50 degrees; only the Pegasus cluster lies south of +30 degrees.

It is immediately apparent that the difference seen in the surface density distribution of galaxies is reflected also in the redshift distribution. The ridge seen in the sky distribution of CGCG galaxies extending from Pegasus through Pisces to Perseus is a real three-dimensional structure; the major dynamical units are embedded in a more loosely distributed but connecting enhancement in the surrounding galaxy distribution. Particularly striking in Figure 2 is the general absence of galaxies in the foreground of Pisces-Perseus.

While Figure 2 is dominated by the Perseus cluster and the Pisces-Perseus supercluster ridge, Figure 3 shows a wealth of connective structures which cannot be explained by the dynamical "finger of God" effect. In combination with the surface density maps, the redshift diagrams present a picture of intersecting structures surrounding volumes that are nearly empty. More detailed examination of finer slices of the redshift distribution shows that features seen in separate declination windows often correlate, but are slightly offset in velocity, as if they represent sheets inclined to the line of sight. Evident in Figure 3 is a

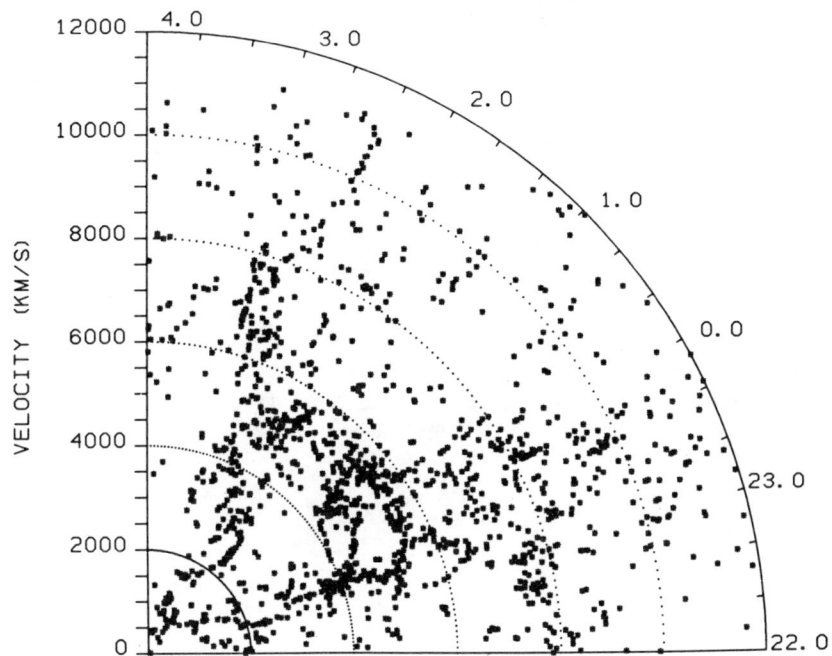

Figure 3. Cone diagram showing the redshift distribution of galaxies included in Figure 1 but which have declinations in the range 0 to +30 degrees.

possible connection to the Local Supercluster around the edges of the noticeable void at +3200 km/s so apparent in Figure 1.

III. CONCLUSION

The preliminary analysis of the three-dimensional structure of the galaxy distribution in the region of the Pisces-Perseus supercluster reveals a wealth of detail. The galaxies are generally found in interconnecting three-dimensional features which can be either sheets or filaments and which surround nearly empty volumes. The complexity of the distribution is only too obvious; the combination of the Hubble expansion, dynamical effects and random motions further confound the contruction of the true galaxy distribution. However, connectiveness appears to play a dominant role in the local universe. Our current understanding is perhaps reminiscent of the difficulties encountered by early attempts to discern the signature of our galaxy's spiral structure.

Acknowledgements. It is a pleasure to thank Guido Chincarini for his collaboration with many phases of this project. The National Astronomy and Ionosphere Center is operated by Cornell University under contract with the National Science Foundation. The National Radio Astronomy Observatory is operated by Associated Universities, Inc. under contract with the National Science Foundation.

References

Einasto, J., Joeveer, M., Saar, E. 1980, M.N.R.A.S. 193, 353.
Giovanelli, R., Haynes, M.P. 1985, A.J. 90, 2445.
Giovanelli, R., Haynes, M.P., Chincarini, G.L. 1986, Ap.J. (in press).
Gregory, S.A., Thompson, L.A., Tifft, W.G. 1981, Ap. J. 243, 411.
Huchra, J., Davis, M., Latham, D., Tonry, J. 1983, Ap. J. Suppl. 52, 89.
Zwicky, F., Herzog, E. Karpowicz, M., Kowal, C.T., Wild, P. 1961-68, Catalog of Galaxies and Clusters of Galaxies, (California Institute of Technology, Pasadena), in six volumes (CGCG).

ELLIPTICAL GALAXIES AND NON-UNIFORMITIES IN THE HUBBLE FLOW

David Burstein, Dept. of Physics, Arizona State University
Roger L. Davies, Kitt Peak National Observatory, NOAO
Alan Dressler, Mt. Wilson and Las Campanas Observatories
S.M. Faber, Lick Observatory, U.C. Santa Cruz
Donald Lynden-Bell, Institute of Astronomy, Cambridge, England
Roberto Terlevich, Royal Greenwich Observatory
Gary Wegner, Dept. of Physics and Astronomy, Dartmouth College

ABSTRACT. We have obtained spectroscopic and photometric data for 400 elliptical galaxies that permit us to predict distances to individual galaxies with accuracies of $\pm 23\%$. Systematic velocities relative to a smooth Hubble flow are observed that are most straightforwardly interpreted as a bulk motion of approximately 700 km/s towards $l=299°$, $b=+1°$ for galaxies within 60 h^{-1} Mpc of the Local Group. A preliminary estimate for the rms motion of elliptical galaxy gruops relative to the bulk motion is 300 km/s. Spiral galaxies studied by Aaronson et al. and by Rubin et al. yield similar results for both the bulk and the rms motion.

1. THE SAMPLE OF ELLIPTICAL GALAXIES

Measures of central velocity dispersion (σ), central absorption line strength, luminosity profile and group/cluster membership have been made for 390 elliptical galaxies evenly distributed on the sky. The sample consists of 100 galaxies in rich clusters, 150 in poor groups and clusters and 140 other galaxies, including 268 out of the 322 ellipticals brighter than $B_T=13.0$. Total magnitudes, B_T, and circular effective radii, A_e, have been determined from an internally consistent set of 3500 photoelectric aperture measures, together with CCD luminosity profiles for 70 galaxies.

Within the three-dimensional space defined by the parameters σ, effective radius and surface brightness, the properties of elliptical galaxies define a plane that is canted to all three axes. We have developed a magnitude-related parameter, D_Σ, that, along with σ, transforms this three-dimensional relationship into a two-dimensional, linear relationship. D_Σ is the diameter within which an elliptical galaxy reaches a mean surface brightness of 20.75 B mag arcsec^{-2}. D_Σ has the observational advantage in that it can be interpolated accurately from photoelectric aperture data.

We assume that deviations from the $D_\Sigma - \sigma$ relation are due to both distance errors and intrinsic scatter. From this, we find a distance accuracy of ±0.10 dex for 100 galaxies in rich clusters (Coma, Virgo, Perseus, A2199, Fornax and DC2345-32), ±0.08 dex for 150 galaxies in poor clusters, and ±0.09 dex for the sample as a whole. We adopt a mean error per galaxy of ±0.09 dex (= ±0.45 mag = ±23% in distance), comparable to the accuracy quoted for the IR-Tully-Fisher (IR-TF) relation for spirals (Aaronson et al. 1986). (A separate correlation between σ and absorption line strength indicates that much of this quoted error is intrinsic to the properties of elliptical galaxies.)

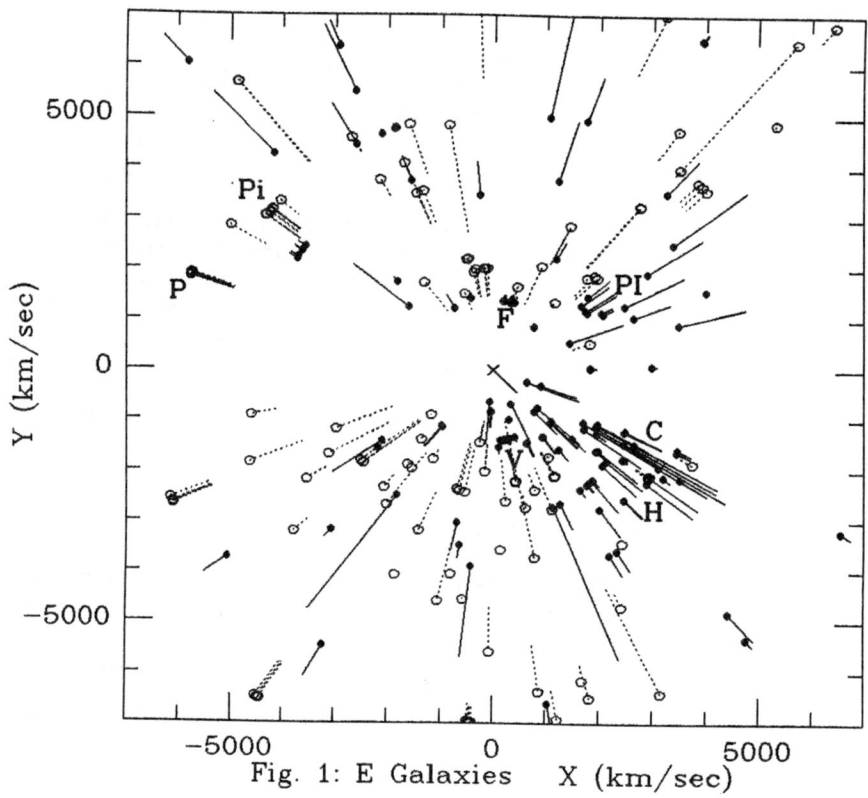

Fig. 1: E Galaxies   X (km/sec)

2. SYSTEMATIC MOTIONS AND PECULIAR MOTIONS OF ELLIPTICAL GALAXIES

The peculiar motions inferred from the $D_\Sigma - \sigma$ relation for 300 elliptical galaxies are shown in Fig. 1 for galaxies with a Supergalactic latitude $<|45|°$. [These motions are projected onto the Supergalactic plane, as we have found that galaxies within 5000 km/s tend to congregate near this plane (Dressler et al. 1986; Tully, this conference)]. Predicted distances have been corrected for Malmquist bias as described by Lynden-Bell et al. (1986). All distances are expressed in units of km/s, with the zero point of velocity set to yield zero peculiar velocity for the Coma cluster. (As there are no nearby elliptical galaxies to act as fundamental distance calibrators, these data cannot currently be used to predict a Hubble constant.)

An important feature of Fig. 1 differentiates it from other similar figures: the velocity reference frame is that of the microwave background (MWB). The motions shown are therefore presumably the motions of galaxies relative to the fundamental cosmic rest frame. The Local Group motion with respect to the MWB is assumed to be 607 km/s towards $l=267°$, $b=27°$.

The estimated distance of each galaxy projected onto the Supergalactic plane is plotted as a point, and the peculiar motion of each galaxy is represented as either a solid line (a radial velocity vector pointed away from the position of the Local Group) or a dashed line (a radial velocity vector pointed towards the Local Group). Galaxies in clusters are plotted with the median peculiar velocity of the group, and have smaller errors. Vectors for individual galaxies in clusters are sometimes overplotted, so the number of galaxies is therefore more than the apparent number of vectors. Positions of prominent groups and clusters are marked (V=Virgo; F=Fornax; P=Perseus; Pi=Pisces; H=Hydra; C=Centaurus; PI=Pavo-Indus). The vector representing the total space velocity of the Local Group relative to the MWB is the solid line emanating from the cross at the origin.

It is apparent that galaxies in the HCPI region and galaxies in the Perseus-Pisces region are moving relative to the MWB in a direction similar to that of the Local Group. In particular, the 44 galaxies in the HC region have a radial velocity component of 700 $\pm$ 150 km/s relative to the MWB reference frame. The most straightforward interpretation of all the data is that of an average bulk motion of galaxies within a sphere of radius $60h^{-1}$ Mpc from the Local Group. A global solution for the bulk motion of the 249 elliptical galaxies in this sphere yields a velocity vector of 700 km/s towards $l=299°$, $b=+1°$. This direction is parallel to the X-axis of Fig. 1. The motion of the Local Group is generally consistent with participation in the same bulk motion.

In addition to bulk motion, we also observe significant peculiar motions of individual galaxies, groups and clusters that are larger than the observational errors. The mean rms velocity 'noise' for 57 groups and clusters relative to the MWB is 500-600 km/s and is of order 300 km/s relative to the bulk motion. (Participants in the Hawaii conference will note that this latter estimate of rms noise value is smaller than the rms noise quoted at that time. This is due to revisions in the analysis that affected primarily the calculation of this quantity.)

The idea of a large bulk motion of galaxies in a large volume of space is reminiscent of a similar such motion found by Rubin et al. in 1976. That result has remained controversial and, for example, appears inconsistent with the peculiar motions of spirals based on the infrared Tully-Fisher (IR-TF) relationship (Aaronson et al. 1982; 1986; Mould, this volume). However, in light of the new elliptical galaxy results, at least some of these discrepancies can plausibly be ascribed to differences in the velocity and spatial distributions of different samples. Some comparisons are shown in the next sections.

3. GALAXIES IN THE ARECIBO DECLINATION RANGE

The most recent measure of the peculiar velocity of the Local Group using the IR-TF method is that of Aaronson et al. (1986; as summarized by Mould, this volume). The 10 galaxy clusters used in that analysis are restricted to declinations accessible to Arecibo; the actual galaxies observed span a range in declination from $+4°$ to $+30°$.

The inferred motion of the Local Group relative to this sample quoted by Mould is 780 ± 190 km/s towards $l=255° \pm 17°$, $b=18° \pm 13°$. The elliptical sample restricted to the same range in declination yield a Local Group motion of 655 km/s towards $l=206°$, $b=27°$, based on 96 galaxies. Our errors are not yet precisely known, but we believe that the two vectors are statistically consistent (see Dressler et al. 1986 for further discussion). The rms velocity noise of <350 km/s estimated by Mould for these clusters is similar to the revised value of 300 km/s calculated for the elliptical galaxy sample (Sec. 2).

We also note that the relative distance moduli between the Virgo and Coma clusters derived from the $D_\Sigma - \sigma$ relation is within 0.06 mag (= 3% in distance) of that quoted by Aaronson et al. (1986). This agreement is actually better than the individual errors predict.

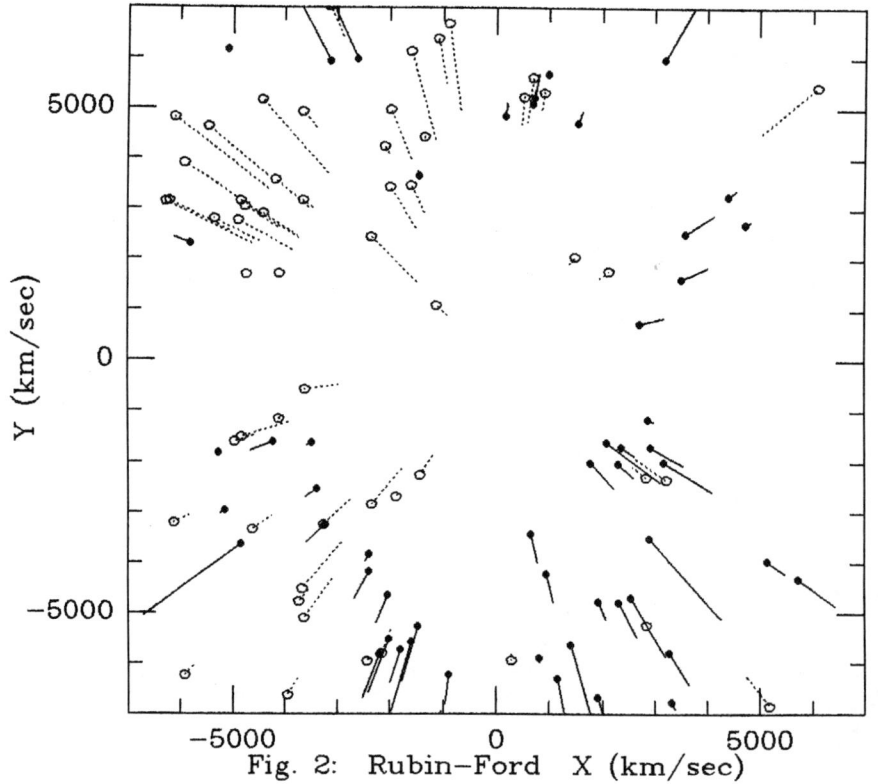

Fig. 2: Rubin-Ford X (km/sec)

4. THE RUBIN-FORD EFFECT

Rubin et al. (1976) derived a then-large motion for the Local Group of approximately 500 km/s towards the direction of $l=163°$, $b=-11°$, relative to 96 Sc I galaxies located in a shell at 3500-6500 km/s. In Fig. 2, we show a diagram similar to Fig. 1 for the Rubin-Ford galaxies within $\pm 45°$ of the Supergalactic plane. The original Rubin-Ford data have been updated with recent photometry (Peterson and Baumgart 1986) and have been numerically corrected for Malmquist bias using a formula similar to that of Eqn. 4-54 of Mihalas and Binney (1981, pg. 241).

From Fig. 2, the peculiar velocities of the Rubin-Ford galaxies, relative to the MWB reference frame, show an apparent bulk motion similar to that seen in the elliptical galaxy sample. A bulk motion vector of 600 km/s towards $l=290°$, $b=+33°$ is derived for all Rubin-Ford galaxies with radial velocities less than 8000 km/s. Both the derived velocity and the longitude are close to that of the elliptical galaxy sample. The difference of $32°$ in galactic latitude probably reflects the fact that the Rubin-Ford spirals sample systematically different regions of space than our elliptical galaxies, a point discussed further below.

5. IR-TF RESULTS FOR SPIRAL GALAXIES IN THE LOCAL SUPERCLUSTER

The apparent motions of 300 spiral galaxies in the Aaronson et al. (1982; AHMST) sample, again relative to the MWB, are shown in Fig. 3 (note the enlarged scale from the previous figures, due to the limitation of velocities to less than 3000 km/s in this sample).

There are two differences in the derivation of IR-TF distances in Fig. 3 and those given by AHMST: a) The new quadratic correlation between IR luminosity and HI linewidth given by Aaronson et al. (1986) has been used. This systematically decreases the estimated distances for low-luminosity spirals, which are common in this nearby sample.
b) IR luminosities have been estimated using the procedures of Burstein (1982). Tests show the latter revision has no systematic effect on the gross features of Fig. 3. A Malmquist correction of the kind used by Mould and Aaronson (1983) has also been applied to these data.

The projected distribution in the Supergalactic plane of the AHMST spirals is quite different from that of the elliptical sample: Only 3% of the AHMST galaxies are located in the HCPI region, whereas nearly 1/3 of the elliptical galaxies with V_{obs} < 3000 km/s are located there. Nearly half of the AHMST sample but only 1/3 of the elliptical sample is located within $45°$ of the core of the Virgo cluster.

This difference in distribution leads to superficially different impressions of the motions of galaxies over large regions of space in Figs. 1 and 3. For example, many of the AHMST spirals in the Ursa Major-Virgo complex in Fig. 3 (negative Y coordinate) appear to be approaching the Local Group. The same region as traced out by the ellipticals appears to show a net motion away. On closer inspection, however, one sees that in both diagrams there is the same dichotomy: galaxies to the left of Virgo (in Ursa Major) approach us, whereas galaxies to the right (in HC) recede. The different global impression is due to the very different distributions of the two samples in the

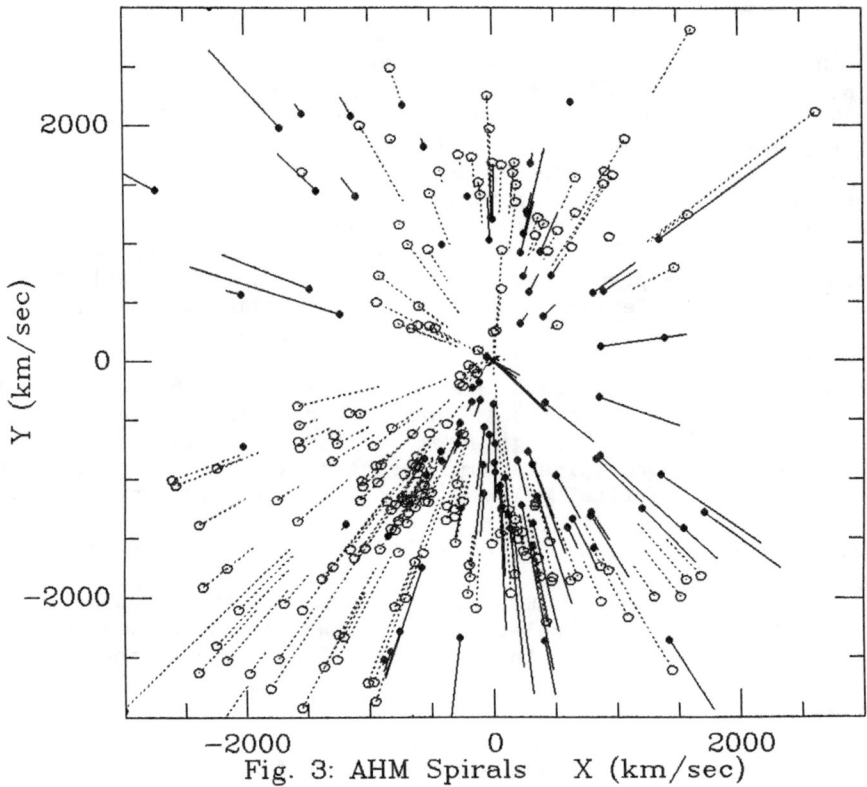

Fig. 3: AHM Spirals   X (km/sec)

figures. Comparison of other regions in Figs. 1 and 3 leads to a similar conclusion: by and large the motions of galaxies in similar regions of the figures are similar; it is the distribution of galaxies that is different. Part the latter difference is due to the tendency of spirals and ellipticals to populate different regions of space, and part due to real sample differences over the sky.

A solution for the bulk motion of 295 spirals in the AHMST sample, relative to the MWB yields a vector with velocity 410 km/sec towards direction $l=286°$, $b=12°$. A solution for the bulk motion of 164 elliptical galaxies in the same velocity range yields a vector with velocity 760 km/sec towards $l=305°$, $b=+1°$, close to that derived in Sec. 2 for elliptical galaxies spanning twice the distance. The bulk motion vectors derived from the AHMST sample and the nearby ellipticals agree reasonably well in direction, but not as well in size. We believe that this difference is not unexpected, given the fact that most of the AHMST sample is distributed on the sky in directions largely perpendicular to the bulk motion, and thus has decreased sensitivity to it.

An rms velocity noise of 300 km/s relative to the bulk flow is derived for this sample of spirals, similar to the value obtained for elliptical groups. (This rough equality is another result of the change in outlook since this meeting, described in Sec. 2). An rms noise of 300 km/s for spirals is larger than the values of 150 km/s or less quoted by AHMST and by Lilje et al. (1986). However, we have not removed a general Virgo infall component from our velocities as did these authors, nor have we included a quadrupole term, as did Lilje et al. Inclusion of these terms could significantly reduce the residual scatter.

On the other hand, both Virgo infall and a quadrupole term vary significantly over short distances, say 2000 km/s. The elliptical galaxy motions we are modelling cover much larger regions (14,000 km/s in diameter), and no detailed model exists to fit small-scale galaxy motions over such a large volume. Thus, a crude model for nearby spirals that removes only their bulk motion is a valid manner of comparing the rms velocity dispersions of these two samples.

6. DISCUSSION

We have shown that, when analyzed in similar fashion, the IR-TF spirals, the Rubin-Ford spirals and the ellipticals all give evidence of a large-scale bulk motion of galaxies in a sphere out to 6000 km/s of $\sim$600-700 km/s in the direction $l\sim 290°$, $b\sim 0°$. The three samples do not agree perfectly, and their motions when expressed in Local Group coordinates appear quite disparate. However, once transformed to MWB coordinates, the motions are relatively consistent. This agreement emerges as a result of the coordinate transformation, which requires the addition of the Local Group vector motion relative to the MWB. This vector is large (600 km/s) and dwarfs the other shorter, but more discrepant vectors. In a real sense, the conclusion of a large bulk motion stems primarily from the dominance of this Local Group motion relative to the MWB. Since this is by far the most accurately known quantity, it thus is hard to argue away the existence of a significant bulk motion of nearby galaxies in terms of observational errors alone.

In addition to a bulk flow, Figs. 1-3 strongly suggest a patchy, correlated behavior in peculiar motions that extends over regions some 1000-2000 km/s in size. Ursa Major, the HCPI region, the Pisces region and the NGC 1600-1700 groups (not shown) are examples. This fact, coupled with spatial segregation of ellipticals and spirals, could lead to apparent large differences in the bulk motions of different samples of galaxies. Once this patchiness is appreciated, it is no longer necessary to insist that the AHMST spiral sample in a small volume and the Rubin-Ford sample in a much larger volume yield similar solar motions. Nor is it necessary that the ellipticals in either volume should give similar results. In other words, the real errors in solar motion solutions are systematic, highly dependent on samples, and are considerably larger than has been thought in the past.

Large scale motions raise interesting cosmological questions. An obvious point is whether a large bulk motion of the magnitude found here is consistent with known large-scale density fluctuations. Another question relates to the value of H_o: How large a volume of space must be sampled in order to obtain a global estimate of H_o? These and other related issues are explored in Dressler et al. (1986) and in Lynden-Bell et al. (1986).

In conclusion, we note that the above interpretation of the observations rests on two crucial assumptions: 1) The MWB radiation defines an absolute cosmological rest frame, and 2) the intrinsic properties of galaxies (both spirals and ellipticals) are everywhere constant. To question either of these assumptions would open up puzzles of equal or greater magnitude than those posed by large-scale motions.

7. ACKNOWLEDGEMENTS

In a survey this large, it is impossible to thank all the people who have helped us. We do wish to thank Rem Stone, Bob Jedrzejewski, Cristina Morea Dalle-Ore, Jesus Gonzalez, Bob McMahan and Ron Marzke for their computing and logistic support over the years. This research was supported in part by NSF Grant AST82-11551, a NATO Travel Grant, an ASU Grant in Aid, a UCSC Faculty Research Grant, Dartmount College, the Carnegie Institution and the Institute of Astronomy. DB thanks Brent Tully and the organizers of this conference for their gracious hospitality.

REFERENCES

Aaronson,M., Bothun,G., Mould,J., Huchra,J., Schommer,R.A. and Cornell,M.E. 1986, Ap. J., in press.

Aaronson,M., Huchra,J., Mould,J., Schechter,P. and Tully,R.B 1982, Ap. J. 258, 64.

Aaronson,M. and Mould, J. 1983, Ap. J. 265, 1.

Burstein, D. 1982, Ap. J. 253, 539.

Dressler,A, Burstein,D., Davies,R.L., Faber,S.M., Lynden-Bell,D., Terlevich,R. and Wegner,G. 1986, in preparation.

Lilje, P. et al. 1986, as quoted at this meeting.

Lynden-Bell, D., Burstein,D., Davies,R.L., Dressler,A.,Faber,S.M., Terlevich,R. and Wegner,G. 1986, in preparation.

Mihalas,D. and Binney,J. 1981, <u>Galactic Astronomy: Structure and Kinematics</u>, (Freeman: San Francisco), pg. 241.

Mould,J. 1986, this conference proceedings.

Peterson,C.J. and Baumgart,C.W. 1986, Ap. J. in press.

Rubin,V.C., Ford,W.K., Thonnard,N. and Roberts,M.S. 1976, Astron. J. 81, 719.

Tully, R.B. 1986, this conference proceedings.

LARGE-SCALE ANISOTROPY IN THE HUBBLE FLOW

C. A. Collins, R. D. Joseph, & N. A. Robertson
Blackett Laboratory, Imperial College, London SW7, England

ABSTRACT. The <u>apparent</u> Local Group motion inferred from infrared observations of an all-sky sample of spiral galaxies is interpreted in terms of a large-scale streaming motion for these galaxies on a scale of 50 h^{-1} Mpc. If this streaming is induced by large-scale density fluctuations, the fluctuations required are compatible with the observed distribution of luminous matter, but only in a closed universe. In the context of current speculation regarding non-baryonic particle species which may dominate the mass of the universe, this result seems to be inconsistent with cold dark matter models.

1. INTRODUCTION

A classic study of the Local Group motion was carried out by Rubin et al. (1976a, 1976b). They identified a sample of 96 spiral galaxies in the redshift interval 3500-6500 km s^{-1} (mean redshift 5100 km s^{-1}) which were all thought to belong to a narrowly-defined luminosity class (ScI-II) and could therefore be used as standard candles. Using the measured redshifts and optical magnitudes for these galaxies Rubin et al. inferred an apparent Local Group motion of 454 ± 127 km s^{-1} toward $\ell = 163°$, $b = -11°$.

This result was criticised on a number of points, and was not confirmed by later studies using other galaxy samples (Hart & Davies 1982, de Vaucouleurs & Peters 1984). We are attempting to resolve these disagreements by re-observing the Rubin et al. sample in the near-IR. In addition to greatly reduced extinction, the IR data permits use of two powerful luminosity indicators using independent third variables, the IR Tully-Fisher relation (Aaronson et al. 1979) and the optical-IR colour-magnitude relation (Tully et al. 1982), to determine distances.

2. OBSERVATIONS AND ANALYSIS

We have so far measured 45 of the 96 ScI-II galaxies comprising the Rubin et al. "minimum bias subset" at J, H, and K in a 35 arcsec aperture using the 1.5 m Infrared Flux Collector in Tenerife. These 45 galaxies are well-distributed around the sky, as shown in Fig. 1, and have a redshift distribution representative of the entire sample.

We have carried out three different solutions for the apparent Local Group motion using these H magnitudes. For the first solution the H magnitudes were corrected for effects of both Malmquist bias and use of a fixed aperture. A "composite" correction for both of these effects was obtained from the slope of a plot of absolute magnitude vs. redshift. Hubble velocities were then derived from the corrected H magnitudes, assuming all the galaxies have the same absolute magnitude. The difference between the observed redshift and the derived Hubble velocity, i.e. $v_{obs} - v_H$, and a putative systematic dipole velocity, v_D, is minimised in the least squares sense, to derive a value for v_D. Since the redshifts have been corrected for the solar motion with respect to the Local Group (300 km s^{-1} toward $\ell = 90°$, b = 0°), v_D is the velocity of the Local Group relative to the frame defined by these galaxies. The result is shown in the first line of Table 1.

In the second solution we have used the HI 21 cm velocity widths to construct the IR Tully-Fisher relation. Application of this relation to correct for luminosity differences among the galaxies gives the solution in line 2 of Table 1.

For a third solution, given in line 3 of Table 1, we have used the B - H vs. H_{abs} colour-magnitude relation for this sample as a luminosity indicator. We find a slope of 1.85 ± 0.3 for this relation, consistent with slopes determined for other galaxy samples (Tully et al. 1982).

We have also used 100 μm flux densities from the IRAS Point Source Catalogue, which are available for for 64 of the 96 galaxies. The distribution of these galaxies around the sky is also shown in Fig. 1. We have corrected this data for Malmquist biasing using a procedure similar to the "composite" correction for the H magnitudes described above. The resulting solution is shown in line 4 of Table 1.

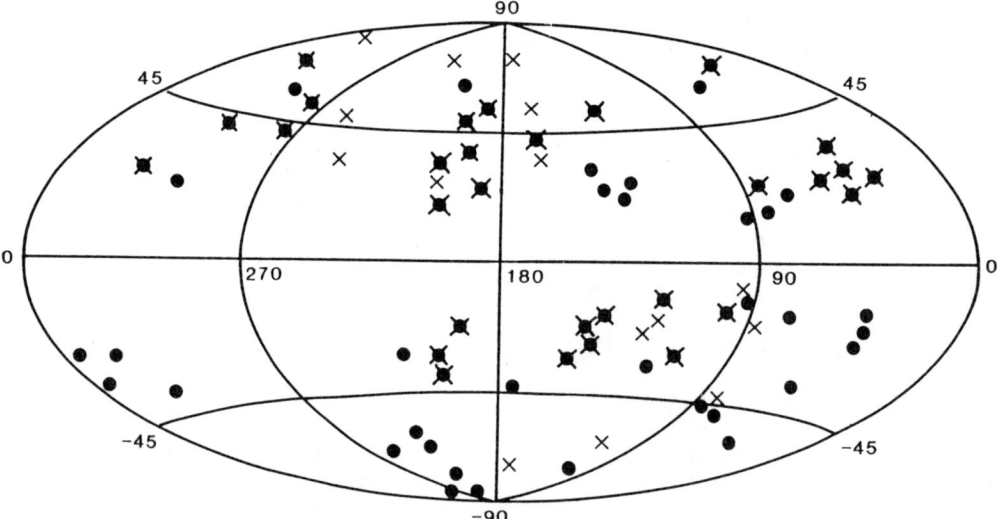

Fig. 1. Positions of galaxies observed in galactic coordinates. Galaxies observed at JHK (**X**), galaxies detected by IRAS (●).

In line 5 of Table 1 we have given the solution Rubin et al. obtained (1976b) for the entire sample, using the blue magnitudes <u>without</u> their diameter correction. To within the errors the results are independent of the photometric data used and inclusion of an independent luminosity indicator in the analysis does not change the solution. For comparison we have also listed the dipole anisotropy in the 3K cosmic background radiation (CBR) (Fixen et al. 1983, Lubin et al. 1983).

TABLE 1 Solutions for the apparent Local Group motion

Solution	v_D(km s^{-1})	ℓ(deg)	b(deg)
H magnitudes, 45 galaxies	621 ± 300	186 ± 30	-03 ± 29
H mags. T-F relation, 40 galaxies	846 ± 334	214 ± 40	-33 ± 24
H mags. C-M relation, 45 galaxies	680 ± 330	184 ± 35	-36 ± 30
IRAS 100 μm, 64 galaxies	662 ± 220	190 ± 20	-06 ± 18
B magnitudes, 96 galaxies	586 ± 200	202 ± 19	-11 ± 17
Dipole anisotropy in the CBR	600 ± 50	265	+35

3. DISCUSSION

The fact that the apparent Local Group motion we find for this sample of galaxies is so different from that inferred from the CBR, leads us to interpret our result in terms of a large-scale streaming motion induced by correspondingly large-scale density fluctuations. Subtracting our apparent Local Group motion, v_D, from the velocity of the Local Group as inferred from the CBR dipole anisotropy gives a galaxy streaming velocity of 970 ± 300 km s^{-1} toward $\ell = 305°$, b = 47°.

Recent studies by Davies & Staveley-Smith (1985) and Aaronson et al. (1986) suggest a motion of the Local Supercluster toward $\ell = 300°$, b = 30°. Whereas the above authors interpret this motion as dynamical attraction toward the Hydra-Centaurus Supercluster, the agreement with the direction of the streaming motion of our galaxy sample suggests instead a consistent picture in which the galaxies within a radius of 50 h^{-1} Mpc seem to be participating in the same overall streaming motion with respect to the fundamental cosmic reference frame.

The net streaming motion for elliptical galaxies out to redshifts of ~ 6000 km s^{-1}, discussed by Burstein et al. in this volume, seems to provide strong support for this picture.

There are two general approaches to using the peculiar velocities of galaxies to constrain models for the material content of the universe. The first is to assume that light traces mass and use linear perturbation theory to constrain the density parameter, Ω_0. Clutton-Brock & Peebles (1981) have calculated the RMS peculiar velocity of a spherical shell of galaxies in the redshift range of this galaxy sample, due to density fluctuations consistent with the galaxy two-point correlation function. They find

$$\langle v^2 \rangle^{1/2} = 300 \Omega_0^{0.6} \text{ km s}^{-1},$$

This shows that consistency between the two-point correlation function and our large streaming velocity requires a high-density universe, and excludes cosmologies with $\Omega_0 \ll 1$.

The second approach is to ask what types of particles can produce density fluctuations of the amplitude required to induce streaming velocities on the observed spatial scales. The mass of the universe is currently thought to be dominated by "cold" or "hot" dark matter composed of some exotic particle species. One difference between these models is the resulting density fluctuation on large scales. Vittorio & Silk (1985) predict a streaming velocity on the scale of the Rubin et al. sample of $\leqslant 160$ km s^{-1} in models dominated by cold dark matter. The streaming velocity we find therefore seems to rule out cold dark matter models. If the dark matter is composed of hot particles, on the other hand, the mass fluctuation spectrum will have greater amplitude at large spatial scale, and will therefore tend to produce higher streaming velocities (cf. Kaiser 1983).

ACKNOWLEDGEMENTS

It is a pleasure to thank Mike Selby, who provided the JHK photometer, Dave King, who measured the galaxy positons, and Gillian Wright, Philip Andrews, Jack Abolins, Carlos Martinez, and the late Carlos Sanchez-Magro, who assisted with some of these observations. We are grateful to the UK PATT and Spanish CAT for allocation of observing time on the 1.5 m IRFC at the Observatorio del Teide, which is operated by the Instituto de Astrofisica de Canarias in Tenerife.

REFERENCES

Aaronson, M., Huchra, J. & Mould, J. 1979, Astrophys. J. **229**, 1.
Aaronson, M., Bothun, G., Mould, J., Huchra, J., Schommer, R.A. & Cornell, M.E. 1986, Preprint.
Clutton-Brock, M. & Peebles, P.J.E. 1981, Astr. J. **86**, 1115.
Davies, R. D. & Staveley-Smith, L. 1985, Proc. ESO Workshop on the Virgo Cluster eds. Richter, O.-G. & Binggeli, B. (Garching: European Southern Observatory), p. 391.
Fixen, D.J., Cheng, E.S. & Wilkinson, D.T. 1983, Phys. Rev. Lett. **50**, 620.
Hart, L. & Davies, R.D. 1982, Nature **297**, 191.
Kaiser, N. 1983, Astrophys. J. Lett. **273**, L17.
Lubin, P.M., Epstein, G.C. & Smoot, G.F. 1983, Phys. Rev. Lett. **50**, 616.
Rubin, V.C., Ford, W.K., Jr., Thonnard, N., Roberts, M.S. & Graham, J.A. 1976a, Astr. J. **81**, 687.
Rubin, V.C., Thonnard, N., Ford, W.K., Jr. & Roberts, M.S. 1976b, Astr. J. **81**, 719.
Schechter, P.L. 1977, Astr. J. **82**, 569.
Tully R.B., Mould, J.R. & Aaronson, M. 1982, Astrophys. J. **257**, 527.
Vaucouleurs, G. de & Peters, W.L. 1984, Astrophys. J. **287**, 1.
Vittorio, N. & Silk, J. 1985, Astrophys. J. Lett. **293**, L1.

AN IMPROVED DISTANCE INDICATOR FOR ELLIPTICAL GALAXIES

S. Djorgovski[1] and Marc Davis[2]
[1]) Harvard-Smithsonian Center for Astrophysics,
60 Garden St., Cambridge, MA 02138.
[2]) Departments of Astronomy and Physics,
University of California, Berkeley, CA 94720.

ABSTRACT. We present a new relation between velocity dispersion, surface brightness, and radius (or luminosity) for elliptical galaxies. Relative distances of galaxies can be estimated to within 0.13 in log D. Most or all of that uncertainty is due to the measurement errors, and it may be reduced further. Our analysis shows that the elliptical galaxies are at least a three-parameter family. We discuss briefly the implications for theories of galaxy formation.

The final step of the Hubble distance ladder is use of correlations of distance-dependent and distance-independent properties of galaxies. The best known and most widely used is the Tully-Fisher relation for spirals. A few such relations for early-type galaxies have been known so far: the luminosity − velocity dispersion (Faber & Jackson 1976), luminosity − line strength (or color), and radius − surface brightness relation [see the review by Kormendy (1982) for other references]. Attempts have been made to use these relations for the mapping of the Local Supercluster velocity field (Tonry & Davis 1981ab; de Vaucouleurs & Olson 1982; Dressler 1984). One outstanding problem which limited such investigations is that the scatter in these relations largely exceeds the error-bars. This effect was attributed to the existence of a "second parameter" (other than the luminosity, or mass) among the fundamental properties of elliptical and S0 galaxies (cf. Terlevich et al. 1981; Tonry & Davis 1981; etc). Thus, the problem of an optimal distance indicator for early-type galaxies is closely related to a wider and more fundamental problem: the manifold of E/S0 galaxies, or the minimal number and identity of fundamental variables which determine the properties of these galaxies.

In order to investigate this question further, we completed a large CCD surface photometry survey of E/S0 galaxies (some 260 objects). Description of the survey, the data, and all other relevant information is presented by Djorgovski (1985). For the present study, we use the surface photometry data on 123 ellipticals (we exclude S0's, and the low surface brightness objects, such as NGC 205), for which we have velocity dispersions from Whitmore et al. (1985), and good-quality, good-seeing surface photometry; for a subset of 108 galaxies we also have the line strength index W_λ (cf. Tonry & Davis 1981ab).

From the surface brightness profiles, we derive "standard radii" (actually, semi-major axes), by using, e.g., Petrosian's (1976) η function (see also Djorgovski et al. 1985). It turns out that the relations explored here are not sensitive to the particular radial scale used; we should only emphasize that the radial scales which we use are *independent* of the mean or zero-point surface brightness for each galaxy. Magnitudes can be measured within these standard radii.

We adopt the following method for this investigation: we perform least-squares fits of radius or absolute magnitude vs. a distance-independent quantity, by assuming a simple virgocentric infall model with $v_{INF} \sim 1/r_{VIRGO}$, and solving simultaneously for the Local Group infall velocity. The least-squares fits minimize error-rescaled perpendicular distances of points from the best fit lines. We then correlate fit residuals with other quantities.

It was immediately apparent that the residuals of luminosity-or-radius vs. velocity dispersion relations correlate well with the surface brightness, and that the residuals of luminosity-or-radius vs. surface brightness relations correlate well with the velocity dispersion. On the other hand, *there is no correlation whatsoever between the velocity dispersion and surface brightness*. This is quite remarkable, as it implies that the dynamics of stars is almost completely independent of the projected density of luminous material in these galaxies!

Indeed, the logarithmic fits of luminosity or radius vs. a linear combination of velocity dispersion and surface brightness produced much improved correlations, with the residual scatter almost completely accountable by the error-bars. The preliminary solutions (pertaining to the $\eta = 0.4$ radius) are:

$$L \sim \sigma^{3.25} \langle SB \rangle^{-0.13} \tag{1}$$

$$L \sim \sigma^{3.35} \langle SB \rangle^{-0.13} W_\lambda^{-0.35} \tag{2}$$

$$R \sim \sigma^{1.45} \langle SB \rangle^{-0.14} \tag{3}$$

$$R \sim \sigma^{1.5} \langle SB \rangle^{-0.15} W_\lambda^{-0.3} \tag{4}$$

where $\langle SB \rangle$ denotes the mean surface brightness in linear flux units. Table 1 lists some of the fit parameters for the old and new distance-indicator relations. Relation (2) looks best, and it is shown in Figure 1.

Apparently, adding the line strength W_λ does not help the fits much. Metallicity is subject to differences in evolution of stellar populations, and there may be a saturation effect in the initial enrichment process; furthermore, the relation of metallicity and line strength (or color) is subject to interpretative difficulties. Thus, we suggest that Eq. (3) be adopted as the optimal distance indicator for elliptical galaxies.

The r.m.s. of log radius residuals in this "fundamental equation" of elliptical galaxies is 0.13, corresponding to distance errors of about 35%, all of which is accountable by the measurement and calibration errors (and possibly inadequacies of our infall model as well). Better data, and possibly also introduction of other parameters in the fit, may reduce the residual scatter even further. This is quite comparable to the Tully-Fisher relation for spirals.

The fundamental relation between the radius, surface brightness, and velocity dispersion was discovered independently by Burstein et al. (1986). Note that their D_Σ incorporates both the radius and surface brightness terms.

Our preliminary solutions for the Virgocentric infall give $v_{INF} \simeq 280$ km s^{-1}, with values with typical 1-σ uncertainity of ~ 120 km s^{-1}.

We can also state that elliptical galaxies are at least a three-parameter family: (1) velocity dispersion, which reflects the depth of the potential well (or the total mass) of a galaxy; (2) projected density of the luminous material (surface brightness); (3+) a variety of shape parameters (ellipticity, ellipticity gradient, isophotal twist rate, surface brightness slope), all of which seem to be mutually uncorrelated. Parameters (1) and (2) determine the luminosity and radius of a galaxy.

The existence of a tight correlation of fundamental properties of elliptical galaxies reflects some important regularity in the recipe for galaxy formation. There are also intriguing similarities with the Tully-Fisher relation for spirals, and indications that inclusion of surface brightness may reduce the scatter there as well. It is quite remarkable that this fundamental relation for elliptical (and other?) galaxies is fairly independent of the details of light distribution. This independence, and the lack of correlation between the velocity dispersion and surface brightness suggest that the fundamental structural and dynamical properties of galaxies are determined by their dark haloes.

It is possible that the zero-point and/or the slope of this fundamental relation vary with the large-scale enviroment, and that they may be different from cluster to cluster. If such effects do exist, they may account in part for the apparent large streaming motions found by Burstein et al. (1986).

REFERENCES:

Burstein, D., Davies, R., Dressler, A., Faber, S., Lynden-Bell, D., Terlevich, R., and Wagner, M. 1986, this volume.
de Vaucouleurs, G., and Olson, D. 1982, *Astrophys. J.* **256**, 346.
Djorgovski, S., Davis, M., and Kent, S. 1985, in *New Aspects of Galaxy Photometry*, J.-L. Nieto (ed.), Lectures in Physics **232**, p. 257, Springer Verlag.
Djorgovski, S. 1985, Ph. D. Thesis, U. C. Berkeley.
Dressler, A. 1984, *Astrophys. J.* **281**, 512.
Faber, S., and Jackson, R. 1976, *Astrophys. J.* **204**, 668.
Kormendy, J. 1982, in J. Binney, J. Kormendy, and S. White: *Morphology and Dynamics of Galaxies*, Saas-Fee Lectures, Geneva Observatory.
Petrosian, V. 1976, *Astrophys. J. Lett.* **209**, L1.
Terlevich, R., Davies, R., Faber, S., and Burstein, D., 1981, *M.N.R.A.S.* **196**, 381.
Tonry, J., and Davis, M. 1981a, *Astrophys. J.* **246**, 666.
Tonry, J., and Davis, M. 1981b, *Astrophys. J.* **246**, 680.
Whitmore, B., McElroy, D., and Tonry, J. 1985, *Astrophys. J. Suppl. Ser.* **59**, 1.

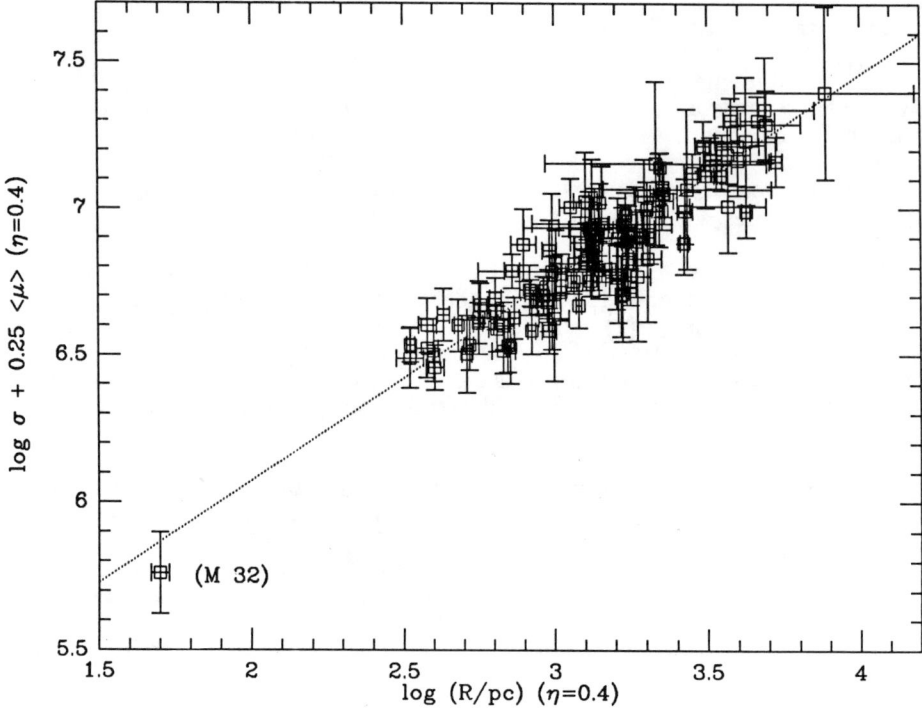

Figure 1. The new relation between log radius, surface brightness, and velocity dispersion (Eq. 3). Fitting procedure and the virgocentric infall model are described in the text. There is no absolute calibration: $H_0 = 100$ km s^{-1} Mpc^{-1} was assumed for convenience. The best-fit infall velocity (Local Group to Virgo) is 278 km s^{-1} for this data set. Statistics for this and other fits are listed in Table 1 below.

- - - -

Table 1. Statistics for distance indicator relations.

Relation	Spearman Rank	Linear Regr.	χ^2/DOF
$L - \sigma$	0.735	0.864	5.36
$L - \mu$	0.326	0.605	31.1
$R - \sigma$	0.562	0.849	7.72
$R - \mu$	0.677	0.886	4.14
$L - \sigma - \mu$	0.832	0.971	1.82
$L - \sigma - \mu - W_\lambda$	0.835	0.964	2.26
$R - \sigma - \mu$	0.897	0.983	0.12
$R - \sigma - \mu - W_\lambda$	0.900	0.984	1.07

A QUADRUPOLAR COMPONENT IN THE VELOCITY FIELD OF THE LOCAL SUPERCLUSTER

P.B. Lilje[1], A. Yahil[1,2,4], and B.J.T. Jones[3]

[1] Institute of Astronomy, Madingley Road, Cambridge CB3 OHA
[2] Astronomy Program, SUNY-Stony Brook, ESS Building,
Stony Brook, NY 11794
[3] NORDITA, Blegdamsvej 17, DK-2100 København Ø, Denmark
[4] John Simon Guggenheim fellow

ABSTRACT. We have detected a significant quadrupolar component in the velocity field of the Local Supercluster. This component is due to the tidal gravitational field from density structures a few tens of Mpc away from it. It is not possible to single out one single source for this field, but the Hydra-Centaurus supercluster seems to give the major contribution. Subtracting the velocity of the Local Group relative to the Virgo cluster due to this effect and the standard Virgocentric infall from the velocity of the Local Group relative to the Microwave Background Radiation, we find that the Local Supercluster is moving towards the Hydra-Centaurus supercluster with a velocity of 500 km/s.

1. INTRODUCTION

This paper is a summary of a paper submitted to the Astrophysical Journal (Lilje, Yahil, and Jones 1986). For further details of our methods and results, see that paper.

For several years, there have been a widespread interest in the velocity field of the Local Supercluster (LSC). It is now generally agreed that the density enhancement of the LSC retards the local Hubble expansion, and that this infall velocity at the distance of the Local Group (LG) is in the range 200-350 km/s. However, the velocity of the LG relative to the Microwave Background Radiation (MBR) is 600 km/s (Lubin, Epstein, and Smoot 1983; Fixsen, Cheng, and Wilkinson 1983), and this velocity is pointing 45 degrees away from Virgo. This discrepancy can best be understood as due to the bulk motion of the LSC.

Because of this bulk motion, one must only use galaxies which are members of the LSC to determine the velocity field inside the LSC. When this has been done, one may use standards outside the LSC to determine the bulk motion. Such a procedure was followed by Aaronson et al.(1982a, AHMST). To get a good fit, they added a random velocity of the LG relative to nearby galaxies to their spherical infall model. This random velocity they found to be 190 ± 45 km/s, much larger than found observationally by Tammann, Sandage and Yahil (1979). It seems evident

that there is another systematic component in the velocity field in the
LSC in addition to the infall. AHMST suggest differential rotation,
but it is difficult to understand how high rotational velocities could
be acquired. Nonradial collapse has also been suggested, but such
anisotropic infall models do not take into account the high central
concentration of the LSC (Yahil 1985). We suggest that this other
systematic component is a quadrupolar component which is the natural
consequence of the tidal forces from nearby structures in the Universe.

2. MODEL

In the regions of the LSC at some distance from the Virg cluster, the
growth of density perturbations are not too different from the expressions
from linear theory. The peculiar flow velocity will then be parallell
with and proportional to the local gravitational acceleration. The
gravitational field is the sum of the one due to the LSC and the tidal
field, and the total peculiar velocity can therefore be well approximated
as the sum of the peculiar velocities due to each of the two fields.

In linear theory, the tidal field from a point mass at distance R
will set up a peculiar tidal velocity field

$$\underline{u}_t = \underline{\underline{\sigma}} \cdot \underline{r} \tag{1}$$

where $\underline{\underline{\sigma}}$ is the symmetric, traceless, shear matrix

$$\underline{\underline{\sigma}} = \begin{pmatrix} -a & 0 & 0 \\ 0 & -a & 0 \\ 0 & 0 & 2a \end{pmatrix} \tag{2}$$

where the z-axis points towards m. The attraction from m will also give
the centre of mass of our system a bulk velocity of

$$u_{bulk} \sim \frac{r}{R} a. \tag{3}$$

In general, when the tidal field is induced by several masses, equation
(1) is still valid, but $\underline{\underline{\sigma}}$ is then a general shear matrix which in the
principial axis system is given by

$$\underline{\underline{\sigma}} = \begin{pmatrix} -a & 0 & 0 \\ 0 & -b & 0 \\ 0 & 0 & a+b \end{pmatrix} \tag{4}$$

We will use the eigenvalues a and b, together with the three Euler
angles which determine the directions of the principial axes as free
parameters to be determined from the data.

We approximate the Virgocentric infall by the linear expression plus
an empirical nonlinear correction (Yahil 1985)

$$\underline{u}_i = -\frac{1}{3} H_o r \Omega_o^{.6} \delta (1+\delta)^{-1/4}. \tag{5}$$

Here, δ is the average value of $\delta\rho/\rho$ inside a shell of radius r. We use
the generally accepted expression $\delta \propto r^{-2}$ and set δ interior to the LG to

3. Different authors find values from 2 to 3 for this quantity. Using 2 instead will only change our value for Ω_o and not our determined velocity field. We keep Ω_o as a free parameter whose value is to be determined from the data.

We proceed close to the manner of AHMST. For a set of parameters Ω_o, a, b and the three Euler angles, we compute the distance of a galaxy from its redshift and angular position by solving equations (1), (4) and (5). From this and the observed H-magnitude, we predict a 21 cm line-width by the IR Tully-Fisher relation:

$$\log\Delta V = e(10-(m-5 \log d))+f \qquad (6)$$

where e and f are free parameters, whose values are to be determined from the data, and d is the distance of the galaxy in units of the distance to the Virgo cluster. We then use a least square method to minimize the difference between the predicted and observed linewidths by varying the 8 parameters.

3. RESULTS

We have done this least square fitting for the galaxies in the catalogue of Aaronson et al. (1982b). However, we have used a somewhat different exclusion zone than AHMST, letting 230 galaxies remain in our fit.

Our results are:

$$\begin{aligned}
u_i &= 315 \pm 49 \text{ km/s} \\
\Omega_o^i &= 0.16 \pm 0.03 \\
a &= 170 \pm 55 \text{ km/s} \\
b &= 26 \pm 43 \text{ km/s} \\
a+b &= 196 \pm 57 \text{ km/s} \\
\text{x-axis: } & l = 65 \pm 38°, b = 55 \pm 18° \\
\text{y-axis: } & l = 214 \pm 16°, b = 32 \pm 20° \\
\text{z-axis: } & l = 308 \pm 16°, b = 13 \pm 9°
\end{aligned} \qquad (7)$$

The three last lines show the direction of the principal axes in galactic coordinates. The tidally induced velocity of the LG relative to the Virgo cluster in the reference system of AHMST is

$$\begin{aligned}
u_x &= -126 \pm 45 \text{ km/s} \\
u_y &= -58 \pm 57 \text{ km/s} \\
u_z &= 46 \pm 70 \text{ km/s}.
\end{aligned} \qquad (8)$$

The significant part of this is lying in the supergalactic plane. The total peculiar velocity of the Local Group, due to both infall and the tidal field, has a magnitude of 387 ± 81 km/s and is pointing towards $l = 206 \pm 26°$, $b = 72 \pm 8°$.

4. DISCUSSION

We have detected a significant quadrupolar term. The main effect of this is an expansion along the z-axis in equation (7) and a contraction along the x-axis. The axis of expansion is pointing towards the Hydra-Centaurus supercluster (Chincarini and Rood 1979; Hopp and Materne 1985). If we subtract the peculiar velocity of the LG from the velocity of the LG relative to the MBR, we find that the bulk velocity of the LSC relative to the MBR is 503 ± 75 km/s and pointing towards $l = 288 \pm 9°$, $b = -9 \pm 10°$. This also is in the same general direction of the sky as the Hydra-Centaurus supercluster.

The shear matrix is different from what one would get from a single attractor, and the expansion axis and the direction of bulk motion point $27 \pm 15°$ away from each other. Therefore, the tidal field can not only be due to one single supercluster. However, we can estimate the distance to the perturbers from equation (3) using the r.m.s. of the eigenvalues instead of a. In units of the distance between LG and Virgo, we find $R \sim 3$, which agrees well with the distance of the Hydra-Centaurus supercluster. We conclude that density structures on the scale of a few tens of Mpc, and predominantly the Hydra-Centaurus supercluster, are responsible for the external contribution to the gravitational field in the LSC.

REFERENCES

Aaronson,M., Huchra,J., Mould,J., Schechter,P.L., and Tully, R.B. 1982a Ap.J.,258,64 (AHMST)
Aaronson et al. 1982b, Ap.J.Suppl.,50,241.
Chincarini,G., and Rood,H.J. 1979, Ap.J.,230, 648.
Fixsen,D.J., Cheng,E.S., and Wilkinson,D.T. 1983, Phys.Rev.Letters,50,620.
Hopp,U. and Materne,J. 1985, Astr.Ap.Suppl., 61, 93.
Lilje,P.B., Yahil,A., and Jones, B.J.T. 1986, submitted to Ap.J.
Lubin,P.M., Epstein,G.L., and Smoot,G.F. 1983, Phys.Rev.Letters, 50, 616.
Tammann,G.A., Sandage,A., and Yahil,A. 1979, Phys.Scr., 21, 630.
Yahil,A. 1985, The Virgo Cluster of Galaxies, eds. O.G. Richter and B. Binggeli, (Garching: ESO) p. 359.

IS THE CRITICISM OF THE VIRGOCENTRIC FLOW MODEL JUSTIFIED?

Amos Yahil
Astronomy Program
State University of New York
Stony Brook, NY 11794-2100, USA

ABSTRACT. Tests of the Virgocentric flow model by numerical simulations must follow strict ground rules in order realistically to mimic the real Virgo Supercluster. An hypothesis is made as to what such "realistic" simulations will find, and a challenge is issued to prove or disprove this hypothesis.

1. INTRODUCTION

For the last decade or so, the Virgocentric flow model (VFM) has continued to be one of the major methods of evaluating the cosmological density parameter Ω_o (e.g., the recent review by Yahil 1985). The model assumes that the peculiar gravitational field generated by the Virgo Supercluster (VSC) is dominated by the monopole term, and there is some observational evidence for this (Yahil et al. 1980). The model further assumes that the peculiar velocity field in the VSC is primarily the radial growing mode generated over the age of the universe by this monopole field.

Note that the model does *not* assume that the distribution of matter in the VSC is perfectly spherically symmetric, only that the gravitational field can be approximated by the monopole term. Equipotentials are always more spherical than isodensity surfaces, so a fair amount of asphericity can be tolerated in the density distribution, without violating the basic assumption of the monopole approximation. Moreover, the flattening in the supergalactic plane is predominant in the region of higher density, $\delta\rho/\rho \gtrsim 10$, and not where the bulk of the mass of the supercluster lies, at densities $\delta\rho/\rho \sim 1$ (Yahil 1985).

Recently the VFM has come under attack on two grounds, which may be related. First, a tidal velocity has been detected in the VSC (Lilje *et al.* 1986; and in this conference). The form of this quadrupolar velocity field, and its magnitude, imply that, while the potential due to material *inside* the VSC may be dominated by the monopole term, other nearby superclusters can generate comparable tidal fields. By mere coincidence, the component of the tidal peculiar velocity in the direction of Virgo is small, and the determination of Ω_o is not seriously altered by the inclusion of the tidal field.

The second attack has come from numerical simulations of the expanding universe, in which superclusters are identified, and then evaluated for the accuracy of the VFM (Bushouse et al. 1985; Villumsen and Davis 1986; Lee et al.1986; and papers in this conference). These models have generally found a wide scatter, as well as bias, in the ratio between the value of Ω_o determined from the VFM and the underlying Ω_o of the simulation. The simulations have so far not included the tidal field in the VFM, so it is not known how much of the discrepancy is due to its exclusion.

This paper is primarily a response to the second criticism. Some pitfalls of the numerical simulations are pointed out in § 2, and possible ground rules are suggested for a more realistic appraisal of the VFM. What should be seen by simulations which follow these ground rules is hypothesized in § 3, and a challenge is issued to the numerical simulators to check this hypothesis.

2. SUGGESTED RULES FOR "REALISTIC" NUMERICAL SIMULATIONS

Numerical simulations may fail properly to test the VFM if they either generate simulations that are significantly different from the real Virgo supercluster, or if the determination of Ω_o from the simulation is sufficiently different from that employed for the real data.

I would suggest the following rules for "realistic" simulations:
- The simulation should mimic the present day structure and velocity field on scales $\lesssim$20-30 Mpc. This means that the galaxy correlation functions, the galaxian velocity dispersion, and the systematic flow velocities should all resemble the observed ones.
- Superclusters chosen for analysis should, just like the VSC, have strong central concentrations.
- They should also be isolated from other superclusters, in the sense that the sphere within which $\langle\delta\rho\rangle/\rho=2-3$ has a radius that is ~3 times smaller than the distance to the nearest comparable supercluster.
- On the other hand the tidal field due to nearby superclusters may not be neglected in the comparison of the VFM with the simulation.

In determining the "observed" Ω_o from the "data" in the simulation, it would be useful to recall the following factors governing the real determination of Ω_o in the VSC:
- One method of determining the peculiar velocity of the Local Group (LG) relative to Virgo is to compare the ratio of the radial velocities of Virgo and more distant clusters with the ratio of their true (i.e., photometric) distances.
- Another method is to compare the radial velocities of galaxies *within* the VSC with their photometric distances. This determination is most sensitive to galaxies infalling into Virgo from locations in the VSC that are (almost) opposite to that of the LG.
- A third method, discussed in this conference by Richter, is to confine the investigation to galaxies whose distance from the LG is small compared with the distance of Virgo, and study the *local* shear field.

- All three methods of determining the peculiar velocity of the LG relative to Virgo give the same result.
- All of them are also sensitive to the special location of the LG in the VSC, from which the real observations are made. But the LG is in a quiet (supersonic) part of the Virgocentric flow, with $u/\sigma \sim 2$-3, and not an arbitrary galaxy in the supercluster, which may be in more turbulent locations.

3. HYPOTHESIS AND CHALLENGE

It is hypothesized that, if the rules advocated in § 2 are followed, the gravitational field will be found to be dominated by the sum of the monopole component due to matter within the supercluster and the quadrupole components due to material outside the supercluster. (There is also a dipole component due to the material outside the supercluster, which imparts a bulk motion to the center of mass of the supercluster. This motion can not be detected by measurements within the supercluster, and requires an external reference frame, such as the microwave background radiation, or distant galaxies).

To be specific, consider the LG to be somewhere in the spherical shell confined between the two surfaces in which the mean density is $\langle\delta\rho\rangle/\rho=2$ and $\langle\delta\rho\rangle/\rho=3$. Take the radius of the supercluster to be twice the radius of the $\langle\delta\rho\rangle/\rho=2$ surface, i.e., everything within this radius is considered "within" the supercluster, and everything else is "outside".

It is further hypothesized that the peculiar velocity field can be written as a similar sum of a monopole velocity field due to material within the supercluster, and dipole and quadrupole fields due to material outside it.

The challenge is to prove or disprove the hypothesis.

4. ACKNOWLEDGMENT

This research was supported in part by USDOE grant DE-AC02-80ER10719 at the State University of New York.

5. REFERENCES

Bushouse, H., Melott, A., Centrella, J., and Gallagher, J. 1985, *M. N. R. A. S.*, **217**, 7p.
Lee, H., Hoffman, Y., and Ftaclas, C. 1986, preprint.
Lilje, P. B., Yahil, A., and Jones, B. J. T. 1986, *Ap. J.*, in press.
Villumsen, J. V., and Davis, M. 1986, preprint.
Yahil, A. 1985, in *The Virgo Cluster of Galaxies*, eds. O. G. Richter and B. Binggeli, (Garching: ESO), p. 359.
Yahil, A., Sandage, A., and Tammann, G. A. 1980, *Ap. J.*, **242**, 448.

ANISOTROPY OF THE GALAXIES DETECTED BY *IRAS*

Avery Meiksin[1] and Marc Davis[1,2]
[1] Physics Department, U.C. Berkeley
[2] Astronomy Department, U.C. Berkeley
Dept. of Astronomy
601 Campbell Hall
University of California
Berkeley, CA 94720

ABSTRACT. We generate a catalogue of 6730 galaxies uniformly selected over most of the sky (> 9.5sr) using the IRAS point source catalogue. Our catalogue reveals a small (4–7%), but robust dipole anisotropy in the galaxy distribution that points towards $l = 235°$, $b = 45°$, within 30° of the microwave dipole anisotropy. Angular correlation analysis suggests that subsets of the catalogue have characteristic distances $D_* \simeq 50 - 100 h^{-1}$ Mpc. The good agreement in direction of these dipoles suggests that the IRAS galaxies roughly trace the large scale mass distribution, that the microwave dipole velocity is mostly induced by the local supercluster ($cz < 3000 \ km \ s^{-1}$), and that the indicated cosmological density is high ($\Omega \approx 0.5$). A full sky redshift survey is in progress to obtain a more precise estimate.

1. INTRODUCTION

A question of great interest in cosmology is the homogeneity and isotropy in the distribution of galaxies. The dipole anisotropy of the cosmic microwave background is presumably due to the net gravitational acceleration of the galaxies around us. Recent observations by Aaronson *et al.* (1986) show a consistency in amplitude and direction of the microwave dipole velocity with the velocity inferred from the Hubble anisotropy of the local group relative to 10 galaxy clusters at distances ranging from 40-100 h^{-1} Mpc ($H_0 = 100 h \ km \ s^{-1} Mpc^{-1}$).

Davis and Huchra (1982) found an anisotropy in the galaxy distribution using the CfA and Shapley-Ames surveys. The dipole points near Virgo, 45° from the microwave result. Using the whole-sky Shapley-Ames catalog, Yahil *et al.* (1980) found an anisotropy similarly dominated by Virgo. The catalogue, however, does not extend beyond Virgo.

To measure the peculiar acceleration induced by the galaxy distribution, ideally one would like a catalogue uniformly selected over the entire sky. Optically selected catalogues fall far short of this goal, chiefly because of Galactic extinction. Moreover, except for the Shapley-Ames catalogue, no catalogue has uniform coverage or photometry between North and South.

With the advent of the IRAS Point Source Catalog (PSC) database, it has become possible to come close to achieving whole-sky coverage. The IRAS database is unaffected by reddening. Near the Galactic plane ($|b| < 10°$), however,

IRAS data is affected by an abundance of Galactic cirrus and is confusion limited. We restrict the galaxy search to $|b| > 10°$, more than doubling the sky coverage over optical catalogues.

We have begun a program to obtain redshifts of a subset of the galaxies. Upon completion, we would then have not only a list free of any contamination by Galactic objects, but information from which we could reliably determine the true depth of the anisotropy result, the two-point spatial correlation function at large separations, the density parameter Ω of the universe, and perhaps even the spectrum of primordial fluctuations.

A more complete description of our work is given in Meiksin and Davis (1986). Our anisotropy results are essentially identical to those reached by Yahil et al. (1985) in a separate analysis of the IRAS data, although our methods are considerably different, as discussed below.

2. SAMPLE DEFINITION

The PSC contains fluxes in bands at 12μ, 25μ, 60μ, and 100μ. The precise meaning of the fluxes is provided in the IRAS Explanatory Supplement (1984). Our primary basis for distinguishing galaxies from stars in the PSC was to require $f(60\mu)/f(12\mu) > 3$, where $f(x) \equiv$ flux in x band. We found that of the 23959 sources with $|b| > 10°$ in the PSC associated with SAO stars, only 0.5% passed this color criterion. By comparison, 73% of the 5501 PSC sources with $|b| > 10°$ associated with Zwicky galaxies passed the color test. We felt it was better to have a strictly selected smaller catalogue rather than to have a large contaminated sample. In addition, the signal had to be of the highest flux quality (flux quality 3) in the 60μ band and had to have a 60μ flux of at least 0.6Jy (the cut-off was based on a logN–logS plot).

Several known HII regions and Local Group galaxies appeared in our map. We eliminated them by cutting out strips around the systems. Two additional blank regions occur due to the small incompleteness in the IRAS coverage. The total remaining solid angle of sky was then 9.55 sr. The strips cover enough of the sky that they must be filled uniformly with density equal to the average density of sources over the remainder of the sky in order not to distort severely the anisotropy result. Equal area galactic projections of the distribution are given by Meiksin and Davis (1986) and Davis (1985).

3. RESULTS

In order to compute the anisotropy, we filled in the blank regions uniformly with sources of the average surface density as those in our sample. Adding the homogeneously filled strips to the 6730 PSC sources brings the effective number of sources to 8854. The vector dipole **D** is then simply computed as the average of the unit direction vectors of the galaxies, giving each galaxy equal weight.

The number weighted result for the 8854 sources is $|\mathbf{D}| = 4.1\%$ directed towards $l = 235°$, $b = 43.5°$. The flux weighted result is similar after deleting the several unusually brightest galaxies from the list. As a measure of the contribution of different depths to the anisotropy, we divided the sources into quartiles in number on the basis of 60μ flux. We refer to Table 1 of Meiksin and Davis (1986) for the results. The anisotropies of the three brightest quartiles agree excellently. The dimmest quartile may be contaminated by weak Galactic sources.

The total dipole vector we measure lies 29° from the microwave result, and the dipole for the brightest quartile lies 22° away. The statistical uncertainty of the microwave anisotropy direction is $\simeq 5°$. Based on the small scatter in direction ($< 11°$) of the top three IRAS quartiles, the misalignment of the microwave and IRAS directions appears to be significant. Perhaps the best measure of the directional uncertainty of **D** is the scatter of directions seen in the top 3 quartiles.

We computed the angular correlation function of the full sample and of the flux quartiles individually to obtain an estimate of the depth of the sample. We found the angular correlations for $\theta < 10°$ to be adequately described by a power law, $w(\theta) = 0.26\theta^{-0.56}$, with a correlation length of $\theta_{corr} = 0.09°$ for the full sample. Table 2 of Meiksin and Davis lists the results for all the analyses. The brightest quartile has angular correlation properties close to those of the spiral-spiral correlations of the UGC catalogue (Davis and Geller 1976), which is not surprising since our sample is largely late type spirals. The characteristic depth of the samples can be estimated by scaling from the baseline of $D = 42h^{-1}\ Mpc$ from the UGC spirals. Because the IRAS luminosity function is likely to be much broader than the optical luminosity function, this is a precarious procedure. The question of the depth of the sample is obviously quite important; improved estimates will have to await more redshifts.

4. POSSIBLE SYSTEMATIC ERRORS

Because of the small amplitude of the anisotropy, we must guard carefully against any anisotropies introduced by systematic effects in the IRAS coverage. Two particular hazards are hysteresis and underestimation of the flux of small extended sources.

There is a known hysteresis in low latitude measured fluxes relative to the scan direction through the galactic plane. For the 60μ detectors, this is typically a 2-4 percent sensitivity enhancement at 20 degrees galactic latitude after a plane crossing, which has not been removed from the PSC fluxes. Further study to accurately calibrate and remove all hysteresis effects is very important, although we believe their systematic effect on the dipole anisotropy to be minor.

The IRAS fluxes of many nearby galaxies will be underestimated since they are extended (>40 arcsec in size), but many of the objects are concentrated in the Virgo direction. The brightest quartile sample will have to be reconstituted with the small extended source catalog data, once available, and the anisotropy direction and amplitude are likely to change.

5. DISCUSSION

Our anisotropy analysis was completed but not submitted prior to the submission of the Yahil et al. (1985) report. The principal differences between their work and ours is the choice of color discrimination and the attention paid to cirrus. Our color choice requires the 60μ flux to exceed the 12μ flux by a factor of 84 above that expected for Rayleigh-Jeans emission; their criterion requires the 60μ flux to exceed the 25μ flux by a factor of 2.0 above Rayleigh-Jeans emission. Thus our sample galaxies are on average much dustier than those of Yahil et al.

Yahil et al. were very conservative regarding cirrus correction, and only 35-55% of the sky was judged sufficiently free of cirrus to be trustworthy. Our approach was more naive, and we kept fully 76% of the sky for our analysis. In spite of these substantial differences of analysis, our anisotropy results agree very well with theirs and are similarly robust.

We have tacitly assumed the IRAS anisotropy is proportional to our peculiar gravitational acceleration. The close alignment of the microwave dipole and the IRAS dipole suggests that in fact the galaxy distribution does trace the mass distribution, although the tracing may still be biased. The observed misalignment between acceleration and velocity is precisely the typical amount seen in the cosmological n-body simulations of cluster formation, where by construction the particles trace the mass (Villumsen and Davis 1985).

The IRAS dipole anisotropy is not determined solely by prominent clusters and voids; in fact it points toward a local minimum of the galaxy distribution. The dipole is truly the vector sum of all the galaxies, with the large low density regions playing as important a role to repel the dipole vector as the observed clusters do to attract the dipole. Given the distance estimates plus the visual identification and known redshifts of the most prominent clusters, it is apparent that the anisotropy is dominated by clustering within $50h^{-1}$ Mpc of us. The CfA catalogue probed well beyond this distance at high galactic latitude, and in that sample the local supercluster ($cz < 3000$ km s^{-1}) completely dominated the gravity anomaly. The expected peculiar velocity found for the CfA sample was $v_p = 670\Omega^{0.6}$ km s^{-1}, (Davis and Huchra 1982). The IRAS catalog shows as expected that the anisotropy is not confined to high latitudes, but an accurate estimate of the expected peculiar velocity must await redshifts of a complete subset of the galaxies (e.g. the brightest quartile).

Given the angular proximity of the earlier CfA anisotropy to the IRAS anisotropy, it is reasonable to expect that the same clustering dominates both samples. The additional anisotropy detected by the IRAS sample will probably raise the expected peculiar velocity to $v_p \approx 900\Omega^{0.6}$ km s^{-1}. This number should be compared directly to the microwave dipole velocity of 600 ± 50 km s^{-1}, and may imply a high value of mean density, $\Omega \approx 0.5$. This will be the largest scale over which we can conceivably determine Ω and therefore the measurement holds considerable interest.

ACKNOWLEDGMENTS

We thank Tom Chester, Martin Cohen, and Frank Low for advice on the IRAS database. This work was supported in part by NSF Grant AST-8419910.

REFERENCES

Aaronson,M., Bothun, G., Mould,J., Huchra,J., Schommer, R.A., and Cornell, M.E., 1986,Ap. J., in press.
Davis,M., in Dark Matter in the Universe, IAU Symp. 117, 1986, (ed. J. Knapp and J. Kormendy; D. Reidel), in press.
Davis,M., and Geller,M.J., 1976,Ap.J., 208;13.
Davis,M., and Huchra,J., 1982,Ap.J., 254;437.
IRAS Catalogs and Atlases,1984,Explanatory Supplement.
Meiksin,A., and Davis,M., 1986,A.J., in press.
Villumsen,J.V., and Davis,M., 1985,in preparation.
Yahil,A., Walker,D., and Rowan-Robinson,M., 1985, preprint.
Yahil,A., Sandage,A., and Tammann,G.A., 1980,Ap.J., 242;448.

THE LOCAL GRAVITATIONAL FIELD

Amos Yahil
Astronomy Program
State University of New York
Stony Brook, NY 11794-2100, USA

ABSTRACT. In the redshift range $v < 3000$ km s^{-1}, the density distributions of the infrared selected IRAS galaxies and the optically selected CfA galaxies are shown to be identical. But the cosmological baseline density level of the IRAS galaxies appears to be higher by a factor ~2 than the one determined for the optical galaxies. If correct, this difference could account for the discrepancy between the determinations of Ω_o from the Virgocentric infall and from the IRAS dipole anisotropy. There are, however, a number of problems with this simple explanation.

1. INTRODUCTION

The main motivation for searching for density inhomogeneities in the distribution of galaxies outside the Virgo Supercluster (VSC) is the discrepancy between the 250 ± 50 km s^{-1} Virgocentric infall velocity of the Local Group (LG) in the direction of Virgo (several papers in this conference, and a previous review by Yahil 1985), and the velocity, $u_\mu = 600$ km s^{-1}, of the LG relative to the microwave background radiation (MBR), in a direction ~45° away from Virgo (Lubin et al. 1983; Fixsen et al. 1983; Lubin, this conference). The difference between the two velocities is most easily understood as the bulk motion of the VSC, induced by the cumulative pull of density inhomogeneities on scales larger than the VSC. The results presented in this conference by Burstein on large scale coherent motions, suggest that the scale of these inhomogeneities may be very large indeed.

A major step in the search for distant inhomogeneities was the discovery of a dipole anisotropy in the distribution of IRAS galaxies on the sky, which is aligned, within the errors, with u_μ (Yahil, Walker, and Rowan-Robinson 1986, henceforth YWR; Meiksin and Davis 1986, henceforth MD). Using depth information provided by the luminosity function of IRAS galaxies (Rowan-Robinson et al. 1986; Lawrence et al. 1986), YWR have been able to obtain an estimate for the cosmological density parameter, $\Omega_o \approx 1$. This high value of Ω_o is at variance with the determination of Ω_o from the Virgocentric infall model and the cosmic virial theorem (e.g., the review by Yahil 1985).

This paper is an initial attempt to address this discrepancy. The main results from the IRAS anisotropy are summarized in § 2. New results of a comparison between the local ($v<3000$ km s^{-1}) density distributions of IRAS and CfA galaxies are presented in § 3. A brief discussion follows in § 4.

2. SUMMARY OF IRAS DIPOLE ANISOTROPY

The IRAS catalog, containing a total ~250,000 sources, offers an opportunity for identifying a complete sample of galaxies, which are calibrated homogeneously over almost the entire sky, and are unaffected by extinction. It is very easy spectrally to discriminate against the hotter stellar sources. YWR apply the condition $S_{25}<3S_{60}$ to the IRAS point sources with high quality detection in the 60μ band, and find that all but a few of the sources identified with stars are eliminated. Lawrence et al. (1986) apply the same criterion in the area of their redshift survey (b>60, 0<l<110). Of 496 sources satisfying the inclusion criteria only 3 have no candidate to the POSS plate limit, and a few can not be distinguished from stars on the plates, but appear diffuse on the integrating TV system at the Isaac Newton Telescope. All the rest are galaxies. The sky average surface density of these galaxies (for a flux limit of 0.5 Jy) is 0.50 $\square^{-1}$, for a total ~20,000 galaxies over the entire sky. MD use the more conservative criterion $3S_{12}<S_{60}$. This eliminates many of the hotter galaxies, resulting in a surface density of 0.28 $\square^{-1}$ (to the same limiting flux).

The main difficulty with the IRAS catalog is contamination by infrared "cirrus" emission from interstellar dust in our own Galaxy, which is spectrally similar to the emission of external galaxies. In the absence of identifications, the solution is to mask parts of the sky. MD exclude a zone of avoidance $|b|\leq 10$, and some known HII regions and prominent LG galaxies, leaving 9.5 str. YWR apply a more conservative mask, based on the CIRR1 flag of the IRAS catalog itself, leaving 5.9 str.

It should be emphasized that the cirrus contamination, and the consequent masking, are only a result of lack of identifications. Once these become available, most of the sky can be included in the dataset. Only a strip of a few degrees around the plane of the Galaxy, and some minor other patches, need to be excluded, due to confusion, contamination of flux measurements, and shadowing effects in the detection system.

Initial identification and redshift surveys have already been performed (Rowan-Robinson et al. 1986; Lawrence et al. 1986). They show that the IRAS galaxies range in distance far beyond the limits of previous redshift surveys. The eighty percentile of the redshift distribution (for a flux limit of 0.5 Jy) is at 20,000 km s^{-1}, and there is a long tail, which allows adequate sampling even beyond that range. Major projects are now underway, both in the U.K. and the U.S., to identify and measure the redshifts of all IRAS sources above 2 Jy, and a fraction of those at fainter fluxes.

An important result to emerge from the study of the IRAS galaxies is the detection of a dipole component in their surface distribution, which is aligned, within the errors, with u_μ. Furthermore, the direction of the anisotropy appears to be independent of flux, and is the same in both the investigations of YWR and MD, despite different sampling and masking criteria.

The interpretation of the IRAS dipole anisotropy requires depth information. In the absence of individual redshifts, this can only be obtained in a statistical fashion from the luminosity function. The studies of Rowan-Robinson et al. (1986) and Lawrence et al. (1986) show that the luminosity function of the galaxies sampled according to YWR is well described by the function

$$\Phi(L) = CL^{-2}(1 + L/\beta L_*)^{-\beta} \tag{1}$$

(the limit $\beta \to \infty$ is a Schechter function). Lawrence et al. (1986) give $C=(11.5\pm0.4)\times10^6 h\, L_\odot\, \mathrm{Mpc}^{-3}$, for a fit of the form of eq. (1), where $L=\nu L_\nu(60\mu)/L_\odot$ (at 60μ 1 Jy$=1.25\times10^5\, L_\odot\, \mathrm{Mpc}^{-2}$). As the area which they study has a source density which is 18% higher than that for the whole sky, their value should be corrected to $C=(9.7\pm0.3)\times10^6 h\, L_\odot\, \mathrm{Mpc}^{-3}$.

For the luminosity function in eq. (1), the product of the flux and the dipole moment of the surface brightness, hereafter loosely referred to simply as the dipole moment, is given by

$$4\pi S\sigma(S) = 12\pi S^2 \Delta S^{-1} \sum \hat{r}_i = \frac{3C}{4\pi} \int D(\mathbf{r})(\mathbf{r}/r^3)(1+r^2/\beta r_*^2)^{-\beta} d^3r \tag{2}$$

Here the dipole moment is *differential* in flux, the sum over the unit vectors in eq. (2) being only over the sources in a flux bin ΔS; $D(\mathbf{r})$ is the local relative density function, $D=1$ corresponding to the cosmological baseline density level (mean density of the universe); and $r_* = \sqrt{(L_*/4\pi S)} = 180\sqrt{(\mathrm{Jy}/S)}\, h^{-1}\, \mathrm{Mpc}$ is the distance at which a source with luminosity $L_* = (5.0\pm0.9)\times10^{10} h^{-2}\, L_\odot$ is seen with flux S.

The point made by YWR is that the dipole moment, as defined by eq. (2), appears to be independent of flux in the measured range. They conclude that the inhomogeneities giving rise to the dipole moment are nearer to us than r_*, and the cutoff term on the right hand side of the equation has no effect. If that is the case, then, apart from a normalization constant, the dipole moment measures the peculiar gravitational acceleration due to the matter which the IRAS galaxies trace.

The peculiar gravitational acceleration results in a peculiar velocity (Peebles 1980)

$$\mathbf{u} = \Omega_o^{0.6}(H_o/4\pi) \int D(\mathbf{r})(\mathbf{r}/r^3) d^3r\, (1 + \langle D \rangle)^{-0.25} \tag{3}$$

where the last term on the right hand side is an approximate nonlinear correction (Yahil 1985). YWR use eqs. (2)-(3) to determine Ω_o. To do that they assume that u_μ is caused by density inhomogeneities which are faithfully traced by the IRAS galaxies. The density moment needed in eq. (3) is then obtained from eq. (2) by dividing the dipole moment $4\pi S\sigma$ by the constant C, taken from the luminosity function determined by Lawrence et al. (1986).

The resulting linear theory value is $\Omega_o = 0.85 \pm 0.16$ (*statistical* IRAS errors only), and the nonlinear correction is estimated at +0.15, bringing Ω_o up to unity. While in accord with some theoretical predictions, particularly in the inflationary scenario, $\Omega \approx 1$ is at variance with dynamical estimates of $\Omega_o = 0.1-0.2$ from both the Virgocentric infall and the cosmic virial theorem (e.g., the review by Yahil 1985).

When a clear paradox develops, at least one of the previous precepts has to be modified. There are various possibilities in this case:

- The measurement of the dipole moment might be wrong, due to instrumental effects that have not been properly corrected, or the underestimate of the flux of extended nearby galaxies. Work is underway to improve the database in these respects.
- It is also possible that the current determination of the IRAS luminosity function is in error, but it is hard to imagine that the error is greater than a factor of two, which is what is required to bring the value of Ω_o down to the level deduced from the Virgocentric infall and the cosmic virial theorem. In any case, the matter needs to be checked by future redshift surveys. One would also like to see a density-independent determination of the luminosity function, followed by an evaluation of the mean density of the surveyed region (cf. Yahil et al. 1980; Davis and Huchra 1982).
- Perhaps the peculiar velocity induced by the inhomogeneities inside the volume encompassed by the IRAS galaxies accounts for only part of u_μ. The discovery of large scale currents, as reported in this conference by Burstein, makes this an interesting possibility. Note, however, that the volume over which coherent motion has been found is still smaller than that covered by IRAS, and there remains the question of the directional coincidence between the anisotropies of the IRAS galaxies and the MBR.
- It is possible that the Virgocentric infall model is not as reliable as has been thought (Bushouse et al. 1985; Villumsen and Davis 1986; Lee et al. 1986; papers in this conference). As I point out in another paper in this conference, however, it is not clear that the analyses of numerical simulations have so far reproduced structures that are reliably similar to the VSC for this judgement to be made. In any case, the argument does not apply to the cosmic virial theorem, which has been shown to hold in N-body models (Evrard and Yahil 1985).
- Finally, it is possible that galaxy formation really is biased. In that case, either the optically selected galaxies, or the IRAS galaxies, or both, might not faithfully trace the total mass-energy, even on scales greater than 10 Mpc, and the determination of Ω_o would be biased. Again, the directional coincidence of the IRAS and MBR anisotropies may be problematic, because if the dipole moment of the IRAS galaxies is corrected for biasing, it may no longer point in the same direction as u_μ. This test of the principle of superposition can not be applied with present data, but will be possible once a complete redshift survey becomes available.

THE LOCAL GRAVITATIONAL FIELD

3. IRAS GALAXIES IN THE VIRGO SUPERCLUSTER

It is very difficult observationally to test for biased galaxy formation, because what one sees is supposedly not what is really there. A possible future test of the principle of superposition is mentioned in § 2. An indirect test is to compare the density distributions deduced from different tracers. If they show distinctly different density profiles, then at least one of them is biased, and possibly both are. On the other hand, if they show the same density structure, then they are at least biased in the same way, and perhaps this is weak evidence for their being faithful tracers of the total mass-energy.

There is concern that the change of galaxy type as a function of density (Dressler 1980), might affect the determination of density profiles. Yahil et al. (1980), however, find that the Virgocentric density profiles, determined separately using the early and late types galaxies in the RSA catalog, are very similar, except at the central Virgo cluster, where the early type galaxies outnumber the late type ones by a ratio of 2:1, the inverse of the ratio found elsewhere. Thus, both tracers appear to be equivalent for the largescale density determination in the supercluster.

A similar test is performed here by comparing the IRAS galaxies with the CfA catalog. The absence of extensive redshift data for the IRAS galaxies precludes an extensive comparison, but a limited one is possible. In fig. 1a of Lawrence et al. (1986) it can be seen that in their main sample, which is complete to 0.85 Jy, there are 46 galaxies in the redshift range $v<3000$ km s^{-1}. Cross correlation with the CfA catalog shows that 41 of the 46 galaxies, or ~90%, are already in the CfA catalog, and therefore have known redshifts. If this result can be generalized to the entire area covered by the CfA survey, then redshifts for IRAS galaxies are already 90% complete for redshifts $v<3000$ km s^{-1}, and their density structure in this region may be studied. This density distribution can then be compared with that of the entire CfA catalog, which contains more than twice as many galaxies in the same region.

Fig. 1 shows the surface density of the IRAS and CfA galaxies with $v<3000$ km s^{-1} in annuli centered on the Virgo cluster (M87 to be precise), with the inner 6° of Virgo excluded. The IRAS galaxies, which are restricted to fluxes above 0.85 Jy, are marked by the solid histogram, with the ordinate on the left, and those of the entire CfA catalog by the dashed histogram, with the right ordinate. The scales are adjusted to be proportional to the total count. If the IRAS galaxies are equivalent tracers, then the distributions should be similar. They are almost identical, except in the innermost bin.

Strictly speaking, the comparison in fig. 1 is incorrect, because the infrared luminosity function of the IRAS galaxies is different from the optical one of the CfA galaxies. Hence the *expected* distributions are somewhat different. A comprehensive study of the distribution is underway, that takes these effects into account. Meantime, it is possible to perform a quick test, by eliminating all galaxies which could not be seen if placed at the edge of the surveyed region, at $v=3000$ km s^{-1}. In such a sample, each galaxy carries equal weight, and the two distributions can be compared directly. This is shown in Fig. 2

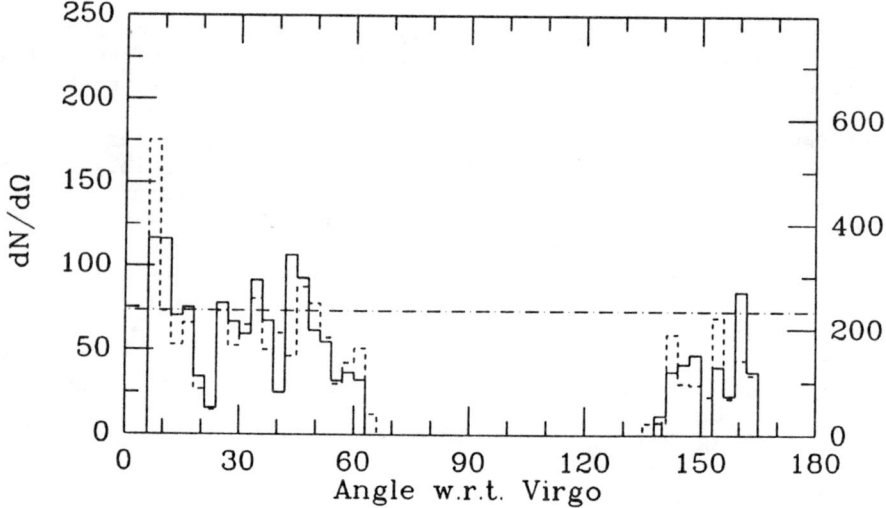

Fig. 2 Same as fig. 1, but all galaxies which would not be included if at the survey limit of v=3000 km s^{-1} are excluded. In this plot all the galaxies have equal weight.

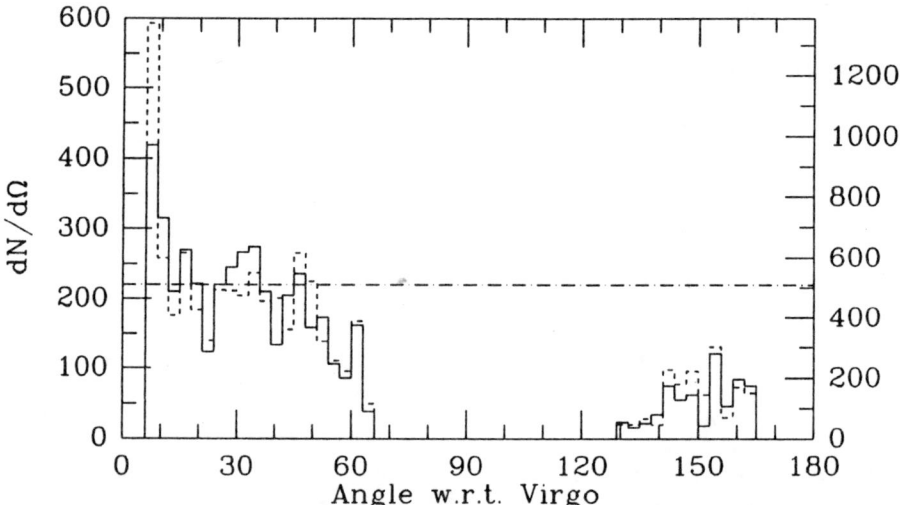

Fig. 1 Surface density of IRAS and CfA galaxies with v<3000 km s^{-1} in annuli centered on Virgo (inner 6° excluded). Solid histogram and left ordinate -- IRAS galaxies with S_{60}>0.85; dashed histogram and right ordinate -- all CfA galaxies; dot-dashed line -- IRAS cosmological baseline density level. Ordinate scales are adjusted to be proportional to the total count.

(no correction has been made for peculiar velocities). The statistics is somewhat poorer, because only a third of the galaxies survive, but the similarity of the IRAS and CfA distributions is unmistakable. It is concluded that the infrared selected IRAS sources and the optically selected CfA galaxies are equivalent tracers in the VSC.

Having established that the Virgocentric density distributions of the IRAS and CfA galaxies are essentially identical, it is of interest to consider the cosmological baseline density level, supposedly the mean density of the universe. The optical baseline, as determined both for the RSA catalog (Yahil et al. 1980) and the CfA catalog (Davis and Huchra 1982), is approximately at the level of the density in the South Galactic Hemisphere (Virgocentric angles larger than $90°$ in figs. 1-2). The IRAS baseline of Lawrence et al. (1986) is shown in figs. 1-2 as dash-dotted lines, and is seen to be about twice higher. This difference in baseline could completely explain away the difference in the determination of Ω_o between the Virgocentric infall model and the IRAS anisotropy.

4. DISCUSSION

One possible interpretation of the results of § 3 is that the optical surveys have used an erroneous cosmological baseline density level. One is reminded of this possibility here in Hawaii, where one is told that Mauna Kea is the highest mountain in the world... as measured from the sea floor. Has the overdensity of the VSC simply been measured relative to a local baseline, that is lower by a factor ~2 than the global one?

This explanation of the difference between the Virgocentric and the IRAS determinations of Ω_o, which would support the higher value of Ω_o, is very tempting. There are, however, several, possibly serious, problems:

- It can be argued that the determination of the baseline from the RSA catalog (Yahil et al. 1980), could have been flawed, because density is only determined out to a redshift of 4000 km s^{-1}, and confidently perhaps only out to ~3000 km s^{-1}. The density of the CfA catalog, however, is determined to a redshift of 8000 km s^{-1} (Davis and Huchra 1982), and one would think that there are enough superclusters and voids in this volume to qualify it as a fair sample of the universe.
- Yahil et al. (1980), use their admittedly local baseline density to predict the number count of galaxies as a function of magnitude, and compare it with the much deeper survey of Kirshner et al. (1979). They find good agreement, and certainly no discrepancy of a factor of 2.
- It is possible that the comparison of the IRAS and CfA catalogs within the redshift range v<3000 km s^{-1} does not tell the entire story. Most of the IRAS galaxies in this range are "normal" galaxies. In fact, they have been known before, because 90% of them are in the CfA catalog. The galaxies which are not in the CfA, and which account for an increasing fraction of the IRAS galaxies at higher redshifts, are mainly galaxies undergoing intensive star formation (star bursters), or active galaxies. If their distribution is more homogeneous than

that of "normal" galaxies, they could contribute to the IRAS baseline density, without affecting the density inhomogeneities.

Additional difficulties are surely also possible, and the simple explanation via baseline differences may well turn out to be wrong. Be that as it may, it is clear where the work ahead lies. The effort should be directed at obtaining a complete IRAS redshift catalog which covers most of the sky. This will allow a firmer determination of the IRAS luminosity function. More importantly, the depth information for individual galaxies will allow a detailed study of the density distribution of these galaxies, and a firmer determination of the gravitational field which they induce. It will be possible not only to obtain the dipole moment, but also the quadrupole one, which may be compared with the recent detection of a quadrupole tidal velocity field inside the VSC (Lilje et al. 1986, and this conference)

5. ACKNOWLEDGMENT

This research was supported in part by USDOE grant DE-AC02-80ER10719 at the State University of New York.

6. REFERENCES

Aaronson, M., Huchra, J., Mould, J., Schechter, P. L., and Tully, R. B. 1982, *Ap. J.*, **258**, 64.
Bushouse, H., Melott, A., Centrella, J., and Gallagher, J. 1985, *M. N. R. A. S.*, **217**, 7p.
Davis, M., and Huchra, J. 1982, *Ap. J.*, **254**, 437.
Dressler, A. 1980, *Ap. J.*, **236**, 351.
Evrard, A. E., and Yahil, A. 1985, *Ap. J.*, **296**, 310.
Fixsen, D. J., Cheng, E. S., and Wilkinson, D. T. 1983, *Phys. Rev. Letters*, **50**, 620.
Kirshner, R. P., Oemler, A., and Schechter, P. 1979, *Ap. J.*, **84**, 951.
Lawrence, A., Walker, M., Rowan-Robinson, M., Leech, K. J., and Penston, M. V. 1986, *M. N. R. A. S.*, in press.
Lee, H., Hoffman, Y., and Ftaclas, C. 1986, preprint.
Lilje, P. B., Yahil, A., and Jones, B. J. T. 1986, *Ap. J.*, in press.
Lubin, P. M., Epstein, G. L., and Smoot, G. F. 1983, *Phys. Rev. Lett.*, **50**, 616.
Meiksin, A., and Davis, M. 1986, *A. J.*, in press (**MD**).
Peebles, P. J. E. 1980, *The Large-Scale Structure of the Universe*, (Princeton: Princeton University Press), § 14.
Rowan-Robinson, M., Chester, T., Soifer, T., Walker, D., and Fairclough, J. 1986, *M. N. R. A. S.*, in press.
Villumsen, J. V., and Davis, M. 1986, preprint.
Yahil, A. 1985, in *The Virgo Cluster of Galaxies*, eds. O. G. Richter and B. Binggeli, (Garching: ESO), p. 359.
Yahil, A., Sandage, A., and Tammann, G. A. 1980, *Ap. J.*, **242**, 448.
Yahil, A., Walker, D., and Rowan-Robinson, M. 1986, *Ap. J. (Letters)*, in press (**YWR**).

MASS SEGREGATION, HI DEFICIENCY AND DWARF IRREGULAR GALAXIES IN THE
VIRGO CLUSTER

E. E. Salpeter
Cornell University, Ithaca, N.Y.
USA

ABSTRACT. Numerical simulations show that moderate mass segregation could already be produced for the Virgo cluster during its "violent collapse" period, but only if most of the mass of the cluster is in individual galaxies. Dwarf irregular galaxies should be much less massive than spirals, but the observational data do not show any systematic evidence for dwarf-spiral segregation. Some minor "antisegregation" artifacts are probably due to high-velocity dwarfs losing their gas completely through ram pressure stripping.

1. MASS SEGREGATION AND RAM PRESSURE STRIPPING

For a cluster of point-masses gravitational collisions slowly transfer kinetic energy from the heavy particles to the light ones. The orbits of the light particles then (a) tend to be at larger distances from the cluster center and (b) should have larger dispersions σ_V of systemic velocity about the cluster center. Point (b) can get obscured observationally by the tendency for clusters formed by gravitational collapse and violent relaxation to have velocity dispersion $\sigma_V(r)$ decrease with increasing distance r from the center: The light, "faster" particles move out to where velocities are lower, so that the overall σ_V for light and heavy particles is almost the same and $\sigma_V(r)$ for light and heavies combined varies little with r (a similar phenomenon is apparent for spiral versus elliptical galaxies in the Virgo cluster, even though the dynamical differences were introduced by causes other than mass segregation). However, a sufficiently large data base can overcome this difficulty by comparing $\sigma_V(r)$ for light and heavies within a moderately small range of values of r. Furthermore, one can sharpen up the contrast in orbital radii by considering only particles with $|v| \le <\sigma_V>$, say: There are fewer light particles at small r and a smaller fraction of them have small velocities. For point-particles with a "Schechter function" mass distribution (N^* particles more mass than the "critical mass" in the exponent), a time t after the era of collapse and violent relaxation, the degree of segregation is of the order of

$$f(6/N^*)[3.5 + (t/T_{Dyn})]. \tag{1}$$

In this approximate empirical relation, based on numerical N-body calculations (Farouki et al 1983), f is the fraction of the total mass of the cluster associated with galaxies (the fraction 1-f is assumed to be distributed in bodies <u>less</u> massive than any galaxy) and T_{Dyn} is a dynamical timescale for the cluster. The factor 3.5 in eqn. (1) represents an enhancement during the violent relaxation period and is important for the Virgo cluster where the outer parts are still "falling in" (Tully and Shaya 1984) so that t/T_{Dyn} is quite small (with $N^* \sim 35$, the expression in eqn. (1) is $\sim 0.6f$).

Ram pressure stripping of the interstellar gas from a galaxy moving through intracluster gas has been calculated by a number of authors [Gisler (1976); Lea and DeYoung (1976); Shaviv and Salpeter (1982)], who give the dependence of the effect on the density n_∞ of the intracluster gas. We are also interested in the dependence on the mass of the galaxy, and on the velocity V_∞ relative to the intracluster gas, which is close to the galaxy's systemic velocity. Recent numerical work by Gaetz (1985) has shown that this dependence is quite strong, the relevant parameter being

$$(n_\infty/\sigma_g)(V_\infty/V_{esc})^{2.4} \tag{2}$$

where σ_g is a typical column density of the galaxy's interstellar gas.

2. THE DATA BASE FOR DWARF IRREGULARS

A systematic data base is available, inside the Virgo cluster proper (which I define as a circle of radius 5°) and for its surrounding, for galaxies down to medium-faint levels ($B_T \sim 14.5$ to 15.5). For extending the data base to fainter levels ($B_T \sim 17$ to 18, to include true dwarfs) a systematic optical catalog is now available (Binggeli, Sandage and Tammann 1985), but without redshifts. HI data for dwarf irregulars can extend the "dynamic range" in studies of (a) systemic velocities V, (b) the Tully-Fisher method via velocity widths ΔV and (c) HI deficiency in disk systems. My colleagues and I (Hoffman et al 1985) are well along in a HI detection survey at Arecibo for most of the dwarf irregulars (over 200) in this optical catalog.

Individual mass to light ratios vary appreciably and inferences about mass segregation from an observed absence of luminosity segregation is difficult (Smith 1985), unless a large dynamic range is available. We have a little mapping data and a lot of velocity width data for the dwarf irregulars in Virgo: probably the mean ratio of total gravitational mass M_T to luminosity is comparable to that in spirals; at any rate ΔV tends to be small and there is little doubt that dwarfs are appreciably less massive than most spirals.

The Tully-Fisher relation (ΔV versus apparent magnitude B_T) extends quite smoothly from the spirals to the dwarf irregulars. We subdivide the main range of systemic velocities V in the Virgo supercluster into "normal velocities", 300 km s^{-1} < V < 1900 km s^{-1} (about

± one standard deviation) and "large velocities", $V > 1900$ km s^{-1}; spatially, IN and OUT refers to within and without a 5° radius circle around the Virgo center. For large and small disk galaxies combined, we compare relative T-F distance moduli: For the OUT group the high velocity mean is about 0.9±0.3 magnitudes fainter than the normal, in keeping with the consensus that larger velocity galaxies in the outer supercluster are at larger distances. However, for the IN group we do not find any significant difference in distance modulus between normal and large velocities (the latter are brighter by 0.15±0.35), when the dwarfs are included.

3. EVIDENCE AGAINST MASS SEGREGATION

Most of the outer and some of the inner parts of the Virgo cluster are still "falling in". Nevertheless, much of the 5° "IN group" is likely to already have undergone collapse and violent relaxation (although only recently). The absence of strong differences in distance moduli (Sect. 2) and the presence of HI deficiencies (Sect. 4) argue in favor of this. Eqn. (1) with $t \sim 0$ then predicts a modest but detectable mass segregation difference between spirals and dwarf irregulars if all the cluster mass is in galaxies, but not if more than $\sim 50\%$ of the mass is in distributed (lighter) chunks of matter.

As mentioned in Sect. 1, there are two practical tests: (i) For the IN groups the dwarf irregulars should have a larger dispersion σ_V in systemic velocities than the spirals if there is mass segregation, the same in the absence of segregation. In fact we found $\sigma_V \sim 776$ km s^{-1} for the dwarf irregulars IN, slightly less than $\sigma_V \sim 852$ km s^{-1} for the spirals IN. (ii) Considering only the "normal" velocity range, 300 to 1900 km s^{-1}, we find for the numbers of (detected) dwarf irregulars 48 IN and 60 OUT (ratio 0.80), compared with 41 spirals IN and 42 OUT (ratio 0.98).

The most important result from (i) and (ii) combined is the absence of any significant segregation: If taken at face value, (i) even has the opposite sign to real mass segregation and the effect in (ii) is only $\sim 20\%$. This may be merely statistical fluctuations, but may also be artifacts caused by the environmental effects discussed below.

4. HI DEFICIENCIES

For the ratio of hydrogen mass M_H to luminosity L for spiral galaxies in Virgo, there is general agreement that the IN spirals are deficient in $<\log_{10} M_H/L>$ by about -0.4 compared with the OUT spirals. Furthermore, the deficiency stems from decreased hydrogen radii, rather than smaller central surface density of gas, just as expected from ram pressure stripping. We now have a fairly large data base of Arecibo HI observations on Virgo dwarfs, but in total almost half are non-detections. There is a great disparity in luminosity functions in the Virgo cluster core (Sandage 1985)--there are almost ten times as many dwarf ellipticals than dwarf irregulars. A category in the optical catalog, "dE or DI", represents a negligible contamination for the dE group, but a significant one for the dI group.

For discussing HI deficiencies we now omit all cases which might possibly be dwarf ellipticals from the dwarf irregular list (and restrict to $B_T < 17$ and omit BCD galaxies). We find a hydrogen deficiency IN versus OUT, but quantitatively slightly less than for spirals, compared with a stronger effect predicted theoretically by eqn. (2). My tentative conjecture is that the "all or nothing" nature of ram pressure stripping has stopped star formation in the most strongly affected dwarf irregulars, so they now mimic the optical appearance of a "slightly unusual dE". This could also explain the 20% fewer "true dI's" observed in the core at "normal" velocities (ii in Sect. 3) and the stronger depletion at "large" velocities (i in Sect. 3; see also eqn. (2) for velocity dependence).

I am grateful to Dr. G. L. Hoffman for considerable help in preparing this talk. This work was supported in part by NSF Grant AST84-15162.

REFERENCES

Binggeli, B., Sandage, A. and Tammann, G., A.J. 90, 1681 (1985).
Farouki, R., Hoffman, L. and Salpeter, E., Ap.J. 271, 11 (1983).
Gaetz, T. J., Cornell Univ. Ph.D. Thesis (1985).
Gisler, G. R., A. and Ap. 51, 137 (1976).
Hoffman, L., Helou, G., Salpeter, E. and Sandage, A., Ap.J. 289, L15 (1985).
Lea, S. and DeYoung, D., Ap.J. 210, 647 (1976).
Sandage, A., Binggeli, B. and Tammann, G., A.J. 90, 1759 (1985).
Shaviv, G. and Salpeter, E., A. and Ap. 110, 300 (1982).
Smith, H., Ap.J. 288, 117 (1985).
Tully, B. and Shaya, E., Ap.J. 281, 31 (1984).

EFFECTS OF MASS SEGREGATION IN ROTATING CLUSTERS OF GALAXIES

A. Kashlinsky
Dept. of Astronomy, University of Virginia, and
National Radio Astronomy Observatory
Charlottesville, Virginia

ABSTRACT. Clusters of galaxies are expected to have some rotation due to tidal interactions in the expanding Universe. We calculate the dynamical friction force in rotating systems and evaluate the changes in the cluster density profile, central mass, rotation curve, and measured M/L with time. Our main conclusions can be summarized as follows: (1) Dynamical friction changes density profiles only marginally. There are always enough galaxies that remain in the outer parts of the cluster and, therefore, there is little prospect of seeing effects of mass segregation by measuring mean radii of galaxies of different magnitudes. (2) Because of the velocity anisotropy induced by rotation there appears a secondary peak in the density distribution as a result of the evolution. (3) Dynamical friction leads to an increase in the rotational velocity of the system and rotation curve becomes flatter as time goes on. (4) Mass-to-light ratio changes very slowly with time and dynamical friction cannot lead to appreciable changes in M/L of clusters unless the amount of mass (light) accumulated at the cluster exceeds observational constraints.

I. INTRODUCTION

In order to explain successfully the large-scale structure of the Universe, one has to understand (among other things) the evolution of the distribution of galaxies. Most galaxies are found in clusters and, therefore, it is important to understand dynamical evolution of clusters of galaxies.

Clusters of galaxies have typically $\sim 10^3$ galaxies, crossing times of $\sim 10^9$ years and velocity dispersions $\sigma < 1000$ km/sec. Most of the mass in clusters is in some hidden form, whose substructure is different from that of the luminous matter. Most naturally this structure is explained within the frame of the isothermal picture for galaxy formation, whereby galactic haloes and clusters of galaxies form by gravitational clustering[1,2] of agglomerations of Population III objects formed after recombination, while luminous parts of galaxies form from the collapse of the remaining gas in the potential wells created by galactic haloes[3,4,5].

The small number of objects in clusters makes relaxation effects very efficient; the most important of them would be that due to dynamical friction, provided the mass of the dark component constituent is less than that of a typical galaxy. Previous investigations of dynamical friction evolution of clusters have shown that it may lead to a change in the density profiles of galaxies in clusters[6], an accumulation of mass in the cluster center, which may be the way that cD galaxies form, and a change in the measured M/L ratio of clusters with time[7,8]. The latter effect is particularly significant, since it would affect the value of Ω and detailed calculations of this effect are needed. Furthermore, clusters are likely to possess some rotation (which was ignored in all the previous works), since they are subject to the same tidal torques as galaxies[9]. An overall rotation (however small) would introduce important differences in the evolution of the system: (1) there is, now, systematic velocity v_{rot}: (2) the velocity distribution is no longer isotropic and this affects the redistribution of energy (ε) and angular momentum (J) in the cluster.

II. THEORY

We have recently calculated evolution in rotating clusters[10]. Rotation of the cluster can be parameterized by a dimensionless spin parameter $\lambda \equiv J|E|^{1/2}/GM^{5/2}$ and from tidal torques clusters would acquire $\lambda \sim 0.1$ on the average[11]. In case of an overall solid-body rotation the distribution function of an isothermal (in absence of rotation) cluster would be given by $f \propto \exp(-(\varepsilon-\Omega J_z)/\sigma^2)$ where ε is the objects energy, J_z is the component of its angular momentum along the axis of rotation and Ω is the angular velocity of the cluster. One can evaluate the dynamical friction force F acting on a galaxy of mass M moving at velocity $\underline{v}$ in such a system[10] to obtain:

$$\underline{F} = -4\pi G^2 \ln\Lambda \rho(r) \frac{f(<|\underline{v}-\underline{v}_{rot}|/\sigma)}{|\underline{v}-\underline{v}_{rot}|^3}(\underline{v}-\underline{v}_{rot}) \qquad (1)$$

where $f(<x) = \int x^2 \exp(-x^2/2)dx/\sqrt{2\pi}$. The force given by (1) has interesting effects on (among others) the measured rotational properties of the system, since it affects co-rotating (having $\underline{v} \cdot \underline{v}_{rot} > 0$) galaxies in a different way than the contra-rotating ones ($\underline{v} \cdot \underline{v}_{rot} < 0$). Since the number of objects in a volume $d\varepsilon dJ dJ_z$ is conserved we have at each time t:

$$dN \equiv \Psi(\varepsilon,J,J_z,t)d\varepsilon dJ dJ_z = f(\varepsilon_0) \int dr/\sqrt{(\varepsilon_0 - J_0^2/2r^2 - \Phi)} d\varepsilon_0 dJ_0 dJ_{z0} \qquad (2)$$

where Φ is the cluster potential and the integral is from the object's pericenter radius to its apocenter. Now by integrating (2) coupled with equations of motion for each (ε,J,J_z) under the influence of (1) one can calculate the evolution of the cluster (luminous) mass profile

$M(r,t)$, the rate of accumulation of mass at the center $M(r = 0,t)$, the evolution of the rotation curve $v_{rot}(r)$ and the change in M/L with time (see reference 10 for more details).

III. RESULTS AND CONCLUSIONS

We have made computations for different values of the the total to core radii ratio (R/r_{core}) and λ and for $m\ell n\Lambda/M = 1.55 \cdot 10^{-3}$, where m and M are galaxy and cluster masses, respectively. Some of the results are shown on Figures 1-3.

Figure 1 shows evolution of the rotation curve with time. It is easy to see that the effect of dynamical friction is to increase the rotational velocity and make rotation curve flatter, because the cluster becomes deficient of contra-rotating galaxies, since they sink to the center at a greater rate. This is particularly significant at smaller radii, where dynamical friction operates more efficiently because of the higher density there. As a result of this the anisotropy in velocity distribution increases (see later). Therefore, if mass segregation is efficient in clusters of galaxies, brighter galaxies should have different rotational properties than the less bright ones.

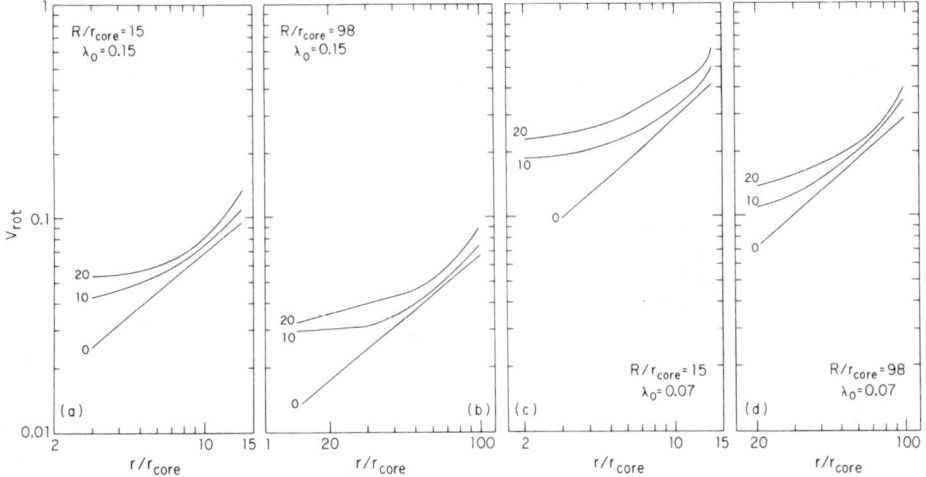

Figure 1. Evolution of rotation curve with time for different models. Time (in units of R/σ) is indicated by numbers near each curve.

The evolution of the luminous mass (i.e., luminosity) profile is shown on Figure 2. It illustrates that dynamical friction makes the mass profile steeper or, consequently, the density profile becomes flatter. This effect is, however, marginal, which indicates that there is little prospect of seeing effects of mass segregation <u>by measuring the mean radii of the distribution of galaxies of different magnitudes which would imply that present observations</u>[12] <u>in no way rule out that</u>

dynamical friction is efficient in clusters of galaxies. As a result of the evolution there appears a secondary peak in the density profile (or where M(r) changes cavity) similar to that found in some clusters[13]. This is a result of the increasing anisotropy in the velocity distribution for galaxies in cluster (because of the overall rotation) and this feature disappears with decreasing λ, as is to be expected in this case.

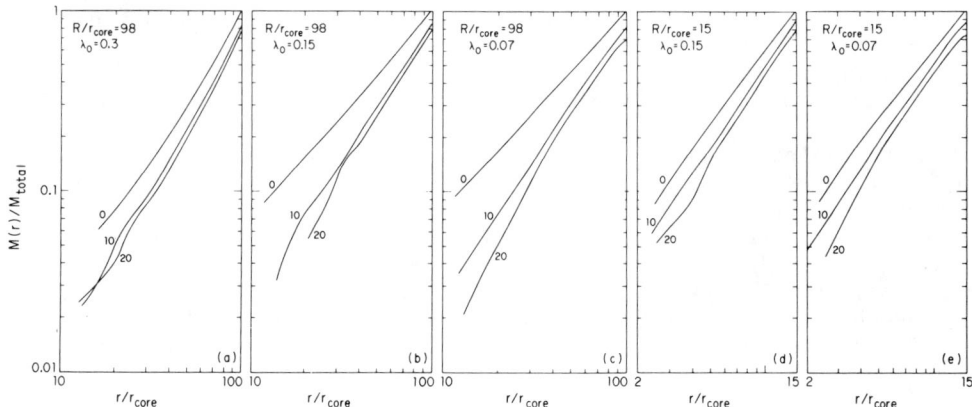

Figure 2. Changes in the luminous mass profile with time for different λ_0 and R/r_{core}. Numbers near the curves show evolution time in units of R/σ.

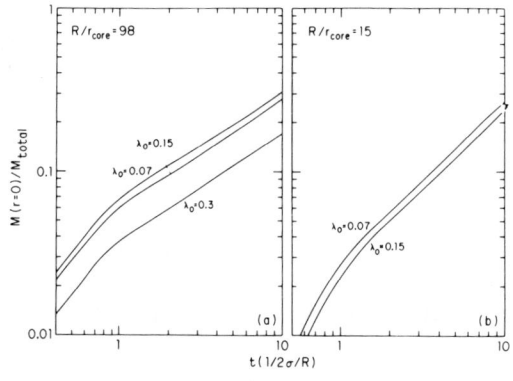
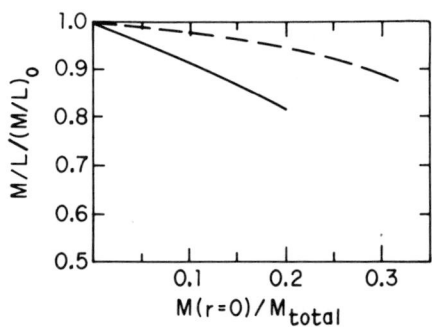

Figure 3

(a) Shows fraction of the total mass that has accumulated at the cluster center as a function of time.

(b) Changes in the measured M/L versus the mass accumulated at the center.

Since different clusters are expected to have different λ, this interpretation of secondary peak produced as a result of the dynamical friction in <u>rotating</u> clusters is consistent with the absence of secondary maxima in some clusters of galaxies[16].

Figure 3 shows how the central mass $M(r = 0)$ increases with time and the evolution of the measured M/L. Because cluster's spin λ determines the elongation of an average orbit and R/r_{core} determines the degree of steepness of the total density profile, the rate of increase in $M(r = 0)$ depends on these parameters. M/L was found to increase slowly with time, but when one plots M/L vs the mass accumulated at the cluster center (Fig. 3c) one can see that this change becomes appreciable when more than half the total luminous mass has accumulated in the cluster. <u>Since we know that central parts of clusters of galaxies do not dominate the total cluster light to such an extent we conclude that variations in clusters' M/L cannot be due to dynamical friction effects.</u> The rate of change of M/L is slow because at most times the galaxies that are the most affected by dynamical friction and contribute to the growth of the central mass are those at inner radii, which least affect M/L, or are on radial orbits, i.e., galaxies that can pass through high density regions where the drag is highest. At any time there remain enough galaxies with high J in the outer parts of clusters to keep the total M/L almost unchanged. The small change in M/L is equivalent to the fact that density profile and, consequently, mean radius change only marginally during cluster evolution. Since dynamical friction cannot be responsible for observed (large) differences in M/L from cluster to cluster, this means that clusters (unlike galaxies) formed with different values of M/L. If the latter is true, that indicates that the ratio of luminous to dark matter masses is not constant, but varies from cluster to cluster.

Acknowledgement. I am grateful to Stefi Baum for reading the manuscript.

REFERENCES

1. Peebles, P.J.E. 1974, Ap. J. Lett., <u>189</u>, L51.
2. Press, W. & Schechter, P. 1974, Ap. J., <u>187</u>, 425.
3. White, S.D.M. & Rees, M. J. 1978, MNRAS, <u>183</u>, 341.
4. Kashlinsky, A. 1982, MNRAS, <u>200</u>, 585.
5. Kashlinsky, A. 1983, in "Clustering in the Universe," eds. Gerbal, D. & Mazure, A., p. 67.
6. White, S.D.M. 1976, MNRAS, <u>174</u>, 467.
7. Hoffman, Y. <u>et al</u>. 1982, Ap. J., <u>262</u>, 413.
8. Barnes, J. 1983, MNRAS, <u>203</u>, 223.
9. Peebles, P.J.E. 1969, Ap. J., <u>155</u>, 393.
10. Kashlinsky, A. 1985, Ap. J., in press.
11. Efstathiou, G. & Jones, B.J.T. 1979, MNRAS, <u>186</u>, 133.
12. Peach, J. V. 1982, in "Clustering in the Universe," eds. Gerbal, D. & Mazure, A., p. 211.
13. Oemler, A. 1974, Ap. J., <u>194</u>, 1.
14. Dressler, A. 1978, Ap. J., <u>223</u>, 765.

MEASUREMENTS OF THE COSMIC BACKGROUND RADIATION

P. LUBIN and T. VILLELA*
Space Sciences Laboratory and Lawrence Berkeley Laboratory
University of California
Berkeley, California 94720

ABSTRACT

Maps of the large scale structure ($\theta > 6$°) of the cosmic background radiation covering 90% of the sky are now available. The data show a very strong 50-100 σ (statistical error) dipole component, interpreted as being due to our motion, with a direction of $\alpha = 11.25 \pm 0.15$ hours, $\delta = -5.6 \pm 2.0$°. The inferred direction of the velocity of our galaxy relative to the cosmic background radiation is $\alpha = 10.6 \pm 0.3$ hours, $\delta = -23 \pm 5$° ($l^{II} = 267$°, $b^{II} = 31$°). This is 44° from the center of the Virgo cluster. After removing the dipole component the data show a Galactic signature but no apparent residual structure. An autocorrelation of the residual data, after subtraction of the Galactic component from a combined Berkeley (3 mm) and Princeton (12 mm) data set, shows no apparent structure from 10° to 180° with a rms of 0.01 mK^2. A 90% confidence level limit of $7x\,10^{-5}$ is placed on a quadrupole component.

INTRODUCTION

It has now been twenty years since the discovery of the cosmic background radiation. In these twenty years numerous experiments have been carried out to study the spectrum, angular structure and polarization. Our current knowledge of the radiation can be summarized by paraphrasing the original 1965 article of Penzias and Wilson namely that the radiation is consistent with a blackbody of temperature of about 3 K, is isotropic and unpolarized. So far, we have learned a great deal but discovered very little. Our prized relic of the early universe is frustratingly simple to describe.

Of course the radiation is slightly anisotropic due to our motion ($\beta \approx 10^{-3}$) but on cosmological scales this is just a local phenomenon. Another local effect is the seasonal difference due to the motion of the Earth around the Sun ($\beta \approx 10^{-4}$) which is observed in both the Berkeley 3 mm and Princeton 12 mm data from flights separated by 6 months.

On scales from arc seconds to a hundred eighty degrees, experiments have been performed with no intrinsic structure being found. The current measurements are summarized in Figure 1. Large scale measurements have reached a sensitivity where the dipole is measurable in real time, as shown in Figure 2, and can be used as a low level calibration source as well as providing us with a convenient way to test if our instrument is working in flight. The dipole has become a "background" to be subtracted away. A much more problematic background is the emission due to our galaxy. With synchrotron and bremsstrahlung radiation decreasing and dust emission increasing with frequency, a natural minimum occurs somewhere around 90 GHz, as shown in

*Also INPE - Departamento de Astrofísica, São José dos Campos, SP, Brazil.

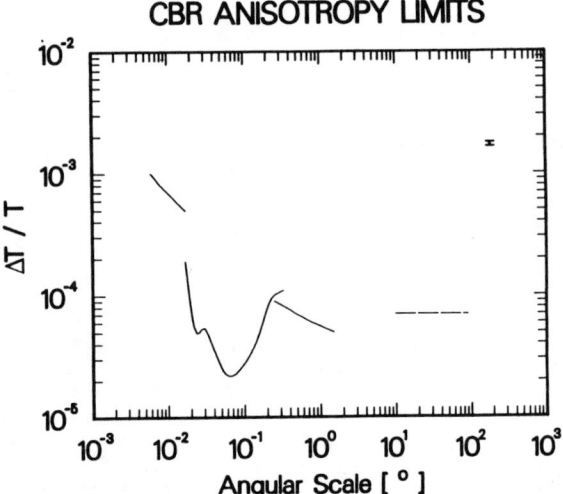

Figure 1 - Anisotropy Measurements at Various Angular Scales.

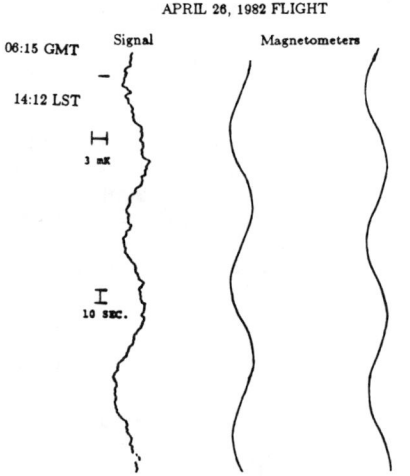

Figure 2 - Real Time Dipole Measurement.

Figure 3, which gives the measured large scale Galactic contribution from recent experiments versus frequency, using a cosecant (b^{II}) Galactic model flattened near the plane. The top plot of Figure 3 gives the ratio of the amplitude of the cosecant (b^{II}) model (pole value) to the dipole amplitude versus frequency. The dipole flux (in antenna temperature) decreases at high frequencies due to the change in slope of the spectrum. At 3 mm wavelength, where we made our measurements, the pole value amplitude is 44 ± 11 μK and hence the Galactic contribution to the dipole ($\approx$3 mK) is at the 1-2% level near the Galactic pole and rises to about 20% at the Galactic plane.

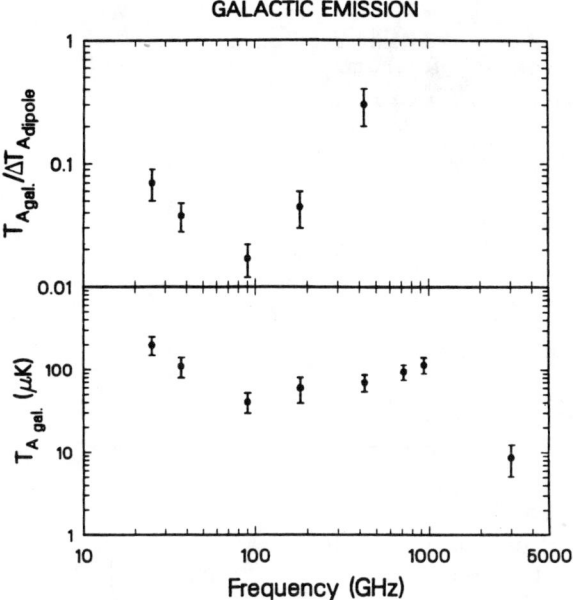

Figure 3 - Measured Galactic Emission Versus Frequency.

A more detailed analysis implies that the pole may be less and the plane more (Lubin and Villela 1985). Inclusion of the Galactic plane only changes the 3 mm dipole and quadrupole fits by 1 σ. In searching for higher order anisotropies (quadrupole, etc.), proper modeling of the Galaxy is crucial since even at 3 mm where the Galactic emission is very small, it is comparable to errors in the quadrupole components, which are 50-80 μK. Future experiments will need to pay particular attention to this problem.

MEASUREMENTS

The data presented here are the result of independent measurements at 12 mm using a maser (Princeton) and 3 mm using a Schottky diode mixer (Berkeley). Both instruments are cooled to liquid helium temperature. The experiments are described in detail in Fixsen, Cheng, and Wilkinson (1983) and Fixsen (1982) for the 12 mm experiment and Lubin, Epstein, and Smoot (1983), Epstein (1983), and Lubin and Villela (1985) for the 3 mm experiment. The combined results we describe here are based on collaborative work to be published (Villela, Walker, Wilkinson, and Lubin 1986). The 12 mm and 3 mm data sets have been combined to cross check the results of each and to produce a combined map which is better connected (see Lubin and Villela 1985) than either, as well as has increased sky coverage. The 12 mm experiment uses a more sensitive detector than the 3 mm experiment but observes in a region of higher Galactic emission, so ultimately both experiments have nearly identical sensitivities to large scale structures (dipole, quadrupole, etc.) as shown in the fits given in Table I. Both experiments have had three data flights, two from Palestine (lat.

TABLE I

COMPARISON BETWEEN 3 MM AND 12 MM FITS			
Values in thermodynamic temperature for T=2.7 K			
FIT	COEFFICIENTS	3 MM[1] DATA	12 MM[2] DATA
Dipole only	T_x	-3.35 ± 0.17	-3.01 ± 0.17
	T_y	0.71 ± 0.10	0.56 ± 0.09
	T_z	-0.35 ± 0.08	-0.28 ± 0.09
Dipole and Quadrupole	T_x	-3.37 ± 0.17	-3.07 ± 0.17
	T_y	0.69 ± 0.09	0.67 ± 0.09
	T_z	-0.42 ± 0.09	-0.45 ± 0.09
	Q_1	0.10 ± 0.09	0.15 ± 0.08
	Q_2	0.15 ± 0.10	0.15 ± 0.11
	Q_3	0.12 ± 0.11	0.14 ± 0.07
	Q_4	-0.09 ± 0.09	0.06 ± 0.11
	Q_5	0.08 ± 0.08	-0.01 ± 0.07

[1] *Excludes data within 5^0 of the galactic plane.*
[2] *Fixsen, Cheng and Wilkinson (1983).*
Errors include calibration.

Table I - Spherical Harmonics Fits.

= 32°), Texas, and one from near São José dos Campos (lat. = -23°), São Paulo, Brazil, for each. Both experiments have nearly identical sky coverage of 85% with a combined sky coverage of 90%.

Corrugated scalar horn antennas with a beam of 6° FWHM and an opening angle of 90° symmetrically placed 45° with respect to the zenith are used in each experiment to measure the temperature difference between the two beams. Both experiments are flown from a balloon platform at an altitude of 25 to 30 km at night. Atmospheric emission is typically 5 to 25 mK but since the instruments only measure temperature differences the atmospheric contribution cancels out to first order. Both instruments have inflight calibration sources which currently limit the error on the dipole amplitude to 5% accuracy while the statistical errors are at the 1% to 2% level.

RESULTS

The dipole and quadrupole fits are summarized in Table 1. Both experiments have dipole directions which agree within 1.6°, with an average of $\alpha = 11.25 \pm 0.15$ hours and $\delta = -5.6 \pm 2.0$ ° for the Solar velocity. The effect of the Earths' orbital velocity around the Sun, which was measured in both experiments, in flights separated by six months, was corrected for. There is still some discrepancy in dipole amplitudes presumably due to calibration errors with a dipole amplitude of 3.45 ± 0.17 mK for the 3 mm data (Lubin, Villela, Epstein, and Smoot 1985) and 3.07 ± 0.17 mK for the 12 mm data (Fixsen, Cheng, and Wilkinson 1983). This calibration error does not affect the dipole direction significantly but does affect the Solar and hence inferred Galactic and Local Group velocities. Taking an average Solar velocity direction of $\alpha = 11.25 \pm 0.15$ hours and $\delta = -5.6 \pm 2.0$ ° with an average dipole amplitude of 3.26 ± 0.23 mK and correcting for the Galactic Solar velocity of 230 km s^{-1} toward $l^{II} = 90°$, $b^{II} = 0°$, gives a Galactic velocity of $\vec{V}_G = 540 \pm 50$ km s^{-1} towards $\alpha = 10.6 \pm 0.3$ hours, $\delta = -23 \pm 5$ ° or $l^{II} = 267 \pm 5°$, $b^{II} = 31 \pm 5°$. This gives an angle of 44° between the center of the Virgo cluster (M87, $l^{II} = 284°$, $b^{II} = 75°$) and $\vec{V}_G$. Assuming that the velocity of the Sun relative to the Local Group is 295 km s^{-1} toward $l^{II} = 97.2°$, $b^{II} = -5.6°$, (Sandage 1986) gives a Local Group velocity of $\vec{V}_{LG} = 610 \pm 50$ km s^{-1} towards $\alpha = 10.8 \pm 0.3$ hours, $\delta = -25 \pm 5°$ or $l^{II} = 272 \pm 5°$, $b^{II} = 30 \pm 5°$. This gives an angle of 45° between the center of the Virgo cluster and $\vec{V}_{LG}$.

The quadrupole limits of both data sets are $7x\,10^{-5}$ as a 90% confidence level upper limit. A combined map made from both data sets is given in Figure 4 and the residual map after subtraction of the dipole is given in Figure 5. An autocorrelation of the residual map shows no obvious structure from 10° to 180° giving a rms of 0.01 mK^2.

Recently, several groups have looked at the IRAS data set and for 60 μm data have found a "dipole" (Yahil, Walker, and Rowan-Robinson 1985). The "dipole" direction is found to be $l^{II} = 248 \pm 9°$, $b^{II} = 40 \pm 8°$ ($\alpha = 10.2 \pm 0.6$ hours, $\delta = -5.4 \pm 8°$). This "dipole" direction is $21 \pm 11°$ from $\vec{V}_{LG}$. It's not clear precisely what significance should be given to this. A "quadrupole" has also been reported in the X-ray background (Fabian, Warwick, and Pye 1980) with an axis direction similar to the dipole direction.

OTHER EXPERIMENTS

Recently, the Soviet Union completed a satellite-borne cosmology mission known as Prognoz 9. Only very preliminary results (Strukov and Skulachev 1984) have been released from this experiment. The instrument consisted of a 8 mm (37 GHz) parametric amplifier and similar corrugated scalar horn antennas to ours, also with a 90° opening angle. Sensitivity should be excellent given the long integration time available for the mission, however Galactic emission, because of the relatively long wavelength, will have to be modeled very carefully. Because neither the Princeton 12 mm data nor the Berkeley 3 mm data are sensitivity (statistics) limited for the dipole measurement, it is unlikely that the Prognoz 9 results will significantly change our understanding of the dipole. Current limitations on the dipole are due to inflight calibration errors ($\approx$5%) and pointing errors ($\approx$1°). The Prognoz 9 data will significantly increase our understanding of the Galactic emission and may provide more sensitive measurements of intrinsic fluctuations. Recent dipole measurements are summarized in Figure 6 as a function of frequency. The 12 mm (25 GHz) and 3 mm (90 GHz) data are also plotted. The broad band higher frequency measurements are from bolometric systems of the M.I.T. (Halpern 1983) and University of Rome (Fabbri, Guidi, Melchiorri, and Natale 1980) groups.

CONCLUSIONS

In the twenty years since the discovery of the background radiation no intrinsic structure has been convincingly measured. The dipole anisotropy, apparently due to our motion, has been measured sufficiently well to determine our direction of motion within 2°. Our galaxy is moving in a direction that is about 44° from the center of the Virgo cluster. The dipole is now a background to be subtracted to get measurements of intrinsic anisotropies. Large scale measurements are now becoming limited by Galactic backgrounds particularly at centimeter wavelengths and even at the best wavelengths around 3 mm an order of magnitude improvement will be difficult without a detailed knowledge of Galactic emission.

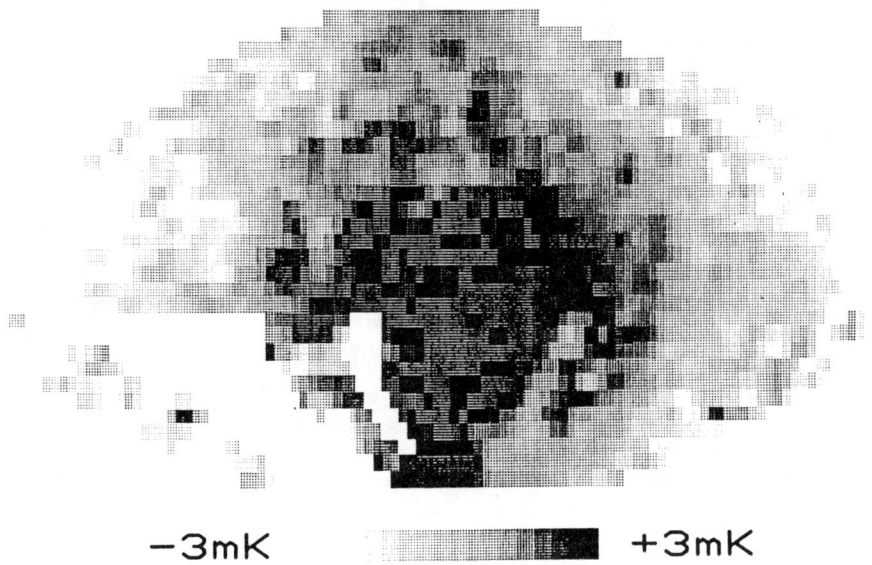

Figure 4 - Combined Dipole Map.

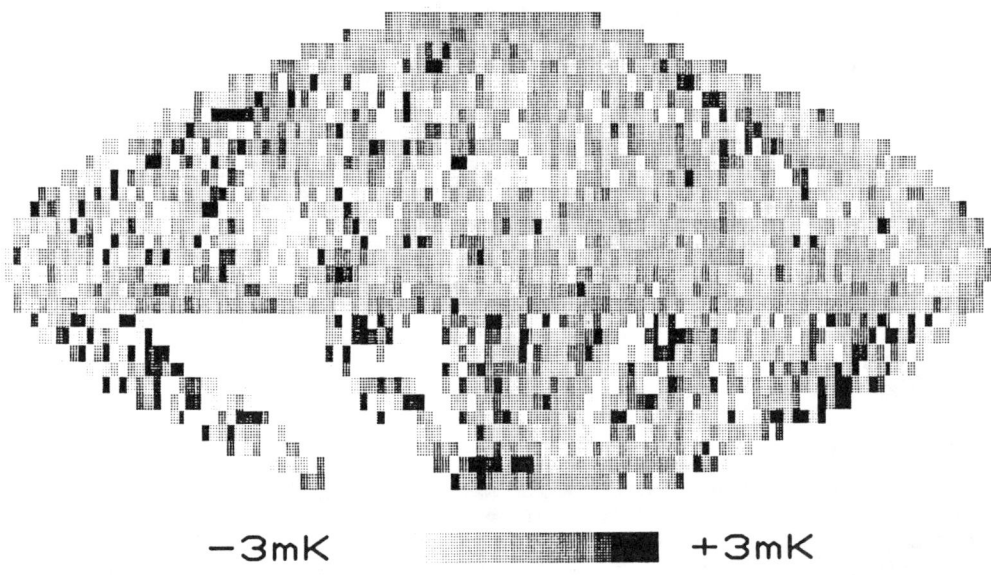

Figure 5 - Combined Residual Map.

DIPOLE MAGNITUDE

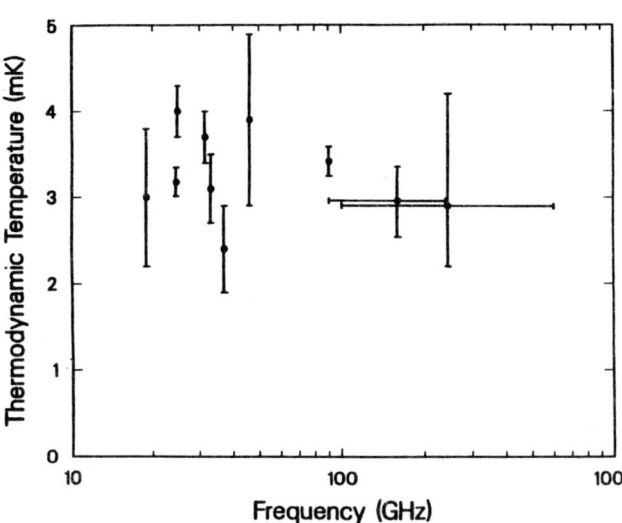

Figure 6 - Recent Dipole Anisotropy Measurements.

This work was supported by the California Space Institute, NASA and by the U. S. Department of Energy. One of us (T. V.) acknowledges support from CNPq and FAPESP.

REFERENCES

Epstein, G. L. 1983, *Ph.D. Thesis*, University of California, Berkeley.
Fabbri, R., Guidi, I., Melchiorri, F., and Natale, V. 1980, *Phys. Rev. Letters*, **44**, 1563.
Fabian, A. C., Warwick, R. S., and Pye, J. P. 1980, *Physica Scripta*, **21**, 650.
Fixsen, D. J. 1982, *Ph.D. Thesis*, Princeton University.
Fixsen, D. J., Cheng, E. S., and Wilkinson, D. T. 1983, *Phys. Rev. Letters*, **50**, 620.
Halpern, M. 1983, *Ph.D. Thesis*, Massachusetts Institute of Technology.
Lubin, P., Epstein, G., and Smoot, G. 1983, *Phys. Rev. Letters*, **50**, 616.
Lubin, P. and Villela, T. 1985, *Società Italiana di Fisica, Conference Proceedings, "The Cosmic Background Radiation and Fundamental Physics"*, **1**, 65.
Lubin, P., Villela, T., Epstein, G., and Smoot, G. 1985, *Ap. J. (Letters)*, **298**, L1.
Penzias, A. A. and Wilson, R. W. 1965, *Ap. J.*, **142**, 419.
Sandage, A. 1986, *Private Communication*.
Strukov, I. A. and Skulachev, D. P. 1984, *Sov. Astron. Letters*, **10**, 1.
Villela, T., Walker, K., Wilkinson, D., and Lubin, P. 1986, *to be published*.
Yahil, A., Walker, D., and Rowan-Robinson, M. 1985, *preprint*.

ON THE MEASURE OF COSMOLOGICAL TIMES

Alvio Renzini
Dipartimento di Astronomia, Università di Bologna
CP 596, I-40100 Bologna
Italy

ABSTRACT. Methods and results of globular cluster dating are synthetically reviewed, and discussed in the cosmological frame. Emphasis is placed on the origin of current uncertainties (particularly in distances), and on the possible observational procedures which can improve both accuracy and reliability of age determinations.

1. INTRODUCTION: THREE COSMOLOGICAL CLOCKS

Cosmological models are described by four main dynamical parameters: H_o, q_o, Ω_o, and Λ. When three among them are specified, the age of the Universe, T_o (the time elapsed since the Big Bang) remains determined by a simple analytical expression. In the simplest possible case ($q_o = \Omega_o = \Lambda = 0$), the age of the Universe is just the reciprocal of the Hubble constant, $T_o = H_o^{-1}$, while for $q_o = 1/2$, $\Omega_o = 1$ (and $\Lambda = 0$), $T_o = 2/3\ H_o^{-1}$. It is also useful to recall that with $H_o = 50$, $H_o^{-1} = 20$ Gyr, and with $H_o = 100$, $H_o^{-1} = 10$ Gyr.

Time being the physical quantity measured by clocks, the most natural clock to use in estimating T_o is then the Universe itself, as just briefly sketched. However, the current estimates of H_o (not for saying of q_o, Ω_o, and Λ) are sufficiently uncertain that it is certainly wise using other, complementary clocks. Indeed, different dating methods involve different sequences of operations, different physical assumptions and biases, and give independent estimates. Ultimately, the intercomparison of the various chronologies may give us a great deal of cosmological and astrophysical information, including (hopefully) a highly constrained model for the Universe (cf. Sandage and Tammann 1984, for a recent discussion).

Two other clocks are currently used in measuring cosmological times, namely: radioactive nuclei, and stars (stellar aggregates), which allow to estimate the age of particular constituent of the Universe (like the oldest atomic nuclei, and the oldest stars) and then provide lower limits to the age of the Universe.

The state of the art in nucleocosmochronology is reviwed by Thielemann at this meeting, while this review deals with the stellar clock, and its application to galactic globular clusters (GCs). The determination of GC ages is doubly intimately interwoven to that of H_o. First, because it is meaningful to compare GC ages to H_o^{-1}, and, second, because from an operational point of view both determinations require accurate distance calibrations.

2. DATING GALACTIC GLOBULAR CLUSTERS

Stars "work" (and can be used) as clocks, because their evolutionary lifetime is a function of their initial mass and composition. This function can be obtained by constructing appropriate model evolutionary sequences, and isochrones in the Hertzsprung-Russell diagram, thus providing a dating method which is most efficient when applied to populous star clusters, in which all stars have presumably the same age and composition. This is indeed the case for (most) galactic GCs, which are believed to be the oldest galactic clusters for which we can estimate the age. However, stars in GCs are certainly not the very first stars which formed in the Galaxy, as they contain heavy elements which must have been manufacted by an even earlier stellar population (sometimes called population III), with primordial, post-Big Bang composition. Moreover, the high degree of chemical homogeneity of (most) GCs ensures that population III stars polluted the protogalactic material before the formation of GCs themselves. They cannot therefore be regarded as primordial objects, but are rather late products of the early organization of cosmic structures.

2.1. Three Dating Procedures

Theoretical isochrones are shapes in the temperature-luminosity plane, hereafter the HR diagram (HRD). The only detail in these shapes which can be used for dating purposes is the section around the so-called main sequence turnoff. More specifically, both turnoff luminosity and temperature are function of age (t) and composition (Y,Z), i.e. $L_{TO}(t,Y,Z)$ and $T_{TO}(t,Y,Z,\alpha)$, where α stands for the adopted ratio between the mixing length parameter and the local pressure scale height, used in modelling the stellar outer convective layers. These theoretical functions are the clock. For instance, using the so-called "Yale isochrones" (Ciardullo and Demarque 1977) one can write:

$$\text{Log } t \simeq 10.423 - 1.025 \text{ Log } L_{TO} - 0.43(Y-0.3) - 0.15(3+\text{Log } Z). \tag{1}$$

More recent isochrones (e.g. VandenBerg 1983, VandenBerg and Bell 1985) give essentially the same relation.

The <u>observables</u> are, in practice, the magnitudes (colors) of the cluster stars, which in turn are used to construct cluster color-magnitude diagrams (CMD). In particular, one can derive the apparent magnitude and color of cluster stars at the main sequence turnoff, e.g. V_{TO} and $(B-V)_{TO}$.

Basically, three distinct procedures have been used in inserting these observables into the theoretical clock, so as to "read" the desired cluster age. All three procedures share the common requirement of an independent estimate of composition, Y and Z. While Z can be obtained spectroscopically, for the helium content Y one generally adopts the value indicated by the so-called R-method ($Y = 0.23\pm0.03$, cf. Buzzoni et al. 1983), or by the Big Bang nucleosynthesis ($Y \simeq 0.24$, cf. Boesgaard and Steigman 1985).

<u>Procedure 1</u>: Turnoff Luminosity. The crucial observable is V_{TO}, to which absorption and bolometric corrections are applied to get the apparent bolometric magnitude of the turnoff, $m_{TO}(bol)$. In the next step, an independent estimate of the cluster distance modulus is used to get $M_{TO}(bol) = m_{TO}(bol) - \text{mod}$, which is then inserted into (1) to sort out the age. In some variety of this method, e.g. when using the magnitude difference between the horizontal branch and turnoff, some of these steps may be either superflous or implicit.

<u>Procedure 2</u>: Turnoff Temperature. Here the crucial observable is the turnoff color $(B-V)_{TO}$. After correcting for reddening, the color is transformed to turnoff temperature, which inserted into the theoretical function $T_{TO}(t,Y,Z,\alpha)$ gives the age. Formally, this procedure does not need the distance modulus to be specified.

<u>Procedure 3</u>: Isochrone Fitting. The isochrones in the theoretical HRD are transformed into isochrones in the CMD, by applying bolometric corrections and temperature-color transformations. The observed cluster CMD is then corrected for reddening, and one looks for the isochrone which "best fits" the cluster locus. Formally, this procedure gives at once both cluster age and modulus. The method implicitly adopts the theoretical zero age main sequence as standard candle in getting the modulus.

It is worth emphasizing that, although all procedures require an independent estimate of composition, uncertainties in composition produce different age uncertainties, depending on the adopted procedure. For instance, in procedure 2, errors in the adopted metallicity generate very large errors in the derived age. This is so because the turnoff temperature is much more sensitive to metallicity than it is to age (cf. Iben and Renzini 1984, Renzini and Buzzoni 1986). Moreover, theoretical temperatures depend on the adopted mixing length parameter, which must be calibrated separately, and theoretical colors obviously depend on the adopted temperature-color transformations. For these reasons procedure 2 is more severely affected by the uncertainties introduced at each of the mentioned steps, and the same disadvantages are present, to some extent, also in the isochrone fitting technique. This leaves procedure 1, as the most promising technique for estimating accurate cluster ages.

With this method the crucial step is the determination of the cluster distance modulus. From Eq. (1) it follows that the relative error in age, $\sigma(t)/t$, is almost exactly equal to the error in cluster distance modulus, $\sigma(\text{mod})$, because the derived age scales approximately as the inverse square of the adopted distance. Incidentally, one can note that H_o^{-1} is linear with respect to the adopted distance, and then $\sigma(\ln H_o^{-1}) = 0.46\sigma(\text{mod})$, where mod is here the distance modulus of galaxies in the Hubble flow. In other words, a 10% accuracy in the distance of galaxies is required to know H_o (and H_o^{-1}) within 10%, while a 5% accuracy in the distance of GCs is needed to know their ages within 10%.

In the frame of procedure 1, three standard candles have been used so far to estimate the distance moduli of GCs: 1) RR Lyrae variables (Sandage 1982), 2) Theoretical horizontal branch models (Iben and Renzini 1984), and 3) local subdwarfs of known parallax and metallicity (Carney 1980). Sandage adopts for RR Lyraes the calibration:

$$M_{bol}^{RR} = M_V^{RR} = 0.8 + 0.35 \, ([Fe/H] + 1.7). \qquad (2)$$

Where the "zero point" (0.8) comes from Baade-Wesselink distances of a few local RR Lyraes with metallicity close to that of the GC M3. The metallicity "slope" (0.35) comes from a chain of semiempirical arguments, globally known as the "period shift effect". On the other hand, Iben and Renzini adopt a calibration based on theoretical horizontal branch (HB) models, which can be written as:

$$M_{bol}^{RR} = 0.5 + 0.18 \, ([Fe/H] + 1.7). \qquad (3)$$

Note that between (2) and (3) there is substantial disagreement in both "zero point" and "slope". The difference in zero point can be taken as an estimate of the current uncertainty affecting HB stars as standard candles. Concerning the slope, the case has been discussed by Iben and Renzini, and more recently by Sweigart, Renzini and Tornambè (1986).

Anyway, Sandage, Iben and Renzini, and Carney, each with their own standard candle, estimate the age of GCs to be 17, 16, and 17 Gyr, respectively, with a typical uncertainty which cannot be smaller than $\pm$ 3.5 Gyr. These estimates agree very well with each other and with the age range indicated by the isochrone-fitting technique, i.e. 15-18 Gyr (cf.

VandenBerg 1983). Note that the first two estimates embarrassingly agree with each other, in spite of the 0.3 mag disagreement in zero point, simply because different bolometric corrections at the turnoff were adopted.

In the last few years many other age estimates for individual GCs have been published, and in some case unrealistically small uncertainties have been claimed. It is then worth emphasizing that GC ages cannot be currently determined with an accuracy better than $\sim$25%, since current uncertainties in the standard candles (subdwarfs and RR Lyraes) imply $\sigma(mod) \sim 0.25$ mag, at least, and $\sigma(t)/t \simeq \sigma(mod)$.

Most of these age estimates agree with the values reported above (e.g. Sandage 1983; Harris et al. 1983; Da Costa et al. 1984; Penny 1984; Buonanno et al 1984a, b, 1986; Sandage and Roques 1984; Fahlman et al. 1985; Richer and Fahlman 1985), while a somewhat lower age, $\sim$13 Gyr, has been derived by Caputo et al. (1984), on the basis of a different choice for the absolute magnitude of RR Lyraes and for the CNO/Fe ratio (cf. Section 4).

The recent advent of CCD detectors has given new momentum to GC studies, mainly because they allow the production of very accurate CMDs, around main sequence turnoff and below (see, for example, Harris and Hesser 1985). Certainly, within a very short time CCD CMDs will become available for most GCs, thus allowing more accurate age determinations. However, it is worth emphasizing that ground based CCD observations can hardly help in improving the absolute calibration of standard candles, RR Lyraes and subdwarfs.

2.2 Perspectives at Improving the Accuracy of GC Age Estimates

The main "bottleneck" in GC dating is the current uncertainty in cluster distances. With $\sigma(mod) \simeq 0.25$ mag, the allowed GC ages range between $\sim$13 and $\sim$20 Gyr. Formally, the situation is only marginally better than for our present knowledge of H_o^{-1}! If we want to know the age of GCs to within 10%, we need to get their distance moduli to within 0.1 mag.

The Hubble Space Telescope can significantly help in better determining the absolute magnitude of traditional distance calibrators, as well as in providing one additional standard candle (see below). Substantial progress appears possible in three directions: 1) Subdwarfs. Accurate parallaxes of quite many local subdwarfs could be obtained, a decisive advantage over the current situation, when one bears in mind that Carney's (1980) study was based on only seven independent stellar parallaxes. Unfortunately, the anticipated astrometric accuracy of HST has recently degraded from 1.7 to 3 milliarcseconds (Fresneau 1985), and therefore the number of usable subdwarfs could be considerably reduced, perhaps by almost a factor of ten. 2) RR Lyraes. Even RR Lyrae itself is too distant for an interestingly accurate parallax to be obtained. However, photometry of GC stars in M31 should provide a direct measure of the metallicity dependence of the HB luminosity (the "slope"), thus greatly helping in understanding the so-called Sandage's period shift effect, which plays a crucial role in GC age estimates (cf. Sandage 1982; Renzini 1983; Iben and Renzini 1984; Sweigart et al. 1986). Still the problem of the RR Lyrae zero point will remain, and could hopefully be solved with ground-based observations, coupled with some improved version of the Baade-Wesselink method. 3) White dwarfs. HST observations of GCs should give access to a completely new standard candle: the white dwarf (WD) cooling sequence (cf. Fusi Pecci and Renzini 1979, Renzini 1986). This approach is methodologically similar to the popular isochrone fitting technique (procedure 3), since also in this case the standard candle is theoretical in nature. However, the WD method has at least two important advantages over procedure 3: unlike theoretical main sequence models, the absolute location in the HRD of the WD cooling sequence is virtually insensitive to both the metallicity and the convection parameter. For this reason I think that the WD method is in principle more reliable than the isochrone fitting technique. The critical step in this procedure is represented by the color-temperature transformations for WDs: in order to get ~ 0.1 mag accuracy in distance modulus (10% accuracy in age) the WD cooling sequence of a GC should be located with $\sim 2\%$ accuracy in temperature, which seems to be within current possibilities (cf. Holberg et al. 1985).

Each of these methods still involves making assumptions which must be verified (or falsified), so that it appears essential to pursue all three methods independently, and to check their consistency.

3. HOW TO CHECK THE CLOCK

The use of a theoretical calibration, e.g. Eq. (1), cannot be eliminated from any attempt at determining GC ages. So, the question of the reliability of the age indicated by the models will remain, even in the ideal case of a perfect knowledge of a cluster's distance and composition. In other words, it is important to put under control not just the internal errors (in distance and composition), but also the systematic errors introduced by the unavoidable use of theoretical models. This require a close "check up" of the models, which can only be accomplished by studying very large and complete samples of GC stars.

It is worth emphasizing that, as a rule of thumb, theoretical effective temperatures are generally much less reliable than theoretical luminosities. This is so because uncertainties in the treatment of convection, and in radiative opacity at intermediate and low temperatures have large effects on model radii (temperatures), but small effects on model luminosities. Procedure 1 does not make direct use of model temperatures, and therefore the crucial test for the models should regard luminosities. As emphasized by Paczynski (1984), what is required are then luminosity functions (LF), for large and complete samples of GC stars. The most massive study in this direction concerns the cluster M3 (Buonanno et al. 1986), where photometry has been obtained for about 10,000 stars, contributing about 10% of the total cluster light. The LF is complete down to ∼2 mag below turnoff. Still, this sample contains only ∼80 stars in the upper 3.5 mag of the red giant branch. According to Rood and Crocker (1985) a ten times more luminous sample (i.e. the whole cluster M3) would be just marginally sufficient for checking the presence of a theoretically predicted bump in the LF. This illustrates the importance of a really complete photometric survey for at least a few among the most populous GCs.

A comparison between observationally complete and theoretical LFs should allow an assessment as to whether in computing evolutionary models: 1) the used classical input physics (e.g. reaction rates, opacity, etc.) is sufficiently correct, 2) it is legitimate to neglect certain (classical) physical phenomena (e.g. rotation, diffusion, non-convective mixing, etc.), and 3) "non-classical" physics may actually play a significant role (e.g. trapped massive neutrinos, cf. Faulkner and Gilliland 1985; production and emission of Nambu-Goldstone bosons, cf. Dearborn, Schramm and Steigman 1985; and the like).

4. CONCLUSIONS

Current estimates for the age of GCs give $t_{GC} = 16\pm3.5$ Gyr. Assuming that the Universe is 10% older one gets $T_o = 18\pm4$ Gyr. Such a value is consistent with a Universe with $H_o = 50$, $\Omega_o \ll 1$ and $\Lambda = 0$ (cf. Sandage and Tammann 1984); it is in marginal (poor) agreement with the popular case $H_o = 50$, $\Omega_o = 1$ ($\Lambda = 0$); and it is clearly at variance with $H_o = 100$, whatever Ω_o, unless $\Lambda \neq 0$.

I am frequently asked the embarrassing question whether I do really believe in 16 Gyr, and what is the strict lower limit to GC ages. In general, people would like to have younger globulars. This desire can be satisfied in various ways, but the most promising one involves a non-solar CNO/Fe ratio, for which there may be good theoretical justifications (e.g. Matteucci and Greggio 1986) and some observational support (Sneden 1985). Indeed, increasing CNO/Fe gives younger ages, and Rood and Crocker (1985) and VandenBerg (1985) estimate a ∼3 Gyr reduction for [CNO/Fe] = 1, over the case [CNO/Fe] = 0. This would give $t_{GC} \simeq 13$ Gyr, in fairly good agreement with e.g. the combination $H_o = 50$ and $\Omega_o = 1$ ($\Lambda = 0$). In my opinion this is all we can reasonably say, for the moment.

Rather than insisting on a particular number, I think it is better trying to reduce the current uncertainties in distance, composition (most importantly CNO), and model ingredients.

REFERENCES

Boesgaard, A.M., Steigman, G. 1985, Ann. Rev. Astron. Ap. 23, 319
Buonanno, R., Corsi, C.E., Fusi Pecci, F., Harris, W.E. 1984a, A.J. 89, 365
Buonanno, R., Corsi, C.E., Fusi Pecci, F., Alcaino, G., Liller, W. 1984b, Ap. J. 277, 220
Buonanno, R., Corsi, C.E., Fusi Pecci, F. 1986, Astron. Ap. (in press)
Bounanno, R., Buzzoni, A., Corsi, C.E., Fusi Pecci, F., Sandage, A. 1986, (in preparation)
Buzzoni, A., Fusi Pecci, F., Buonanno, R., Corsi, C.E. 1983, Astron. Ap. 128, 94
Caputo, F., Castellani, V., Quarta, M.L. 1984, Astron. Ap. 138, 457
Carney, B. 1980, Ap. J. Suppl. 42, 481
Ciardullo, R.B., Demarque, P. 1977, Trans. Yale Univ. Obs., Vol. 33
Da Costa, G.S., Mould, J.R., Ortolani, S. 1984, Ap. J. 282, 125
Dearborn, D.S.P., Schramm, D.S., Steigman, G. 1985 (preprint)
Fahlman, G.G., Richer, H.B., VandenBerg, D.A. 1985, Ap. J. Suppl. 58, 225
Fualkner, J., Gilliland, R.L. 1985, Ap. J. 299, 994
Fresneau, A. 1985, STScI Newsletter, 2, no. 3, 9
Harris, W.E., Hesser, J.E. 1985, Dynamics of Star Clusters, ed. J. Goodman, P. Hut (Dordrecht: Reidel), p. 81
Harris, W.E., Hesser, J.E., Atwood, B. 1983, Ap. J. Letters, 268, L111
Holberg, J.B., Wesemael, F., Wegener, G., Bruhweiler, F.C. 1985, Ap. J. 293, 294
Iben, I.Jr., Renzini, A. 1984, Physics Reports, 105, 329
Matteucci, F., Greggio, L. 1986, Astron. Ap. 154, 279
Penny, A.J. 1984, MNRAS, 208, 559
Renzini, A. 1983, Mem. Soc. Astron. It. 54, 335. 1986 (preprint)
Richer, H.B., Fahlman, G.G. 1985 (preprint)
Rood, R.T., Crocker, D.A. 1985, Production and Distribution of CNO Elements, ed. I.J. Danziger, F. Matteucci, K. Kjär (ESO), p. 61
Sandage, A. 1982, Ap. J. 252, 553. 1983, A. J. 88, 1159
Sandage, A., Roques, P. 1984, A. J. 89, 1166
Sandage, A., Tammann, G. 1984, Large Scale Structure of the Universe, Cosmology and Fundamental Physics, ed. G. Setti, L. Van Hove (CERN)
Sneden, C. 1985, (same as Rood & Crocker), p. 1
Sweigart, A.V., Renzini, A., Tornambè, A. 1986, Ap. J. (submitted)
VandenBerg, D.A. 1983, Ap. J. Suppl. 51, 29
VandenBerg, D.A. 1985, (same as Rood & Crocker), p. 73
VandenBerg, D.A., Bell, R.A. 1985, Ap. J. Suppl. 58, 561

AGES FROM NUCLEAR COSMOCHRONOLOGY

F.-K. Thielemann[*] and J.W. Truran
Department of Astronomy, University of Illinois
Urbana, IL 61801 USA

ABSTRACT. The long-lived radio nuclides ^{187}Re, ^{232}Th, ^{238}U, ^{235}U, and ^{244}Pu can be used to set limits on the galactic age. The interpretation depends on (1) the production ratios of these isotopes in the appropriate nucleosynthesis site, (2) the time-dependent nucleosynthesis rate during galactic evolution, and (3) the isotopic ratios in the matter which condensed in meteorites at the formation of the solar system 4.6 billion years ago. The results range from 11.6 to 24.6 billion years, given the present uncertainties in the three quantities listed above. They are strongly dependent on variations in the production ratios and the amount of initial enrichment from pre-disk populations. A large initial enrichment tends to reduce predicted galactic ages. The strongest constraints on the age come from the ^{232}Th/^{238}U ratio, while the ^{235}U/^{238}U ratio is sensitive to the time-dependence of the nucleosynthesis rate. The inclusion of the ^{187}Re/^{187}Os pair within the same framework allows consistent solutions, but the intrinsic uncertainties in the Re/Os pair insure that it cannot lead to a more precise age determination. Thus, at present, nucleocosmochronology cannot provide age determinations with smaller uncertainties than those obtained from the analysis of globular cluster ages.

1. INTRODUCTION

Previous attempts to determine the age of the Galaxy with the aid of nucleocosmochronlogy made use of long-lived nuclei like ^{187}Re($t_{1/2}$ = 4.5 x 10^{10} y), ^{232}Th(1.405 x 10^{10} y), ^{238}U(4.46 x 10^9 y), and ^{235}U(7.038 x 10^8 y). Nuclei with shorter half-lives like ^{244}Pu(8.26 x 10^7 y) and ^{129}I(1.57 x 10^7 y) have been used to obtain information about nucleosynthesis occurring close to the time of the formation of the solar system.

---

[*]on leave from Max-Planck-Institut für Physik und Astrophysik, Garching b. München, FRG

The general idea of nucleocosmochronology is based on the equation

$$N_A/N_B = f(P_A/P_B, \text{ galactic nucleosynthesis rate}(t)) \qquad (1)$$

where N_A/N_B is the ratio with which two isotopes A and B condensed in meteorites at the time of formation of the solar system. P_A/P_B denotes the production ratio for those nuclei in a nucleosynthesis site. The fact that all of the above mentioned nuclei are products of a single nucleosynthesis process, the r(rapid neutron capture)-process, reduces the problem to knowing only one type of process. The usage of several chronometric (isotopic) pairs, resulting in several equations like eq. (1) can put constraints on galactic evolution and specifically determine the duration of nucleosynthesis and therefore also the age of the Galaxy.

Since at the present time the astrophysical site of the r-process is not clearly identified, no complex chemical evolution model can be applied which makes use of abundance yields as a function of stellar mass. Thus all previous studies were performed with a simple ansatz, the exponential model, which assumes that the time dependence of the nucleosynthesis production follows an exponential behavior (Fowler, Hoyle 1969; Fowler 1972, 1978). A different, "model-independent", approach was pursued by Schramm and Wasserburg (1970; see also Symbalisty and Schramm 1981). The latter determines only mean ages of the elements. An interpretation in terms of a total age of the Galaxy is, however, again somewhat model dependent (Tinsley 1977; 1980 Meyer and Schramm 1986).

We therefore continue to utilize the method of a model ansatz which allows us to determine free paramters in the model. In addition, we consider the possibility of an initial enrichment of metals, which is incorporated in many galactic evolution models to account for pre-disk populations of stars (Pop II and Pop III). It is of special interest to determine whether such an initial enrichment can reduce the relatively large ages obtained with the standard exponential model, when the reduced production ratios for $^{232}Th/^{238}U$ and $^{235}U/^{238}U$ are applied which resulted from the inclusion of β-delayed fission and neutron emission in r-process calculations (Thielemann, Metzinger, Klapdor 1983; Thielemann 1984). These latter results ($t_G = (17.6 \mp 4) \times 10^9$ y) differ significantly from earlier determinations (Fowler 1978, $t_G = (10.9 \mp 2) \times 10^9$ y).

In the following sections we want to concentrate on the three necessary input parameters to studies of nucleocosmochronology:
(a) isotopic production ratios in the r-process,
(b) the time dependence of galactic nucleosynthesis,
(c) abundance ratios at the time of formation of the solar system (meteoritic values).

2. NUCLEOSYNTHESIS PRODUCTION RATIOS

The cosmochronologically important radioactive isotopes ^{187}Re, ^{232}Th, ^{238}U, ^{235}U, and ^{244}Pu, are formed in a single astrophysical process, the rapid neutron capture driven r-process which synthesizes heavy elements

up to Pu. This helps substantially to reduce the uncertainties.
Moreover, any statistical uncertainties associated with the production
of an individual isotope are greatly reduced by the fact that the abundances of the long-lived species listed above (except for ^{187}Re) are
determined by summing over the abundances of short-lived progenitors in
alpha decay-chains.

These are: ^{232}Th: ^{232}Th, ^{236}U, ^{240}Pu, ^{244}Pu, ^{248}Cm, and ^{252}Cf;
^{235}U: ^{235}U, ^{239}Pu, ^{243}Am, ^{247}Cm, ^{251}Cf, and ^{255}Fm; ^{238}U: ^{238}U, ^{242}Pu,
^{246}Cm; ^{244}Pu: ^{244}Pu, ^{248}Cm, ^{252}Cf.

It was mentioned already in Section 1 that the astrophysical site
of r-process nucleosynthesis has not yet been unambiguously determined.
Existing theoretical models may be grouped broadly into two classes:

(1) Sites that involve the expansion of neutronized matter just
outside the cores of exploding (type II) supernovae; for a review see
Hillebrandt (1978).

(2) Sites which operate via neutron capture on pre-existing iron-peak nuclei, due to an intense neutron release in explosive helium
burning environments by ^{22}Ne(α,n)^{25}Mg or ^{13}C(α,n)^{16}O reactions. These
include shock-processed helium zones of supernovae or helium core
flashes (for a detailed discussion see Truran 1984).

At present, class (1) face the problem that Type II supernova calculations are not yet accurate enough for the mass-cut between the
remaining neutron star and the ejected envelope to predict the ejection
of 10^{-6}-10^{-5}M$_0$ per event, which is needed to explain solar r-abundances.
Based on the present knowledge of thermonuclear and beta rates, realistic stellar model conditions for scenario (2) seem not to allow for high
enough neutron fluxes to enable the operation of an r-process (Cowan et
al. 1983; Bosch et al. 1985).

It is, however, possible to simulate neutron fluxes which reproduce
nicely the r-process abundance curve. In the following we will draw on
such calculations and their predictions for production ratios of chronometric pairs. An effect, not included in early calculations, which was
first discussed by Wene and Johansson (1976) is the effect of beta-delayed fission. There are many heavy nuclei with A⩾220 which are
unstable against fission. In the r-process, very neutron-rich heavy
nuclei are formed which will (beta) decay until finally the beta-stable
progenitors, listed above, are populated. Beta decay of a nucleus
proceeds to various excited states of the daughter nucleus. Within the
region of heavy nuclei where fission becomes energetically possible,
excited states have a substantially enhanced fission probability. This
effect of fission after beta-decay (beta-delayed fission) can strongly
influence the population of progenitor nuclei and therefore the chronometer production ratios.

Table Ia gives a short historical summary of production ratios for
r-process calculations performed without the inclusion of beta-delayed
fission. When delayed fission was first introduced, no full r-process
calculations were performed, but the effect was rather estimated by
following the decay of an abundance distribution along the capture-path,
which was assumed to be constant as a function of mass number A. This
is a reasonable first order approximation far from the r-process peaks
at A=130 and 196. Such results are listed in Table Ib. The presently

existing r-process calculation which also includes delayed fission is listed in Table Ic. Hopefully, further calculations will be available soon which can help to provide a reasonable estimate of the uncertainties involved (Cowan et al., in preparation).

TABLE Ia
Production Ratios from R-Process Calculations

$^{232}Th/^{238}U$	$^{235}U/^{238}U$	$^{244}Pu/^{238}U$	Ref.
1.65	1.42	0.90	Seeger, Fowler, Clayton (1965) Fowler (1972)
1.90	1.89	0.96	Seeger, Schramm (1970)
1.80	1.42	0.90	Fowler (1978)

TABLE Ib
Constant Abundances in R-Path and Fission

$^{232}Th/^{238}U$	$^{235}U/^{238}U$	$^{244}Pu/^{238}U$	Ref.
1.70	0.89	0.53	Wene, Johansson (1976)
1.50	1.10	0.40	Krumlinde et al. (1981)
1.63	1.21	0.13	Thielemann et al. (1983ab)
1.60	1.42	0.57	Meyer et al. (1985)

TABLE Ic
R-Process Calculation Including Delayed Fission

$^{232}Th/^{238}U$	$^{235}U/^{238}U$	$^{244}Pu/^{238}U$	Ref.
1.40	1.24	0.12	Thielemann et al. (1983ab)

The pattern presented above (in Table Ia) can be understood in the following way. A rather "flat" mass formula (in terms of nuclear mass excess as a function of N-Z), such as the one by Myers and Swiatecki (1967) used in Seeger and Schramm (1970), yields a relatively flat abundance distribution. The reason is that, in such a case, neutron separation energies of 1.5-2.0 MeV, which determine the location of the r-process path, occur relatively far from the stability line, where very short β-decay half-lives are encountered. This leads to a "flat" (almost constant) abundance distribution with nearly equal abundances for all short-lived radioactive progenitors.

For the case in which equal abundances are assigned to all progenitors, ratios of $^{232}Th/^{238}U = ^{235}U/^{238}U = 2$ and $^{244}Pu/^{238}U = 1$ are expected (because of the occurrence of either 6 or 3 progenitors in the particular α-decay chains), close to the values obtained by Seeger and Schramm (1970). If a "steeper" mass formula is assumed, the abundances are more sensitive to structure effects in β-decay half-lives. In such a case, a slight maximum occurs at A≅238-240. This led to reduced production ratios for ^{232}Th/^{238}U and ^{235}U/^{238}U in the original work of Seeger, Fowler, and Clayton (1965).

Thielemann, Metzinger, and Klapdor (1983a) used a relatively "steep" mass formula (Hilf, von Groote, and Takahashi 1976) and the same effect was evident. A recent overview by Haustein (1984), which compares experimental data and mass formula predictions, comes to the conclusion that steep mass formulae, like e.g. Liran and Zeldes (1976), are to be preferred over "flat" formulae, e.g. Myers (1976). The Hilf et al. (1976) formula is not listed in this comparison but behaves similar to the one by Liran and Zeldes (1976).

The inclusion of β-delayed fission results in a drastic reduction in the abundances for A≥238. This leads to a strong reduction of heavy progenitors; it thus reduces ^{232}Th/^{238}U and has a drastic effect on ^{244}Pu/^{238}U. The calculations of Krumlinde et al. (1981), Thielemann et al. (1983), and Meyer et al. (1985) which incorporate the recent (and only available) fission barrier predictions by Howard and Möller (1980) give similar predictions (at least for the most important ^{232}Th/^{238}U pair), when constant abundances in the r-path are assumed (Table Ib): $1.50 < ^{232}$Th/^{238}U < 1.63, $1.10 < ^{235}$U/^{238}U < 1.42, and $0.12 < ^{244}$Pu/^{238}U < 0.57. The largest uncertainty is in ^{244}Pu/^{238}U which, however, changes drastically in both cases in comparison to the previous value of 0.90.

These two effects, the slight maximum around A=238-240 and the fission losses, add up in an r-process calculation with the inclusion of delayed fission (Table Ic), which reduces ^{232}Th/^{238}U even further to 1.40. While we will use these results in Section 5 to determine galactic ages, it cannot be excluded that the shape of the abundance maximum around A=238-240 is sensitive to the time dependence of the neutron flux in the r-process. As, however, the r-process site is not yet determined unambiguously, we will take the values derived from a constant abundance curve (Table Ib) as a conservative limit.

The results presented are dependent on (1) the fission barrier heights, (2) the β strength function predictions, and (3) the nuclear masses far from stability. All these quantities bear uncertainties; the general effect described above, i.e. the reduction of the chronometer production ratios due to β-delayed fission, should however persist.

3. NUCLEOSYNTHESIS PRODUCTION AND GALACTIC EVOLUTION

The production function of r-process material is dependent upon our knowledge of the star formation history of our Galaxy and the associated contamination of the interstellar gas. A critical uncertainty associated with the production function arises from the fact that the astrophysical site of r-process nucleosynthesis has not yet been clearly identified (see Section 2). Thus, it is not yet possible to perform complex galactic evolution calculations which make us of nucleosynthesis (r-process) yields as a function of stellar mass. The appropriate approach then is a simple galactic evolution model with instantaneous recycling (e.g. Tinsley 1980).

When applying such a model, the abundance of nucleus A in the interstellar gas is governed by the differential equation (Tinsley 1980 or also Thielemann, Metzinger, Klapdor 1983b)

$$\dot{N}_A = -\omega N_A + P_A \Psi \quad (-\lambda_A N_A \text{ if A radioactive}) \tag{2}$$

with P_A describing the production of nucleus A due to stellar evolution and ejection into the interstellar medium and Ψ denoting the star formation rate (the mass of interstellar gas being transformed into stars per unit time). ω is given by

$$\omega = -\dot{m}_g/m_g + f/m_g \, (1-Z_f/Z) \tag{3}$$

and for the case of metal-free infall ($Z_f=0$) ω describes the gas consumption rate due to star formation. Under the assumptions ω = const and $\Psi(t) = \Psi_0 \exp(-\mu t)$ the solution is

$$N_A(t) = P_A \Psi_0 \exp(-\omega t)/(\lambda_R - \lambda_A) \left(\exp(-\lambda_A t) - \exp(-\lambda_R t) \right) \tag{4}$$

with $\lambda_R = \mu - \omega$. This leads to

$$N_A(t)/N_B(t) = P_A/P_B \, f(\lambda_R, \lambda_A, \lambda_B, t) \tag{5}$$

which is only dependent on λ_R or $\exp(-\lambda_R t)$. The latter function has the same time dependence as $\Psi(t)\exp(\omega t)$, under the mentioned assumptions. Therefore it is called the "effective nucleosynthesis rate" and cosmo-chronological studies describe its time dependence rather than the time dependence of the nucleosynthesis production $P_A \Psi(t)$. For a more general derivation see Tinsley (1980).

One of the major constraints on simplified chronometer studies is the need to find solutions which give a time dependence of the effective nucleosynthesis rate (ENR) which is in accordance with limits derived from galactic evolution models and astronomical observations. This means that $\Psi(t)$ and ω have to be known. ω is related to the more commonly used gas consumption time scale τ_g by $\omega = 1/\tau_g$. Miller and Scalo (1979) give limits to the above mentioned quantities $0.125 \leq \Psi(\text{present})/\Psi_{av} \leq 2.5$ and $1 \leq \tau_g/10^9 y \leq 3$, (see also the discussion in Scalo 1986). If those limits are used within an exponential model it translates to $\lambda_R \leq 0$ with $\text{ENR}(t) = \Psi_0 \exp(-\lambda_R t)$. Meyer and Schramm (1986) on the other hand find from their model-independent approach that $\text{ENR}(t)$ has been constant within a factor of 3, i.e. the upper limit is a rising $\text{ENR}(t)$ by a factor of 3. This sets limits for λ_R of $-1.5 \times 10^{-10} \leq \lambda_R \leq 0$.

3.1. Chronometric Equations with Initial Enrichment

The model described above for the evolution of the galactic disk (see e.g. Fowler 1972), contains the parameters Δ (duration of galactic nucleosynthesis) and λ_R (coefficient governing the time dependence of the effective nucleosynthesis rate $\Psi(t)\exp(t/\tau_g)$, with Ψ being the star formation rate and τ_g the gas consumption time scale). Solving the chronometer equations with such a model ansatz, for the recent values of the chronometric production ratios and meteoritic abundance ratios, resulted in relatively large galactic ages (Thielemann, Metzinger, Klapdor 1983ab; Thielemann 1984).

Many galactic evolution models also include an initial enrichment of metals (Truran and Cameron 1971; see also Tinsley 1980 and references therein) due to pre-disk populations (Pop II and Pop III?). The question arises as to whether the inclusion of an initial enrichment can significantly alter these chronological results (Fowler and Meisl 1986). Therefore, Thielemann and Truran (1986) generalized the exponential model to a form which included an initial enrichment of the order S_0. This leads to the generalized equation

$$N_A(\Delta)/N_B(\Delta) = P_A/P_B \; f(\Delta, S_0 \lambda_R; \lambda_A \lambda_B) \tag{6}$$

which relates the (r-process) production ratios P_A/P_B and the meteoritic ratios $N_A(\Delta)/N_B(\Delta)$.

It has to be noted that, in the present context, only the initial enrichment of r-process nuclei is of interest. To obtain an estimate for reasonable values of S_0, it is necessary to seek observational clues to the history of r-process nucleosynthesis in the Galaxy. A important contribution to the understanding of this problem was provided by the observations of Spite and Spite (1978). Their spectroscopic studies of metal deficient stars revealed that the concentration of europium, an element whose abundance in nature is dominated by its r-process component, is essentially solar ([Eu/Fe] ≅ 0) even for stars with [Fe/H] ~ −2.6. This confirms that r-process nucleosynthesis occured in the earliest stellar populations.

Truran (1981) pointed out that that in r-process material the (Sr-Y-Zr)/Eu and Ba/Eu abundance ratios will be less than those of solar system matter, because Sr-Y-Zr and Ba are dominated by s-process nucleosynthesis. The heavy element abundance patterns characteristic of the most iron-deficient stars ([Eu/Fe] ~ 0, [Y/Fe] ~ −0.5, [Ba/Fe] ~ −0.8) are therefore entirely compatible with an r-process origin. This picture has recently been confirmed by the observational results of Sneden and Parthasarathy (1983) and Sneden and Pilachowski (1985).

These observational constraints might be summarized as follows: (1) the r-process contribution dominates the heavy element abundance pattern in stars of low Fe/H, and (2) the abundance ratios r-process/Fe are variable from star to star. It seems to be strongly suggested that a mechanism which is not dependent on iron seed nuclei but produces r-process nuclei as primary nucleosynthesis products is responsible for those observational findings (see Section 2). At least a primary r-process contribution is required to explain the abundances in extreme halo population stars. The fact that the overall levels of r-process and heavy element (carbon-iron) abundances are roughly compatible with solar values indicates that the relative initial enrichment in metals and r-process nuclei is of the same order, i.e. $S_0 = 0.1-0.3$.

4. METEORITIC ABUNDANCE RATIOS

In the following, all ratios which will be given are dated back to the time of meteorite formation, i.e. $t = \Delta$.

The best known ratio is $^{235}U/^{238}U$, as it is not affected by chemical fractionation. The value is 0.317 (Anders and Ebihara 1982) or 0.315 (Cameron 1982). For a literature overview see Anders and Ebihara (1982).

$^{232}Th/^{238}U$ came down from 2.5 (see Fowler 1978, who gives a weighted average over a variety of sources) to present values 2.32 (Anders and Ebihara 1982) or 2.22 (Cameron 1982); these latter authors reject several sources because of probable elemental fractionation effects.

$^{244}Pu/^{238}U$, which can at present only be determined from decay products of ^{244}Pu, has undergone drastic changes: 0.035 (Wasserburg et al. 1969), 0.015 (Podosek 1972), 0.016 (Drozd et al. 1977), 0.0068 ∓ 0.0019 (Hudson et al. 1984).

The large uncertainty for $^{244}Pu/^{238}U$ in the production ratios (see Section 2) as well as in the meteoritic values makes it a unreliable chronometer, and it is not included in the analysis of Section 5.

5. RESULTS

In this section we concentrate on the three long-lived chronometers ^{232}Th, ^{238}U, and ^{235}U to determine the long-term behaviour in galactic r-process nucleosynthesis. The evolution models described in Section 3 have the three parameters S_0, λ_R, and Δ. The three chronometers, listed above, result only in two chronometric pairs, i.e. $^{232}Th/^{238}U$ and $^{235}U/^{238}U$, or two chronometric equations like eq. (6). Therefore we will use λ_R as a free parameter and determine Δ and S_0 from the two equations. This also allows us to explore the possible range in λ_R, derived from astronomical observations (see Section 3).

Fig. 1 shows the results for two cases: (a) the production ratios are taken from Fowler (1978) with values of 1.80 and 1.42 for $^{232}Th/^{238}U$ and $^{235}U/^{238}U$, respectively (see Table Ia), and (b) the corresponding values 1.40 and 1.24 from Thielemann et al. (1983) (see Table Ic) are used. Solutions are displayed as a function of λ_R while its absolute value is also given in the middle frame on a logarithmic scale, to enable a better reading. The S_0-frame contains only values with $S_0 \leq 0.35$, i.e. less than 35% initial metal enrichment. The oldest disk populations show 10-30% of the solar metal content and therefore larger values of S_0 are excluded on the basis of observations.

Solutions with $S_0 = 0$ and the production ratios of 1.80 and 1.42 give a declining effective nucleosynthesis rate ($\lambda_R > 0$), $\Delta = 6.3 \times 10^9$ y, and a galactic age of 10.9×10^9 y, in accordance with the findings of Fowler (1978). It may be recognized that the inclusion of a non-zero initial enrichment even reduces the age further, an effect which was expected (Fowler and Meisl 1986). When applying the more recent production ratios 1.40 and 1.24, which result from the inclusion of β-delayed fission, larger ages result and, in the vicinity of $S_0 = 0$ a drastic increase is noticeable. This strong dependence on S_0 introduces a large uncertainty in Δ and values in the range 11×10^9 y $\leqslant \Delta \leqslant 20 \times 10^9$ y are possible.

It is interesting that the old production ratios allow only for a declining effective nucleosynthesis rate, while the more recent values give a (slightly) increasing nucleosynthesis rate with $-2 \times 10^{-10} y^{-1} \leqslant \lambda_R \leqslant -10^{-10} y^{-1}$, in agreement with the discussion of Section 3.

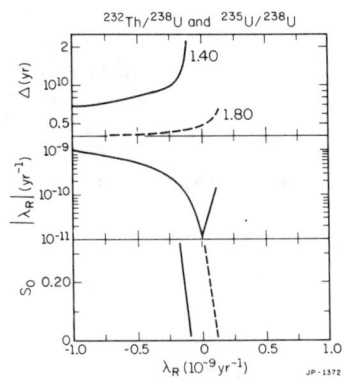

Fig. 1

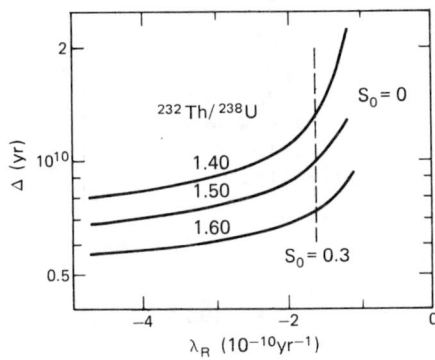

Fig. 2

A sensitivity study with variations of both chronometers showed that the $^{235}U/^{238}U$ ratio mainly affects λ_R, while the values for Δ at a given S_0 are very close, largely independent of the change in λ_R. Therefore we show a final study with $^{235}U/^{238}U = 1.24$ but using the full uncertainty range for $^{232}Th/^{238}U$ (1.40-1.60), where 1.60 is the conservative upper limit discussed in Section 2. The results in Fig. 2 range from 7 to 20×10^9 y for Δ, if initial enrichment values between 0 and 30% are allowed for. If we constrain S_0 to roughly 0.25 according to the observed metal enrichment in metal-poor disk stars, we can set lower limits from 7 to 15×10^9 y.

A more general approach (Thielemann and Truran, 1986) which included also the long-lived $^{187}Os/^{187}Re$ pair showed that all long-lived chronometers might give a consistent picture of galactic evolution, but it is not possible to achieve narrower age limits than given in the previous discussion (see also Yokoi et al. 1983 and Arnould et al. 1984).

6. CONCLUSIONS

The calculations presented in the previous section showed that the present knowledge about cosmochronologies only allows one to draw conclusions concerning the duration of nucleosynthesis in the galactic disk which show large uncertainties [11.6 - 24.6 (19.6) billion years]. These values have to be combined with the period of pre-disk evolution and the time scale for galaxy formation after the big bang to obtain a limit at the age of the universe. The uncertainties have several origins: (a) Nuclear physics: properties of nuclei far from stability (β-decay half-lives, fission barriers, mass formulae). (b) The interpretation relies on a model ansatz for the nucleosynthesis production as a function of time. This model ansatz can introduce additional uncertainties, as it can deviate from the "real" galactic nucleosynthesis history. (c) Meteoritic abundance ratios. Most of the long-lived chronometer ratios used in the present context, seem, however, to be fairly well determined.

Beer and Macklin(1985) have succeeded in determining the radiogenic ^{207}Pb abundance (resulting from ^{235}U-decay). Thus, an additional and maybe more precise chronometer can be used in the future, similar to the Re/Os pair, for age determinations.

We want to thank W.A. Fowler, B. Meyer, D.N. Schramm and K. Takahashi for valuable discussions. This study was supported in part by NSF Grant AST 83-14415.

REFERENCES

Anders, E., Ebihara, M. 1982, Geochim. Cosmochim. Acta **46**, 2363
Arnould, M., Takahashi, K., Yokoi, K. 1984, Astron. Astrophys. **137**, 51
Beer, H., Macklin, R.L. 1985, Phys. Rev. **C32** 738
Bosch, U. et al. 1985, Phys. Lett. **164B**, 22
Cameron, A.G.W. 1982, in Essays in Nuclear Astrophysics, eds. C.A. Barnes, D.D. Clayton, D.N. Schramm (Cambridge Univ. Press) p. 23
Cowan, J.J., Cameron, A.G.W., Truran, J.W. 1983, Ap. J. **265**, 429
Drozd, R.J., Morgan, C.J., Podosek, F.A., Poupeau, G., Shirk, J.R., Taylor, G.J. 1977, Ap. J. **212**, 567
Fowler, W.A. 1972, in Cosmology, Fusion and Other Matters, ed. F. Reines (Colorado Associated University, Boulder) p. 67
Fowler, W.A. 1978, Proceedings of the Welch Foundation Conferences on Chemical Research XXI, ed. W.D. Milligan (Robert A. Welch Foundation, Houston) p. 61
Fowler, W.A., Hoyle, F. 1960, Ann. Phys. **10**, 280
Fowler, W.A., Meisl, C. 1985, Symposium on Cosmogonical Processes, Boulder, Colorado
Haustein, P.E. 1984, Proc. 7th Intl. Conference on Atomic Masses (AMCO-7), ed. O. Klepper, THD Schriftenreihe für Wissenschaft und Technik **26**, p. 413
Hilf, E.R., von Groote, H., Takahashi, K. 1976, CERN 76-13, p. 142
Hillebrandt, W. 1978, Space Sci. Rev. **21**, 639
Howard, W.M., Möller, P. 1980, At. Data Nucl. Data Tables **25**, 219
Hudson, B., Kennedy, B.M., Podosek, F.A., Hohenberg, C.M. 1984, Geochim. Cosmochim. Acta
Krumlinde, J., Möller, P., Wene, C.O., Howard, W.M. 1981, Proc. 4th Intl. Conference on Nuclei far from Stability, CERN 81-09, p. 260
Lederer, C.M., Shirley, V.S. 1978, Table of Isotopes, 7th Edition (John Wiley Sons)
Liran, S., Zeldes, N. 1976, At. Data Nucl. Data Tables **17**, 431

Meyer, B.S., Howard W.M., Mathews, G.J., Möller, P., Takahashi, K. 1985, in Proc. 1985 Meeting of the American Chemical Society, Chicago
Meyer, B.S., Schramm, D.N. 1986, in preparation
Miller, G.E., Scalo, J.M. 1979, Ap. J. Suppl. **41**, 513
Myers, W.D. 1976, At. Data Nucl. Data Tables **17**, 411
Myers, W.D., Swiatecky, W.J. 1967, Ark. Fys. **36**, 343
Podosek, F.A. 1972, Geochim. Cosmochim. Acta **36**, 755
Scalo, J.M. 1986, Fund. of Com. Phys., in press
Schramm, D.N., Wasserburg, G.J. 1970, Ap. J. **162**, 57
Seeger, P.A., Fowler, W.A., Clayton, D.D. 1965, Ap. J. Suppl. **11**, 121
Seeger, P.A., Schramm, D.N. 1970, Ap. J. (Letters) **160**, L157
Seelmann-Eggebert, W., Pfennig, G., Müzel, H., Klewe-Nebenius, H. 1981, Chart of the Nuclides, Kernforschungszentrum Karlsruhe
Sneden, C., Parthasarathy, M. 1983, Ap. J. **267**, 757
Sneden, C., Pilachowski, C.A. 1985, Ap. J. Lett. **288**, L55
Spite, M., Spite, F. 1978, Astron. Astrophys. **67**, 23
Symbalisty, E.M.D., Schramm, D.N. 1981, Rep. Prog. Phys. **44**, 293
Thielemann, F.-K. 1984, in Stellar Nucleosynthesis, eds. C. Chiosi, A. Renzini (Reidel, Dordrecht) p. 389
Thielemann, F.-K., Metzinger, J., Klapdor, H.V. 1983a, Z. Phys. **A309**, 301
Thielemann, F.-K., Metzinger, J., Klapdor, H.V. 1983b, Astron. Astrophys. **123**, 162
Thielemann, F.-K., Truran, J.W. 1986 in Nucleosynthesis and its Implications on Nuclear and Particle Physics, eds. J. Audouze (Reidel, Dordrecht)
Tinsley, B.M. 1977, Ap. J. **216**, 548
Tinsley, B.M. 1980, Fund. Cosmic Phys. **5**, 287
Truran, J.W. 1981, Astron. Astrophys. **97**, 391
Truran, J.W. 1984, Ann. Rev. Nucl. Particle Sci. **34**, 53
Truran, J.W., Cameron, A.G.W. 1971, Astrophys. Space Sci. **14**, 179
Wasserburg, G.J., Hunecke, J.C., Burnett, D.S. 1969, J. Geophys. Res. **74**, 4221
Wene, C.O., Johannson, S.A.E. 1976, Proc. 3rd Intl. Conference on Nuclei far from Stability, CERN 76-13, p. 584
Yokoi, K., Takahashi, K., Arnould, M., Astron. Astrophys. **117**, 65

CONSTRAINTS ON THE AGES OF DISTANT GIANT ELLIPTICAL RADIO GALAXIES

Rogier A. Windhorst
Mount Wilson and Las Campanas Observatories, Pasadena
Carnegie Institution of Washington

David C. Koo
Space Telescope Science Institute, Baltimore

Hyron Spinrad
University of California, Berkeley

1. INTRODUCTION

The oldest observed ages of elliptical galaxies and globular clusters form a logical lower limit to the age of the Universe. To find the earliest possible formation epoch of elliptical galaxies one needs to study candidates at high redshift, which have very red colors and whose spectra consist entirely of starlight. To derive ages from observed spectra or colors one must first understand the stellar populations that compose the gE spectrum, their evolutionary tracks in the HR diagram and the resulting spectral evolution of the entire stellar population with cosmic time.

Complicating factors must be kept in mind. Non-coevally formed stellar populations can contribute significantly to the galaxy spectral energy distribution (SED), affecting the apparent ages of elliptical galaxies (e.g. O80, P85). Also, the contributions of several evolutionary stages to the HR diagram are uncertain, like the Asymptotic Giant Branch (W85) and the Horizontal Branch (RB).

Distant elliptical galaxy candidates have been found by selection of either distant clusters (O84), faint red field galaxies (K85, H85) or faint red radio galaxies (KKW). These authors all conclude that there is little evidence for drastic spectral evolution of gE galaxies at redshifts <1.0. Others (LLA, LE, DS) have argued that at least the radio galaxies in the 3CR catalogue - which are the most powerful radio sources in the Universe - have undergone more active spectral evolution.

In the current project we will constrain the oldest possible ages of elliptical galaxies. We deliberately use the faint, red **radio** galaxies from ultradeep radio surveys, because: 1) the optical-IR SED of weak radio galaxies is entirely due to their stellar population, without having major non-thermal contributions; 2) their redshift distribution is strongly biassed towards high redshifts, because they are radio sources (they are optical luminous galaxies); 3) they have very red colors, hence are most likely very old galaxies.

2. THE NATURE OF THE FAINT, RED RADIO GALAXIES

At the 3CR level the radio source population consist for ~2/3 of gE radio galaxies, ~1/3 of quasars and a few peculiar galaxies. At the mJy level these fractions decrease to 55% gE's and 20% quasars, while the fraction of blue radio galaxies increases to ~25% (WKK, KKW). The latter are blue, morphologically peculiar, often interacting and merging or sometimes very compact galaxies. These blue radio galaxies most likely cause the upturn in the mJy source counts (WMOKK).

The optical luminosities of radio galaxies are best studied from a magnitude redshift diagram. Figure 1 gives the photographic F-magnitude (6100Å) vs. redshift for all radio galaxies from the KKW sample (in the flux range 0.1<S<100 mJy), including some recent redshifts obtained with the KPNO 4m Cryogenic Camera. The figure shows two types of radio galaxies: red, high surface brightness radio galaxies (circles) and blue, low surface brightness radio galaxies (triangles). Filled symbols are extended radio sources, which coincide always with the optical luminous galaxies, which are spectroscopically mostly like the red giant ellipticals.

In Figure 1 several model predictions have been plotted: the passively evolving C-model of B83, the mildly evolving μ=0.7 and the drastically evolving μ=0.4 models. All models have been computed for H_o =50, q_o =0.1, M_F=-22.80 at z=0, and galaxy ages of 15, 14, and 13 Gyr, which corresponds to early galaxy formation (z_f =5.3), galaxy formation at the QSO turn-on epoch (z_f=3.4) and later galaxy formation (z_f=2.4). Section 3 discusses the models and their reliability.

Figure 1 shows that the red galaxy class has a narrow absolute magnitude distribution (M_F=-22.80 for H_o =50, q_o =0.1) even down to mJy levels, consistent with the "standard candle" behavior of gE radio galaxies. The blue radio galaxies have in general lower optical luminosity than the giant ellipticals. The models illustrate that a standard candle gE galaxy with F>22 is most likely at z>0.75 and that a gE galaxy at z=2 can have apparent magnitudes anywhere in the range 22<F<26, depending on its precise star formation history.

3. CONSTRAINTS ON THE AGES OF DISTANT ELLIPTICAL RADIO GALAXIES

The reddest occurring colors of radio galaxies at high redshifts will constrain the earliest possible formation epoch of the gE galaxies. We study this from the J-F (4650-6100Å) color redshift diagram in Figure 2. Bruzual's (1983) passively evolving C-models are plotted for various epochs of galaxy formation. A larger complete sample is given by WKS. Here we summarize the assumptions of B83's models:
1a) The initial mass function (IMF) is independent of cosmic time, galaxy type and age; 1b) Both the Miller-Scalo (1979) and the Scalo (1986) IMF are used and result in very similar predictions, except immediately after the starburst; 1c) The high mass cut-off in the IMF is 75$M_\odot$ and the low mass cut-off is 0.08$M_\odot$. The former affects only the predicted UV spectrum shortly after the starburst, the latter only the predicted near-IR spectrum, but over a longer timescale.
2) The time dependent star formation rate (SFR) are simple analytical

CONSTRAINTS ON THE AGES OF DISTANT GIANT ELLIPTICAL RADIO GALAXIES

Fig. 1. The Hubble diagram for faint radio galaxies. Open symbols are compact radio sources, filled symbols are extended radio sources.

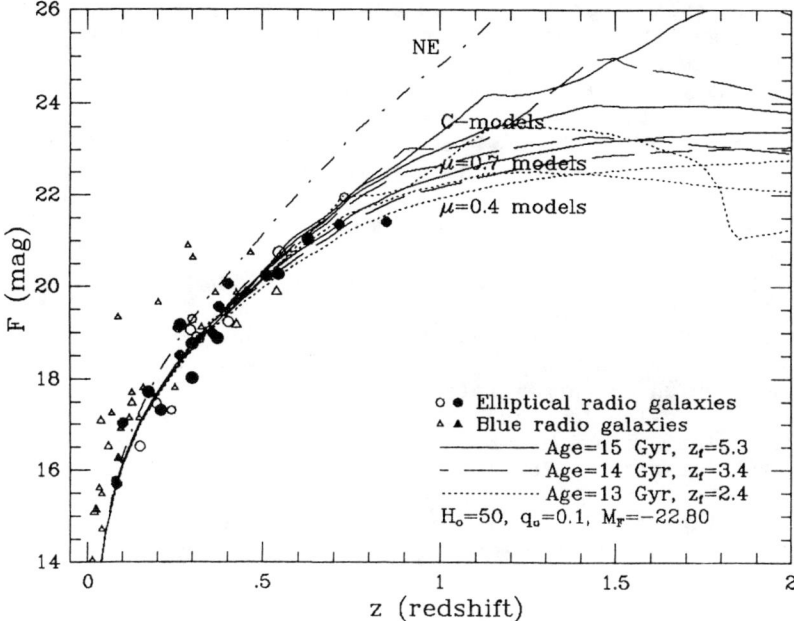

Fig. 2. The J-F color-redshift diagram for faint radio galaxies. Symbols are the same as in Fig. 1, models are described in the text.

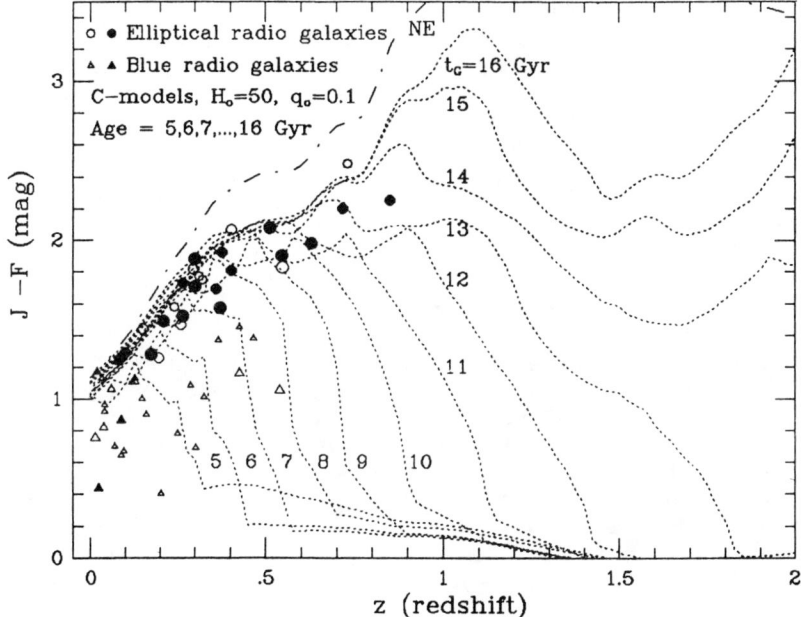

functions: 2a) The passively evolving C-model, which has a constant SFR during the first Gyr after galaxy formation and after that zero SFR; 2b) The evolving μ-models, which have an exponentially with time declining SFR, such that a fraction μ x Mgal forms stars during the first Gyr after galaxy formation (which occurs at z_f); 2c) Random starbursts are simulated by series of C-models at various z_f.
3) The evolutionary tracks in the HR diagram (CD). The models are only valid to the extent in which the understanding of the HR diagram is correct. We expect problems with late stages of stellar evolution: 3a) The contribution of AGB and carbon stars is not well known theoretically and is not properly accounted for. These stellar populations only dominate at $\lambda > 1\mu$ during $10^{7.5}-10^9$ yrs after the starburst (RB, W85); 3b) Horizontal Branch stars are ignored, which may enhance blue/UV part of the spectrum, but might have produced more red light when the stellar population was younger (RB).
4) The optical-IR SED consists of 100% starlight. The following effects are ignored in the models: 4a) Non thermal components and emission lines, which are negligible in most mJy radio sources; 4b) Internal and intergalactic reddening, because these effects are small for giant elliptical galaxies (gE's) at z<1.0. None of our measured color redshift diagrams (3600Å<λ<2.2μ) shows that the predicted SED becomes too bright in the blue, neither with decreasing wavelengths nor with increasing redshift, which should have been the case had dust significantly absorbed the blue/UV light at higher redshifts. Dust absorption might however be significant for the blue starburst galaxies. Intergalactic absorption might be important for z>3, if the (young) galaxies are shrouded by dust.
5) The metallicity of the stellar population is close to solar. Hence the models ignore the unknown chemical evolution of the galaxies, which is justified only if metals form quickly and if we are not witnessing the initial starburst.
6) World models and galaxy ages are necessarily free parameters in the models, since the models are computed in terms of cosmic time and the observables are measured in terms of look-back time. Hence we must iterate the values of H_o, q_o and z_f.

The good match of the reddest models to the red upper boundary of the data in Figure 2 is an a posteriori justification of the models and their assumptions. The validity of the models in the optical is further discussed by Sp86 and WKS. Assuming that the models are indeed applicable, the few very red (J-F~2.4) high redshift radio galaxies in Figure 2 seem to require ages of 14-15 Gyr. The reddest J-F color a radio galaxy ever gets is about 2.4-2.5, as shown with better statistics in Figure 3.

If these gE radio galaxies are indeed 14-15 Gyr old, lower values of H_o would be favored, namely H_o~50 for q_o=0.1, H_o~40 for q_o=0.5, or H_o~60 for q_o=0, while H_o> 60 would be only possible if $\Lambda \neq 0$. These old galaxy ages are in conflict with high values of H_o. A similar problem exists for the apparently very old ages of Galactic globular clusters. In both cases the argument is based on the evolutionary tracks in the HR diagram. Our distant, very red radio galaxies provide <u>independent</u> evidence for this apparent inconsistency, since they must already have been <u>very old</u> at least half a Hubble time ago.

4. CONSTRAINTS FROM ULTRADEEP CCD SAMPLES

According to KKW and OKSW the radio source population at F>22 contains both extended and unresolved radio sources. These most likely consist of both high redshift giant ellipticals (z>0.8) and blue radio galaxies (z>0.5) - in analogy to the brighter sample (see Fig. 1, 2).

A complete sub-sample of 70 radio sources in KKW's Hercules field has VLA positions accurate to 0".1 (OKSW). This sample has been studied by Windhorst and Koo (1986, in preparation), using long integrations with the Hale 200" Four-shooter CCD array. Preliminary results are shown in Figure 3, the J-F vs. F color-magnitude diagram. (The CCD photometry was done in Gunn g and r; g-r is comparable to and can be transformed to J-F). The objects with F<22.7 and J<23.7 form the complete photographic 4-m sample of KKW. The additional fainter objects with r<26.0 and g<26.0 are the complete Four-shooter sample in 1.25 sq.deg., together with a few upper limits, both from the 4m sample and the 200" Four-shooter sample.

The most important point is that the Four-shooter radio source sample is essentially <u>completely</u> identified down to r<26.0. Out of 70 mJy radio sources, 68 (or 97%) have reliable identifications. Because the optical completeness is very close to 100%, selection effects are not important. Only the radio selection is relevant. The almost 100% identification score also means that the highest redshifts expected for mJy radio sources must be present in the Four-shooter sample.

Fig. 3. The J-F vs. F color-magnitude diagram of faint radio galaxies. Symbols and models are as in Figure 1.

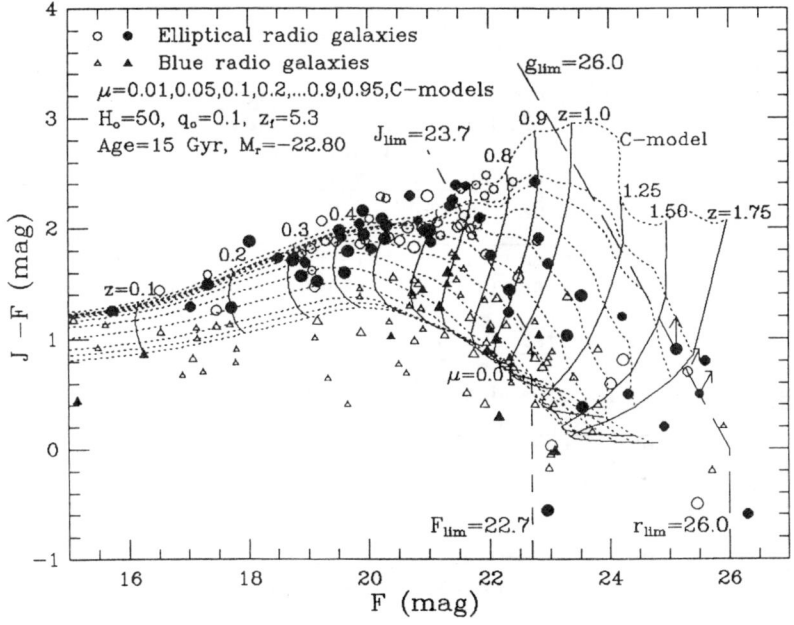

A range of Bruzual's μ-models have been superposed in Figure 3 (dashed lines), computed for the quoted parameters. The solid lines are redshift isochrones for $M_F = -22.80$. For F>22 the extended radio sources show a wide color range, $-0.6 < J-F < +2.4$, which might be indicative for the various star formation histories of the gE radio galaxies. Most of these are consistent with $\mu > 0.4$ models, although the few bluest objects might have some non-thermal contribution.

Figure 3 shows with better statistics than Figure 2 that the reddest color of the candidate gE radio galaxies is J-F~2.4. This supports the conclusion of Section 3 that the oldest possible ages of gE radio galaxies are of the order 14-15 Gyr. No matter what the precise redshifts of these faint radio galaxies (F>22) would be in Figure 2, the reddest occurring colors seem to require ages of at least 14 Gyr, if the models are indeed valid.

A few rather red objects in Figure 4 at F>24 have not even been detected in the blue at g=26.0. These objects might be the reddest and faintest mJy radio galaxies that exist. Once their redshifts have been measured, they will place very strong constraints on the oldest possible ages of elliptical (radio) galaxies. Spectroscopy on these objects is an important challenge to the next generation telescopes.

We thank Harry van der Laan for presenting this contribution on behalf of us at the Kona workshop.

REFERENCES

Bruzual A., G.: 1983, Astrophys. J., **273**, 105 (B83).
Ciardullo, R.B., Demarque, P.: 1977, Trans. Astron. Obs. Yale, **33** (CD).
Djorgovski, S., Spinrad, H.: 1986, Astrophys. J., in press (DS).
Hamilton, D.: 1985, Astrophys. J., **297**, 371 (H85).
Koo, D.C.: 1985, Astron. J., **90**, 418 (K85).
Kron, R.G., Koo, D.C., Windhorst, R.A.: 1985, A. A., **146**, 38 (KKW).
Lebofsky, M.J., Eisenhardt, P.R.M.: 1986, Ap. J. **300**, 151 (LE).
Lilly, S.J., Longair, M.S., Allington-Smith: 1985, MNRAS, **215**, 37 (LLA).
Miller, G.E., Scalo, J.M.: 1979, Ap. J. Suppl., **41**, 513 (MS).
O'Connell, R.W.: 1980, Astrophys. J. **236**, 430 (O80).
Oke, J.B.: 1984, in Clusters and Groups of Galaxies, pg. 99, eds.
F. Mardirossian, G. Giuricin and M. Mezetti (Reidel, Dordrecht), (O84).
Oort, M.J.A., Katgert, P., Steeman, F.W.M, Windhorst, R.A.: 1986, A. A., submitted (OKSW).
Pickles, A.J.: 1985, Astrophys. J. **296**, 340 (P85).
Renzini, A., Buzzoni, A.: 1986, in "The Spectral Evolution of Galaxies", ed. C. Chiosi and A. Renzini (Reidel, Dordrecht) (RB).
Scalo, J.M.: 1986, Fund. Cosm. Phys., in press (Sc86).
Spinrad, H.: 1986, Publ. Astron. Soc. Pac., in press (Sp86).
Windhorst, R.A.: 1984, Ph.D. Thesis, University of Leiden (W84).
Windhorst, R.A., Kron, R.G., Koo, D.C.: 1984, A. A. Suppl., **58**, 39 (WKK).
Windhorst, R.A., Miley, G.K., Owen, F.N., Kron, R.G., Koo, D.C.: 1985, Astrophys. J., **289**, 494 (WMOKK).
Windhorst, R.A., Koo, D.C., Spinrad, H.: 1986, in preparation (WKS).
Wyse, R.F.G.: 1985, Astrophys. J. **299**, 593 (W85).

EXTRAGALACTIC GAS AT HIGH REDSHIFT: A CHRONOGRAPH OF NONLINEAR DEPARTURES FROM HUBBLE FLOW

Paul R. Shapiro[1]
Department of Astronomy
University of Texas at Austin
Austin, TX 78712
USA

ABSTRACT. Studies of the state and evolutionary history of intergalactic and pregalactic gas provide important diagnostics of the conditions which led to the formation and to the present-day clustering and peculiar motions of galaxies. Three topics are discussed here: (1) cosmological H II regions and the photoionization of the IGM; (2) molecular hydrogen and the radiative cooling of pregalactic shocks; and (3) the possible detection of a hot, metal-enriched IGM in X-ray absorption. We find that the observed quasar population is not sufficient to explain the absence of the Gunn-Peterson effect at $z > 3$ by overlapping quasar H II regions. We find that metal-free intergalactic and pregalactic shocks which radiatively cool commonly produce an H_2 fraction $\gtrsim 10^{-3}$ and cool to $T \lesssim 10^2$ K. Such shocks should be detectable in quasar spectra as large-column-density H I Ly α absorption lines as well as H_2 Lyman band absorption, unless the shocks break up by $z \gtrsim 3.5$. Lastly, we note that the recent discovery of an O VIII X-ray absorption trough towards a BL Lac object at $z = .12$ is consistent with earlier predictions of the X-ray opacity of a hot IGM with substantial density and oxygen-enrichment.

1. INTRODUCTION

We have heard many exciting new results reported during this conference concerning the evidence for both large-scale inhomogeneities in the galaxy distribution on scales of at least several 10's of Mpc's and correspondingly large peculiar velocities for the galaxies of the order of several hundreds of km/sec. These inhomogeneities and large-scale departures from the Hubble flow are doubtless telling us something fundamental about the formation and clustering of galaxies as a result of the combined effects of the gravitational growth of density fluctuations and accompanying hydrodynamical phenomena. Thus far, most of the information about these inhomogeneities and departures from Hubble flow is limited to the nearby universe within about 10^4 km/sec on the Hubble scale, a tiny fraction of the observable universe. I will argue today that there is much to be learned about the origins of the present structure and peculiar velocities of the nearby universe by studying the condition and evolutionary history of the extragalactic

---

[1]Alfred P. Sloan Research Fellow

gas at high redshift, thereby sampling a much larger volume and range of look-back times. The term "high redshift" will here refer loosely to the range $z \lesssim 4$ probed by the known quasars. Studying this extragalactic gas -- its density, temperature, composition, and peculiar motions -- provides a chronograph, or time-chart, of the onset of the nonlinear departures from Hubble flow which formed galaxies and/or pregalactic stars.

I will discuss three separate (though related) problems in the history of the extragalactic gas: (1) cosmological H II regions and the photoionization of the IGM; (2) H_2 molecules and the radiative cooling of pregalactic shocks; and (3) the evidence for a hot, metal-enriched IGM.

2. COSMOLOGICAL H II REGIONS AND THE PHOTOIONIZATION OF THE IGM

The well-known absence of Gunn-Peterson H Ly-α absorption troughs in the spectra of quasars requires that any appreciable intergalactic gas distributed smoothly on cosmological scales be highly ionized. It has long been assumed that the quasars themselves are sufficient to photoionize a low-density IGM (with $\Omega_b h^2 \lesssim 0.1$) in order to satisfy the Gunn-Peterson test (e.g. Arons and McCray 1969, Bergeron and Salpeter 1970, Sherman 1981). Recent studies of faint quasars at high redshift, however, which suggest a decline in the number density of quasars for $z > 3$ (e.g. Koo 1985) have motivated us to reconsider this long-held belief. In what follows, I will attempt to answer the question, "Can the H II regions of high-redshift quasars overlap early enough to satisfy the Gunn-Peterson test?" (My collaborators in the work described in this section are Mark Giroux of The University of Texas and Ira Wasserman of Cornell.)

We have generalized the classical H II region problem to the case of a point source of ionizing radiation in a cosmologically expanding gas in a Friedmann-Robertson-Walker universe (Shapiro, Giroux, and Wasserman 1986; Shapiro, Wasserman, and Giroux 1986). We have derived a cosmological generalization of the Strömgren radius r_s which, in comoving coordinates, is given by

$$r_s(t) = [3 N_{ph}/(4\pi \alpha_2 n_{H,i}^2)]^{1/3} a(t) \equiv r_{s,i} a(t), \tag{1}$$

where N_{ph} is the ionizing photon number luminosity of the central source, the scale factor $a(t) = (1+z_i)/(1+z)$, α_2 is the recombination coefficient to levels $n \geq 2$, $n_{H,i} = n_H^\circ (1+z_i)^3$, n_H° is the present value of the intergalactic H density, and z_i is the redshift of source turn-on. As defined, proper and comoving length have the same value at $z = z_i$. As in the standard interstellar version of this problem, the Strömgren radius defines the volume within which the total number of recombinations per second balances N_{ph}. It provides a useful standard of comparison for the comoving ionization front radius r_I.

The actual H II region expands bounded by an ionization front defined as the surface across which the outgoing ionizing photon flux is balanced by the incoming neutral particle flux. This "jump" condition leads to an ordinary differential equation for the time-evolution of r_I. We have solved this equation numerically and plotted in Figure 1 some illustrative curves for r_I (in units of r_s) and the ionization front peculiar velocity [in units of r_s/t_i, where $t_i = t(z_i)$]. The ionization front fills the time-varying Strömgren

EXTRAGALACTIC GAS AT HIGH REDSHIFT

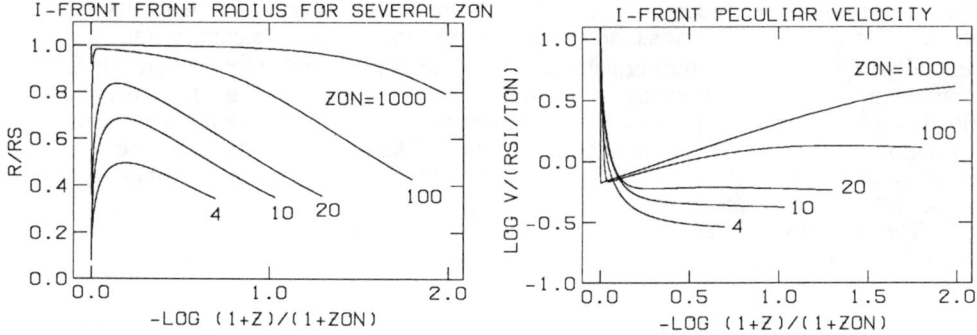

Figure 1 (a) (left) $r_I(t)/r_s(t)$ is plotted against redshift, where "ZON" is z_i, and $\Omega_b = 0.1$, $h = 1$, and $q_0 = 0.5$ are assumed. (b) (right) peculiar velocity $a(t) (dr_I/dt)$ for the same ionization front as shown in (a) is plotted (in units of $r_{s,i}/t_i$) against redshift.

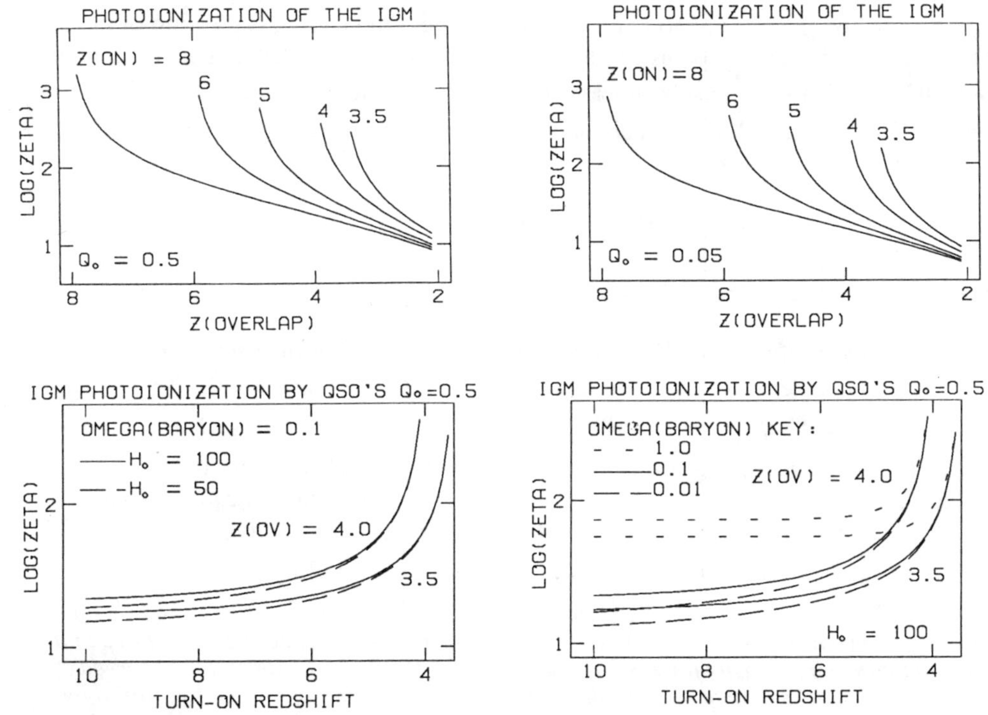

Figure 2 (a) (upper left) ζ_{crit} versus z_{ov} for several values of z(on), or z_i, assuming $\Omega_b = 0.1$, $h = 1$, and $q_0 = 0.5$; (b) (upper right) same as (a), but $q_0 = 0.05$; (c) (lower left) ζ_{crit} versus z_i for $z_{ov} = 3.5$ (lower curves) and $z_{ov} = 4.0$ (upper curves) for different H_0; (d) same as (c) except for different Ω_b, with curves which converge and approach the vertical at $z_i = 4$ corresponding to $z_{ov} = 4$, while others are for $z_{ov} = 3.5$.

radius after roughly one recombination time only for sources which turn on at very high redshift. For quasar H II regions, however, the front radius never reaches r_s and eventually falls further and further behind. This reflects the fact that, at small redshift, the recombination time exceeds the age of the universe. In the meantime, the front peculiar velocity rapidly decreases to a relative minimum and then increases again. For quasar H II regions, this minimum is supersonic. Hence, unlike the interstellar case, the ionization front never slows to the point at which it must make a transition from R-type to D-type, bounded by a shock wave.

The condition that a uniform distribution of point sources of number density n_x photoionize the IGM by their overlapping H II regions is given by

$$(4/3) \pi r_I^3 a^3 n_x = 1. \tag{2}$$

We assume that the sources all turn on at $t_i = t(z_i)$ and, thereafter, $n_x = n_x^\circ (1+z)^3$, where $n_x^\circ = n_{x,i}(1+z_i)^{-3}$. It is useful to define two dimensionless ratios: (1) $\lambda \equiv \alpha_2 n_{H,i} t_i$ (i.e. λ is the ratio of the age of the universe at t_i to the recombination time at t_i; and (2) $\zeta \equiv (2 n_x^\circ N_{ph})/(3 H_o n_H^\circ)$ (i.e. for the Einstein-de Sitter case, ζ is the ratio of the total number of ionizing photons emitted in a Hubble time to the total number of H atoms in the IGM). Clearly, ζ must be at least of order unity in order that the IGM by fully ionized. The critical value of ζ such that the H II regions overlap by a given redshift z_{ov} is given by

$$\zeta_{crit} = (2/3) \lambda (H_o t_i)^{-1} (r_I/r_{s,i})^{-3}, \tag{3}$$

where r_I is evaluated at z_{ov}. We plot the results for this ζ_{crit} for several values of z_i in Figure 2.

In order for the quasar H II regions to overlap by $z \sim 3$ as they must in order to satisfy the Gunn-Peterson test, we find that $10 \lesssim \zeta_{crit} \lesssim 10^2$. For example, if $z_i = 4$ and $z_{ov} = 3.5$, $\zeta_{crit} = 68$ if $q_o = 0.5$, while $\zeta_{crit} = 38$ if $q_o = 0.05$, assuming $\Omega_b = 0.1$ and $h = 1$. (These values of ζ_{crit} are only weakly dependent on Ω_b and h for this low value of Ω_b, as Figure 2 indicates.) Expressed in terms of the fractional volume filling factor f of ionized gas, this same example, with $z_i = 4$ and $q_o = 0.5$, yields a filling factor at $z = 3.5$ of $f = .015 \zeta$. What is the observed value of ζ? The highest redshift data available [for z ~ 3, from Koo (1985), Fig. 6] imply that the observed value is $\zeta_{obs} \lesssim 1$-$3 (\Omega_b h^2/0.1)^{-1}$, or perhaps much less if the QSO number density turns over (i.e. decreases) relative to a constant comoving density for $z > 3$ as suggested by the data. In short, the observed high redshift QSO's cannot be the sole source of the ionization of the IGM required by the Gunn-Peterson test!

The factor $\Omega_b h^2$ cannot be much less than 0.1 for the IGM, so increasing ζ_{obs} by greatly reducing the assumed value of the IGM density is not a permissible way to resolve the discrepancy. For example, Ostriker and Ikeuchi (1983) have demonstrated that the "Ly-α forest" clouds require an ambient IGM with $\Omega_b h^2 \sim 0.1$ in order to pressure-confine them. We are forced to conclude that either the observed turn-over in the QSO number density for z > 3 is not real, or else something else, such as shock waves from explosive galaxy formation or pancake collapse or both, heats the IGM at $z \gtrsim 3.5$ to temperatures $T \gtrsim 10^6$ K in order to collisionally ionize it to the point of avoiding the Gunn-Peterson effect. It is interesting to note that the large peculiar velocities reported at this conference for galaxies in the present universe imply ambient gas dynamical motions of the same order which, if "thermalized," as by shock waves, would be more than enough to ionize the IGM as required. Of course, we have no information about the magnitude of peculiar velocities at higher redshift. Perhaps we should take the results discussed here as an indirect argument in favor of widespread peculiar motions at high redshift (i.e. $z \gtrsim 3.5$) of the order of ~200 km s^{-1} or more.

3. HYDROGEN MOLECULES AND THE RADIATIVE COOLING OF PREGALACTIC SHOCKS

When a pregalactic gas of H^+ and He is heated and ionized, as by a shock wave occurring in the nonlinear gravitational collapse of cosmological density fluctuations or in an explosion in the IGM, the gas cools radiatively and recombines out of equilibrium. The temperature drops faster than the ions can recombine. As a result, when the temperature falls below 10^4 K, the ionized fraction greatly exceeds its ionization equilibrium value at these temperatures. This nonequilibrium enhancement of the proton and electron concentrations makes possible the formation of H^- and H_2^+ ions which, in turn, form H_2 molecules, according to the following reactions:

$$H + e \rightarrow H^- + h\nu$$
$$H + H^- \rightarrow H_2 + e$$

and

$$H^+ + H \rightarrow H_2^+ + h\nu$$
$$H_2^+ + H \rightarrow H_2 + H^+.$$

The presence of H_2 molecules allows the gas to cool further, by rotational-vibrational line excitation and the formation process itself. This extra cooling to a temperature well below the canonical equilibrium value of ~10^4 K for a molecule-free gas without metals has an important effect on the Jeans mass and subsequent fragmentation.

The importance of H^- and H_2^+ as intermediaries in the formation of H_2 molecules in an astrophysical gas of primordial composition was first discussed in a different context by Peebles and Dicke (1968) and Hirasawa (1969) and subsequently by a number of others (Hutchins 1976; Struck-Marcell 1981a,b; Palla, Salpeter, and Stahler 1983; Lepp and Shull 1984). The new results I will summarize in this section are specifically directed at the problem of the radiative cooling of a shock-heated primordial gas. (My collaborator in this work is Hyesung Kang of The University of Texas.)

We have solved the problem of the time-dependent, nonequilibrium radiative cooling of a primordial gas in the presence of a steady-state shock wave (Shapiro and Kang 1986a,b). For each value of the shock velocity and the preshock gas density and ionization state, we have satisfied the shock jump conditions and the hydrodynamical conservation equations for the postshock flow simultaneously with the integration of the rate equations for ionization, recombination, molecule formation and dissociation. We find that, to a very good approximation, the flow behind the shock is isobaric, so the illustrative cases presented here will actually assume that the postshock flow is isobaric. We consider cases with no external radiation source as well as those with an external source of ionizing and photodissociating radiation such as would correspond to a background from quasars or the flux from a nearby quasar or other source. For the cases in which an external radiation source is included, we have taken account of optical depth and self-shielding effects.

An illustrative selection of our results is shown in Figure 3 for a shock velocity of 270 km s^{-1}, which corresponds to a postshock temperature of 10^6 K. Figure 3(a) illustrates the difference between the equilibrium cooling curve, which drops precipitously below 10^4 K, and the actual, time-dependent, nonequilibrium cooling curves for two values of the immediate postshock H density $n_{H,i}$. Figure 3(b) illustrates the fact that a shock-heated gas which, in the absence of molecules, cools to 10^4 K in less than a Hubble time will cool to $T \lesssim 10^2$ K in a wide range of cases when nonequilibrium recombination and H_2 formation are taken into account. The particular value assumed for the immediate postshock H density, $n_{H,i} = 10^{-3}$, corresponds, for example, for $\Omega_b h^2 = 0.1$, to a shock in a uniform IGM at z ~ 6 or a shock at z ~ 2 in a region which is overdense by a factor of 20 or a shock at z ~ 0 in a region which is overdense by a factor of ~300. In the first of these cases, the gas cools as described above, but the cooling time, although shorter than the age of the universe at z = 0, is actually longer than the age at z = 6. In that case, adiabatic expansion cooling would also contribute, and the shock need not remain steady throughout the cooling time. In the second of these cases, the cooling time is comparable to the age of the universe at z = 2, but only in the absence of a quasar-like UV background. With the external radiation included, the cooling time can still be less than the Hubble time at z = 0, but again, adiabatic expansion would also contribute to the cooling, and the shock need not remain steady. In the last of these cases, the cooling occurs in less than a Hubble time even with the external radiation included, even for very high flux levels. [Strictly speaking, for z > 3, for a shock with these characteristics, inverse Compton cooling is important and will reduce the cooling time. However, since the Compton cooling rate is a sensitive function of z, including it requires that the redshift of shock-heating be fixed for a given calculation. The same results could not then be used to describe shocks at different redshifts, so for the purposes of this discussion we have neglected inverse Compton cooling in Figure 3(b).]

In this low density limit with no external radiation, the results for different densities are approximately the same if plotted against the product $t\, n_{H,i}$ rather than time t alone. With an external radiation source of intensity I, this scaling is still roughly true if the quantity $I/n_{H,i}$ is held constant as I and $n_{H,i}$ are varied. The scaling breaks down at low temperatures where the H_2 molecules are important, since the collisional dissociation and the rotational-vibrational excitation rates exhibit extra density dependence. (In fact, if $n_{H,i}$ is a high as ~10^2 cm^{-3}, molecular cooling is greatly suppressed.) As shown in Figures 3(b) and (c), the H_2 fraction approaches ~10^{-3} without external radiation, while a wide range of values of the external radiation flux can considerably enhance the H_2 concentration to values as high as ~$10^{-1.5}$. This latter effect results from the enhancement of the ionized fraction by photoionization and the corresponding increase in the H$^-$ and H_2^+ formation rates. It also depends upon the ability of the flow to gradually shield

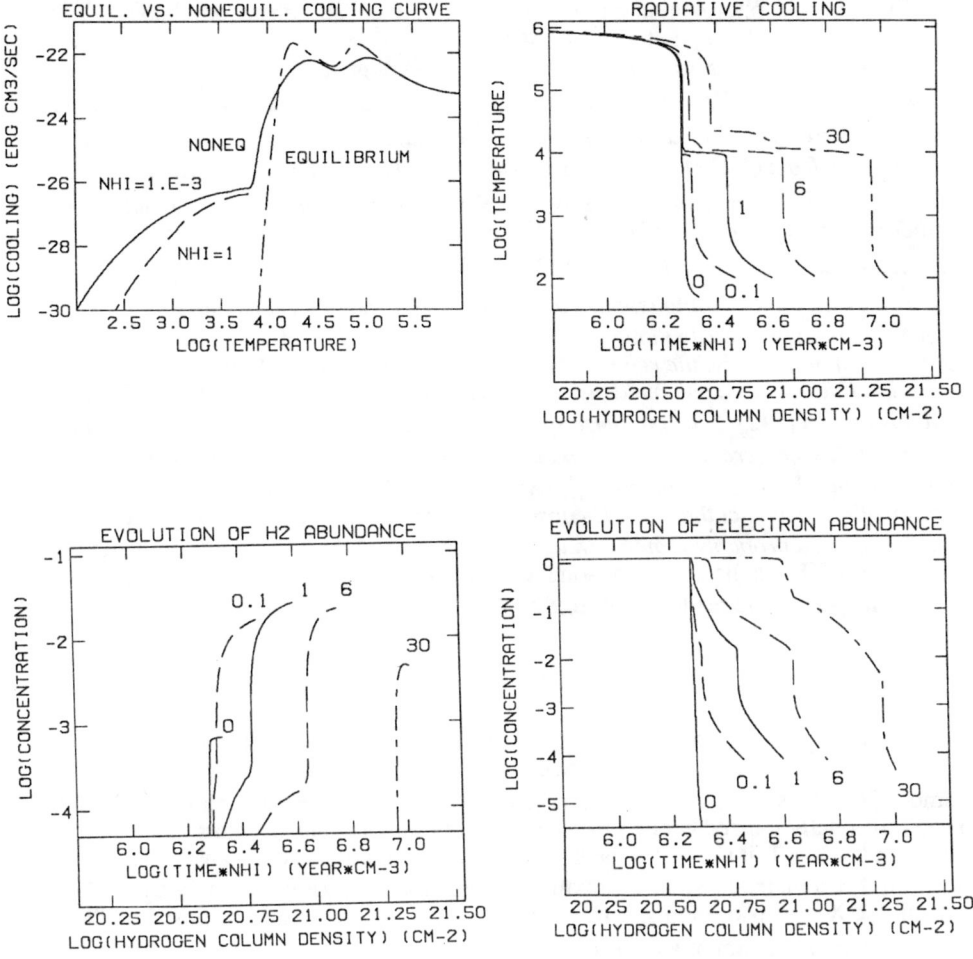

Figure 3 (a) (upper left) cooling rate Λ/n_H^2 versus T (°K) for ionization equilibrium (dash-dot) and the time-dependent, nonequilibrium calculations with $n_{H,i} = 10^{-3}$ cm^{-3} (solid), 1 cm^{-3} (dash); (b) (upper right) T (°K) versus $tn_{H,i}$ and versus N(H), the H column density between shock and fluid element, with $n_{H,i} = 10^{-3}$ cm^{-3} and $v_{shock} = 270$ km s^{-1}. Curves are labelled with the assumed external flux I at 13.6 eV, in units of 10^{-21} erg cm^{-2} sec^{-1} Hz^{-1}, for $I \propto \nu^{-0.7}$, $h\nu > 1.0$ eV; (c) (lower left) $n(H_2)/n_H$ versus $tn_{H,i}$ and versus N(H) for same cases as in (b); (d) (lower right) n_e/n_H versus $tn_{H,i}$ and versus N(H) for same cases as in (b).

itself, first from the photoionizing radiation above 13.6 eV and then from the H_2 photodissociating radiation below 13.6 eV.

In short, pregalactic shocks which are strong enough to ionize the gas and which are able to cool to $T \lesssim 10^4$ K within the age of the universe will quite commonly produce an H_2 concentration $\gtrsim 10^{-3}$ and cool to $T \lesssim 10^2$ K (unless $n_{H,i} \gtrsim 10^2$ cm^{-3}). External radiation (e.g. from QSO's) can, in some cases, actually increase the H_2 fraction to ~$10^{-1.5}$. For a wide range of conditions, therefore, the Jeans mass in the cooled gas is significantly reduced compared to the values predicted for cooling gas without H_2. For example, an explosion in an IGM with $\Omega_b h^2 = 0.1$ which produces a shock at ~200 km s^{-1} at $z \sim 6$ will, after cooling isobarically to $T \lesssim 10^2$ K, have a Jeans mass of $\lesssim 10^5$ $M_\odot$, much lower than previously thought. Finally, as also noted by MacLow and Shull (1986) in a similar study, the remnants of such shocks should result not only in H I absorption lines of large column density in quasar spectra, but in detectable H_2 Lyman band absorption (912-1120Å) in these spectra as well. The H_2 column densities which we have calculated here are, in fact, so large that even _fragments_ of the postshock cooling shells with total H column densities of greater than 10^{16} cm^{-2} should produce detectable H_2 absorption. The fact that such H_2 absorption has not yet been established as common in quasar spectra probably requires that radiatively cooled shock remnants break up by a redshift $z \sim 4$. [For a discussion of some earlier attempts to identify H_2 in quasar absorption spectra, the reader is referred to Aaronson, Black, and McKee (1974).]

4. EVIDENCE FOR A HOT, METAL-ENRICHED IGM?

Several years ago, Shapiro and Bahcall (1980) suggested that the discovery of quasar X-ray sources provided an opportunity to renew earlier, unsuccessful attempts to detect a cosmologically distributed IGM. We noted that, if the IGM were hot enough ($T \gtrsim 10^6$ K) to be collisionally ionized enough to escape detection by the Gunn-Peterson test for H I even with $\Omega_{IGM} \sim 1$, it might still be detectable as a result of its X-ray absorption spectrum if it contained a small admixture of metal ions. Toward this end we calculated in detail the X-ray absorption spectrum above 0.1 keV that would be imposed on the X-ray continuum spectrum of a quasar by an intervening, uniform, hot ($T \gtrsim 10^6$ K) IGM containing C, N, O, Ne, Mg, Si, S, and Fe. This included a total of 56 ionic stages, 156 resonant absorption troughs, and 106 continuum edges. We concluded that, if 10^6 K $\lesssim T \lesssim 10^7$ K, such an IGM should be detectable even with $\Omega_{IGM} < 1$ and elemental abundances below their solar value. An example of the optical depth which such a gas would contribute even toward a nearby quasar like 3C 273 is shown in Figure 4(a), where the metals are assumed to have a solar mix of relative abundances. The most prominent absorption feature on the plot for $T = 10^{6.5}$ K is the O VIII Ly α resonance trough between 654 eV and 654 $(1+z_0)^{-1}$ eV, where $z_0 = 0.15$. At 10^6 K, lower ionization stages of oxygen produce analogous features at adjacent, but lower energies.

Several years after these predictions were made, Canizares and Kruper (1984) reported the discovery of a broad X-ray absorption feature in the spectrum of the BL Lac object PKS 2155-304, one of the strongest extragalactic X-ray sources known, using the objective grating spectrometer on the Einstein Observatory [see Figure 4(b)]. They noted that the feature would be consistent with the predictions for a hot IGM by Shapiro and

EXTRAGALACTIC GAS AT HIGH REDSHIFT

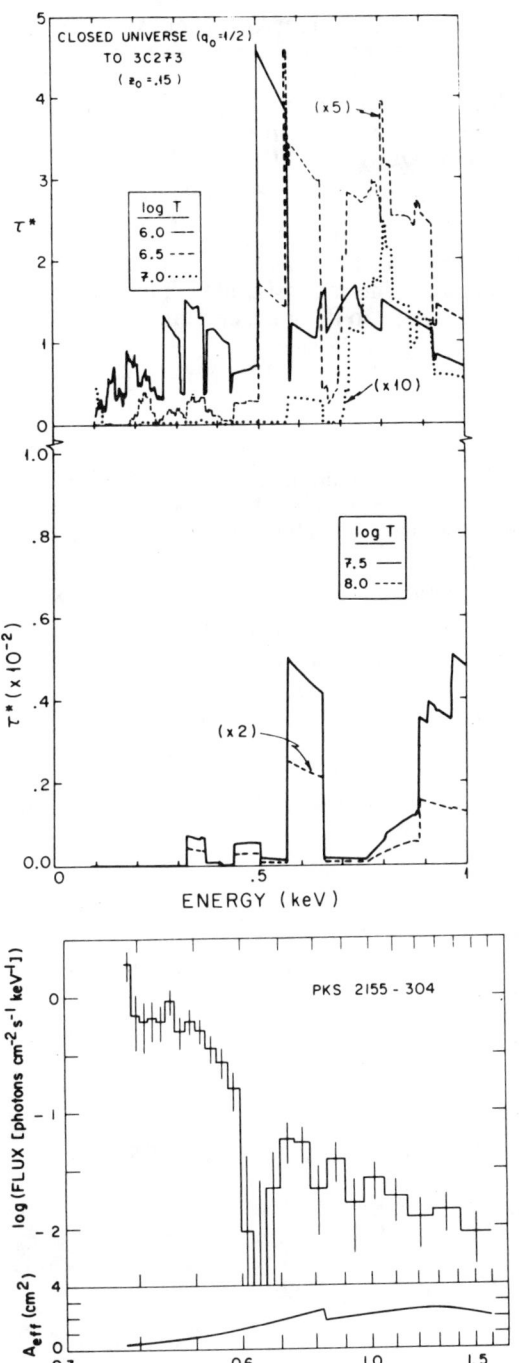

Bahcall (1980) if the temperature of the IGM were $\sim 5 \times 10^6$ K and the redshift of the BL Lac object were $z = 0.12^{+.03}_{-.06}$. Bowyer et al. (1984) subsequently measured the redshift and found $z = .117$, just as required by the interpretation of the X-ray absorption feature in terms of a hot, oxygen-enriched IGM.

A review of the arguments against alternative explanations, many of which are discussed by Canizares and Kruper, is outside the scope of this talk. The only viable alternative is that the hot gas producing the O VIII resonance trough is intrinsic to the BL Lac object. This requires that a somewhat implausible set of special circumstances conspire to mimic the effect of the hot IGM. The gas must be expelled from the BL Lac object in a narrowly collimated jet beamed precisely towards us with the full spread of velocities from zero to 0.1 c required to explain the trough width as the result of Doppler broadening internal to the jet rather than cosmological expansion. The gas in the jet must all be at the same temperature so that no other ionization stages are detectable. It is difficult to imagine a collisionally ionized jet which would accomplish this despite the large velocity spread. For a photoionized gas, on the other hand, this requires an unusual X-ray spectral index and highly collimated X-ray emission along the same direction. No other known object has this kind of very large column density jet of uniformly ionized oxygen.

Figure 4 (a) (upper) Optical depth versus observed photon energy (0.1 - 1.0 keV), where $\tau^* \equiv \tau_\nu/(2\Omega_{IGM}h\xi)$, where ξ is ratio of metal abundances to the solar value. Reprinted from Shapiro and Bahcall (1980). (b) (lower) *Einstein Observatory* OGS X-ray spectrum of PKS 2155-304 with 1σ error bars (upper) corrected for the efficiency of the OGS (lower). Reprinted from Canizares and Kruper (1984) by permission of the authors.

Of course, future X-ray observations will settle this issue by observing other extragalactic X-ray spectra to see if they show trough widths and depths which uniquely support the hot IGM explanation.

In the meantime, *we must take seriously the possibility that a hot, metal-enriched IGM has been discovered. According to Canizares and Kruper, fitting the observed X-ray spectrum of PKS 2155-304 would require a density of O VIII in the IGM of $5^{+9}_{-3.6} \times 10^{-9}$ h cm^{-3} (1σ error limits). Taking the 1σ lower limit, this implies a lower limit to the IGM density which corresponds to $\Omega_b h \gtrsim (1/4) ([O]/[H])_{solar}/([O]/[H])_{IGM}$, at least out to z ~ .1.* Apparently, the [Fe]/[O] ratio must be less than solar in order to explain the absence of other features in the spectrum. This is consistent with the fact that Pop III stars and massive stars, for example, are thought to favor the enrichment of oxygen relative to other elements like Fe.

5. CONCLUSION

(1) The observed QSO's are not numerous or luminous enough to photoionize the IGM in time to satisfy the Gunn-Peterson test for neutral H atoms in the IGM at z > 3. Either the observations are failing to detect the true number density of high redshift quasars or else something else must ionize the IGM at high redshift!

(2) Shock-heated primordial gas which can radiatively cool to $T \lesssim 10^4$ K can cool further to $T \lesssim 10^2$ K under a wide range of conditions, including the presence of quasar UV radiation, by forming an H_2 concentration ~10^{-3} or greater. This substantially reduces the Jeans mass in the cooled gas and alters its fragmentation history. Where are the H_2 absorption bands which such cooled shock remnants should introduce in quasar spectra?

(3) An X-ray absorption trough just like that predicted for a hot IGM enriched by metals has been detected for a nearby (z = .12) source. This may indicate that an IGM of substantial density and nearly solar oxygen abundance exists, at least at small redshift!

Acknowledgements. I would like to thank H. Kang and M. Giroux for their collaboration and for help in preparing several of the figures included here. This work was supported in part by NSF Grant AST 84-01231.

REFERENCES

Aaronson, M., Black, J. H., and McKee, C. F. 1974, *Ap. J. Letters*, **191**, L53.
Arons, J. and McCray, R. 1969, *Astrophys. Letters*, **5**, 123.
Bergeron, J. and Salpeter, E. E. 1970, *Astrophys. Letters*, **7**, 115.
Bowyer, S., Brodie, J., Clarke, J. T., and Henry, J. P. 1984, *Ap. J. Letters*, **278**, L103.
Canizares, C. R. and Kruper, J. 1984, *Ap. J. Letters*, **278**, L99.
Hirasawa, T. 1969, *Prog. Theor. Phys.*, **42**, 523.
Hutchins, J. B. 1976, *Ap. J.*, **205**, 103.

Koo, D. C. 1986, in *Structure and Evolution of Active Galactic Nuclei*, Trieste, April 1985 (Dordrecht: Reidel).
Lepp, S. and Shull, J. M. 1984, *Ap. J.*, **280**, 465.
MacLow, M. M. and Shull, J. M. 1986, preprint.
Ostriker, J. P. and Ikeuchi, S. 1983, *Ap. J. Letters*, **268**, L63.
Palla, F., Salpeter, E. E., and Stahler, S. W. 1983, *Ap. J.*, **271**, 632.
Peebles, P. J. E. and Dicke, R. H. 1968, *Ap. J.*, **154**, 891.
Shapiro, P. R. and Bahcall, J. N. 1980, *Ap. J.*, **241**, 1.
Shapiro, P. R., Giroux, M., and Wasserman, I. 1986, to be submitted to *Ap. J.*
Shapiro, P. R. and Kang, H. 1986a, in *IAU Symposium 117: Dark Matter in the Universe*, ed. G. Knapp (Dordrecht: Reidel), in press.
_____ 1986b, to be submitted to *Ap. J.*
Shapiro, P. R., Wasserman, I., and Giroux, M. 1986, in *IAU Symposium 117: Dark Matter in the Universe*, ed. G. Knapp (Dordrecht: Reidel), in press.
Sherman, R. D. 1981, *Ap. J.*, **246**, 365.
Struck-Marcell, C. 1981a, *Ap. J.*, **259**, 116.
_____. 1981b, *Ap. J.*, **259**, 127.

MICROWAVE AND X-RAY CONSTRAINTS ON LOCAL LARGE-SCALE OVERDENSITIES

L. Danese and G. De Zotti
Istituto di Astronomia
Vicolo dell'Osservatorio, 5
I-35122 Padova
Italy

ABSTRACT. We discuss the small scale temperature fluctuations of the microwave background induced by a local overdensity, of size larger than the Local Supercluster, that may eventually account for a substantial fraction of the observed dipole anisotropy, and briefly mention the new prospects of observing them opened by space-borne experiments currently in project. We also show that useful information on the large scale distribution of galaxies and, in particular, on the correlation function of rich clusters can be elicited from an analysis of the HEAO 1 A-2 all sky survey data, exploiting also the spectral resolution of this experiment.

1. INTRODUCTION

Although non-linear models for the collapse of the Local Supercluster (White & Silk, 1979) may entirely account for the motion of the Local Group through the microwave background (MWB), as inferred from the dipole anisotropy, several lines of evidence seem to imply that the actual infall velocity toward Virgo is too small and in the wrong direction (Tammann & Sandage, 1985). If so, a region of space encompassing at least the Local Supercluster should be moving with a velocity V ≈ 500 km/s relative to the MWB. Sandage & Tammann (1985) also argue that the size of the moving region, while ⩾ 20 Mpc (H_0 = 50 km/s Mpc), is probably ⩽ 50 ÷ 100 Mpc, in agreement with the results of Hart & Davies (1982).
On the other hand Rowan-Robinson et al. (1985) find that the IRAS source counts at 60μ and 100μ show evidence for a 20% overdensity on a scale at least comparable to the characteristic depth of the IRAS survey (≈ 200 Mpc). Yahil et al. (1985), using the 60μ IRAS sources to analyze the local gravitational field in terms of spherical harmonics, find that the direction of its dipole component is pretty close to the direction of the dipole anisotropy in the MWB. Altogether IRAS data then suggest that the size of the inhomogeneity responsible for the motion of the Local Group is ≈ 200 Mpc.
From the above short summary it appears that our knowledge of the

matter distribution beyond the Local Supercluster is still far from satisfactory. The aim of this note is to discuss how (hopefully) forthcoming data on the anisotropy of the MWB and new analyses of the existing maps of the X-ray sky can help to settle this problem.

2. ANISOTROPIES OF THE MWB

Let us assume that a substantial contribution to the observed dipole comes from a large scale perturbation, of which the Local Supercluster may be a part, and whose $\delta\rho/\rho \ll 1$, so that the linear perturbation theory applies.

As first demonstrated by Sachs & Wolfe (1967), gravitational potential fluctuations on a scale R_L induce, in addition to a dipole term, smaller scale anisotropies whose amplitude is lower by a factor $\approx H_0 R_L/c$. For R_L in the range 50 to 200 Mpc a decrease of the MWB temperature of amplitude $|\Delta T/T| \gtrsim 10^{-5}$ is to be expected on an angular scale $\lesssim 1°$, approximately in the direction of the apex of the MWB dipole, as observed in the reference frame of the Local Supercluster.

Measuring such fluctuations is well within the possibilities of the recently developed high sensitivity He-3 cooled bolometers. In fact, Melchiorri et al. (1985) have recently submitted to ESA a proposal for a space-borne experiment dedicated to high sensitivity measurements of microwave anisotropies at mm. wavelengths, where the MWB intensity peaks and, consequently, the fluctuations due to discrete sources or to patchy emission from the Galaxy are minimal. They plan to use a two-meter telescope, attaining an angular resolution of $\approx 10'$, and a detector which can reach $\Delta T \approx 0.2$ mK s$^{-1/2}$ allowing a 3σ detection of a $\Delta T/T \approx 10^{-5}$ in only about 10^m of observing time.

Actually, in still not unreasonably long times (≈ 13 hours), the detector could reach $\Delta T/T \approx 10^{-6}$. This may open the possibility of looking also for the gravitational effect of the overdensity on photons passing through it (Rees & Sciama, 1968; Dyer, 1976; Raine & Thomas, 1981; Kaiser, 1982; Goicoechea & Sanz, 1985). The effect has an amplitude smaller than the dipole by a factor $\approx (H_0 R_L/c)^2$ and occurs only in universes other than Einstein-de Sitter since, as stressed by Peebles (1980, Sect. 93), in a flat universe the gravitational potential for a linear, growing perturbation is a function of space coordinates alone because the density contrast grows proportionally to the scale factor (see, however, Goicoechea & Sanz (1985) for the case of a non-uniform matter distribution within a density perturbation in a $\Omega = 1$ universe). Clearly, this temperature anisotropy is detectable only if the overdensity has a very large size; on the other hand, the shape of the temperature profile and the amplitude of the effect would be informative on the density run inside the perturbation as well as on its size and on the distance of the observer from its center. In addition, this is a local effect, independent of the properties of fluctuations at decoupling.

In addition to looking for anisotropies associated to density perturbations, Melchiorri et al. (1985) also aim to accurately measure the Sunyaev-Zeldovich (1972) effect in as many rich clusters of galaxies as possible, within $z \approx 0.04$ (the limit in z follows from the fact that the signal from distant objects is strongly diluted because of the limited angular resolution of the telescope). Combining these measurements with X-ray observations would give the distances of the clusters (Cavaliere et al., 1977, 1979; Gunn, 1978; Silk & White, 1978), thus providing a suitable reference frame against which to determine the motion of the Local Supercluster.

3. CONSTRAINTS FROM X-RAY DATA

Some evidence for a dipole anisotropy consistent with the microwave results has also been found in the UHURU (Protheroe et al., 1980) and in the HEAO 1 A-2 all sky survey data (Shafer & Fabian, 1983; Boldt, 1985). Although the amplitude of the dipole in X-rays is expected to be a factor $\simeq 3.4$ larger than in microwaves, it turns out to be much harder to measure accurately because of the uncertainties in the correction for the Galactic emission and because the X-ray background is much rougher than the MWB.

On the other hand, the HEAO 1 A-2 data can provide useful additional information on the large scale distribution of galaxies through the angular correlation function of intensity fluctuations in directions separated by an angle α:

$$\Gamma(\alpha) = <\delta I(\theta,\phi) \delta I(\theta+\delta\theta, \phi+\delta\phi)>$$

$$\cos \alpha = \cos \theta \cos(\theta+\delta\theta) + \sin \theta \sin(\theta+\delta\theta) \cos \delta\phi.$$

Neglecting the smearing effect of the detector's response function, assuming that the spatial correlation function is very small on scales comparable to the effective depth of the survey ($\approx$ 350 Mpc), and confining ourselves to angular separations much smaller than one radian, we have (Dautcourt, 1977):

$$\Gamma(\alpha) = \int dz\ F(z)\ G(d_A(z)\ \alpha)$$

where d_A is the angular diameter distance,

$$G(y) = \int dx\ g(\sqrt{x^2 + y^2})$$

and $g(r)$ is the spatial correlation function of X-ray sources.

As is well known, extragalactic sources at the HEAO 1 A-2 flux limit subdivide almost equally between clusters of galaxies and active (mostly Seyfert) galaxies. Thanks to their very different X-ray spectra and to the spectral resolution of the A-2 experiment, it should be possible to analyze separately the angular correlation functions for the two classes of sources. Clusters of galaxies have bremsstrahlung spectra with typical temperatures of $\approx$ 6 keV; thus their contribution above $\approx$ 10 keV is negligible in comparison to that of active galaxies which have power law spectra with energy index $\alpha_x \simeq 0.7$. On the other hand, unless X-ray selected active galaxies are much more strongly clustered than galaxies as a whole (an unlikely possibility since they are predominantly spirals), the dominant contribution to $\Gamma(\alpha)$ below 10 keV is due to cluster-cluster correlations (cf. Bahcall & Soneira, 1983).

REFERENCES

Bahcall, N.A., Soneira, R.M.: 1983, Ap. J. 270, 20
Boldt, E.A.: 1985, Proc. Third Rome Meeting on Astrophysics, in press
Cavaliere, A., Danese, L., De Zotti, G.: 1977, Ap. J. 217, 6
Cavaliere, A., Danese, L., De Zotti, G.: 1979, Astr. Ap. 75, 322
Dautcourt, G.: 1977, Astron. Nachr. 298, 141
Dyer, C.C.: 1976, MNRAS 175, 429
Goicoechea, L.J., Sanz, J.L.: 1985, Ap. J. 293, 17
Gunn, J.E.: 1978, in Observational Cosmology, ed. A. Maeder
Hart, L., Davies, R.D.: 1982, Nature 297, 191
Kaiser, N.: 1982, MNRAS 198, 1033
Melchiorri, F. et al.: 1985, COSP: a Mission Proposal to the European Space Agency
Peebles, P.J.E.: 1980, The Large Scale Structure of the Universe, Princeton University Press, Princeton
Protheroe, R.J., Wolfendale, A.W., Wdowczyk, J.: 1980, MNRAS 192, 445
Raine, D.J., Thomas, E.G.: 1981, MNRAS 195, 649
Rees, M.J., Sciama, D.W.: 1968, Nature 217, 511
Rowan-Robinson, M., Chester, T., Soifer, T., Walker, D., Fairclough, J.: 1985, MNRAS, in press
Sachs, R.K., Wolfe, A.M.: 1967, Ap. J. 147, 73
Shafer, R.A., Fabian, A.C.: 1983, Proc. IAU Symp. No. 104, p.333
Silk, J., White, S.D.M.: 1978, Ap. J. Letters 226, L103
Sunyaev, R.A., Zeldovich, Ya.B.: 1972, Comm. Ap. Space Phys. 4, 173
Tammann, G.A., Sandage, A.: 1985, Ap. J. 294, 81
White, S.D.M., Silk, J.: 1979, Ap. J. 231, 1
Wilson, M.L., Silk, J.: 1981, Ap. J. 243, 14
Yahil, A., Walker, D., Rowan-Robinson, M.: 1985, Ap. J., in press

SUPERCLUSTER INFALL MODELS

Jens Verner Villumsen
California Institute of Technology
Marc Davis
University of California, Berkeley

ABSTRACT. We investigate the structure of the velocity field around large clusters formed in $\Omega = 1$ cosmological N-body simulations that form large scale structure in reasonable agreement with observations. We can test the applicability of spherical infall models commonly used to measure the mass of the Virgo supercluster and the cosmological density parameter. The spherical model is seen to be reasonably accurate for mean overdensity $\delta < 3$, but only when the infall velocity is averaged over 4π steradians. Large asphericities in the flow field are the norm, even at low overdensities. These asphericities result from significant subclustering seen on small scales and from the non-spherical nature of the exterior mass distribution. The simple spherical model for the Virgocentric flow as a measure for Ω is subject to large systematic and random errors. A much better test is to use a model independent, local criterion, noting that the local peculiar force field determines the local peculiar velocity extremely well both in direction and in amplitude. This relationship which must be valid in linear theory can be extended surprisingly well into the non-linear perturbations of the developing clusters.

1. INTRODUCTION

The flowfield around the Virgo supercluster can be used to study the mass of a system on scales that are much larger than is possible with virial analyses of more distant clusters. This is possible because we have redshift-independent distance measures that allow us to determine roughly where a galaxy is relative to us and to Virgo.

In the simple spherical model it is assumed that a single spherical overdensity region is centered on the Virgo cluster, that all peculiar velocities are pointed at Virgo, and that the flow field is spherically symmetric. With these assumptions there is a unique relationship between mean interior overdensity δ and the peculiar velocity V_p normalized by the local Hubble velocity Hr.

$$\frac{V_p}{Hr} \approx \frac{\Omega^{0.6}}{3}\delta(1+\delta/3)^{-1/2}$$

If we can measure peculiar velocities and mean interior overdensities this circles are overdensity shells $\delta_i = 2^{\frac{(i-1)}{2}}$. For each particle the proper velocity $V_{proper} = Ha\mathbf{x} + a\frac{d\mathbf{x}}{dt}$ is shown, a is the cosmological scalefactor and the scaling is arbitrary. It is readily apparent that the clusters are not isolated, and the velocity field is neither radial nor spherically symmetric. There is also significant subclustering that distorts the flowfield. We measure the mean infall velocity in an overdensity shell averaged over 4π. Figure 2 shows the mean infall velocity normalized to the Hubble velocity as a function of overdensity averaged over the clusters. The dispersion between the clusters is also shown. The solid curves are the predictions of the spherical models for various values of Ω. For $\delta < 2$ we would conclude that $\Omega = 0.9 \pm 0.1$ while for $2 < \delta < 3$ we would conclude $\Omega = 0.8 \pm 0.1$. For larger overdensities the discrepancy becomes more pronounced. So the spherical model predicts the infall velocity quite well in the outer parts but becomes rapidly poor near turn around.

We characterize the deviations from spherical symmetry of the velocity field by expanding the velocity field in spherical harmonics up to quadrupole order taking into account that the density distribution is not spherically symmetric. All velocities are measured relative to the center of the cluster. Figure 2 illustrates the importance of the quadrupole contribution to the infall velocity. The open squares and circles denote the infall velocity as seen along the major and minor axes of the velocity ellipsoid averaged over the clusters. It is clear that underestimates of a factor of 3 in Ω are possible if galaxies are not sampled over 4π steradians.

If we scale the typical deviations from spherical symmetry to the Virgo cluster we would see a transverse circulation of $75-150 km/sec$ at our Virgo-centric radius, an irrotational transverse velocity of the same order, the typical dipole term would be about $130 km/sec$ and the radial quadrupole distortion would be nearly twice as large. These amplitudes are by no means negligible and just might explain our transverse velocity within the local supercluster.

We attempted to correlate the deviations from spherical infall with other observables, generally with poor results. There is some correlation between the quadrupole distortions of the radial velocity and radial force field. We see no correlation between the principal axes of the inertia tensor of the interior mass distribution and the principal axes of the radial velocity field and neither with the quadrupole distortion of the radial force field. This indicates that there is a strong contribution to the force field from exterior matter which is ignored in the spherical model. Lilje et al. (1986) find a significant quadrupole velocity term in the Virgo flow field induced by exterior forces.

3. FORCE-VELOCITY CORRELATIONS

Now the good news: In linear theory the peculiar velocity of a particle V_p will be parallel to the peculiar force vector $\mathbf{F}$ and is given by $V_p = \frac{2}{3\Omega^{0.4}H_0}\mathbf{F}$. For the spherical model this equation can be extended into the non-linear regime. We calculate the ratio allows us to measure Ω on scales of our distance to Virgo. This analysis has been done by a number of authors, see review by Davis and Peebles (1983). Newer results are given by Aaronson et al. (1985). The general conclusion is that we are situated at the $\delta = 2.0 \pm 0.2$ surface with an infall velocity of 200-300 km/sec. This leads to an estimate of $\Omega \approx 0.2-0.4$.

The problem with the universe is that there is nothing to compare it to, and we do not have independent evidence for the validity of the spherical model.

It is thus apparent that n-body simulations might allow us to test the spherical and other models for the flow field around large clusters like Virgo. Large-scale N-body simulations from random phase initial conditions naturally produce large clusters with substantial substructure and no particular symmetry surrounded by other clusters and voids. Of course we have complete information. Studies of clusters in such simulations have been presented by Bushouse et al. (1985) and by Villumsen and Davis (1986).

2.SIMULATIONS

Seven simulations were run from random phase initial conditions with power spectra $|\delta_k^2|$ with power law indices of $n = -1, -2$, and with Cold Dark Matter(CDM) spectrum. In some of the CDM runs the fundamental wave in the z-directions was enhanced to force a pancake instability. A total of 19 large clusters were identified by eye and the centers were found by iterating the center of mass on succesively smaller spheres centered on the cluster.

The flow field was analyzed in overdensity shells centered on the cluster ranging from unit overdensity to $\delta \approx 12$ spaced uniformly in log δ.

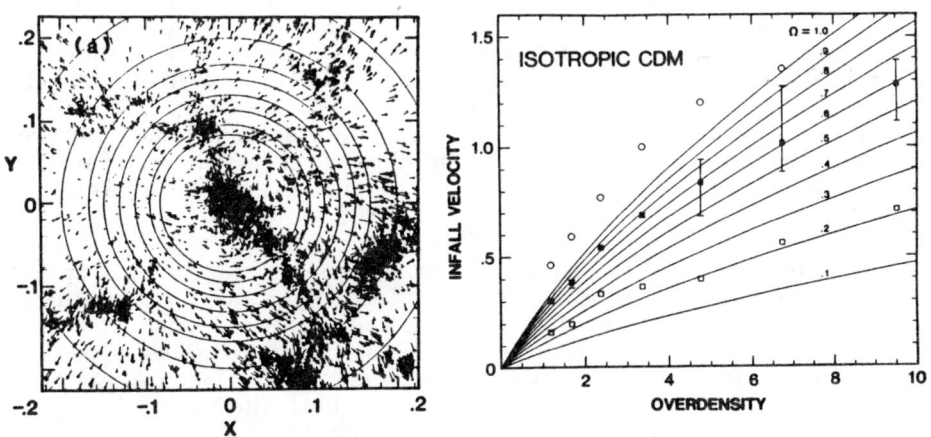

Figure 1: Flowfield near large cluster.   Figure 2: Measured infall velocities.

In figure 1 we show a closeup of a cluster in an $n=-2$ simulation. The

$$\alpha = \frac{2}{3\Omega^{0.4}H_0} \frac{V_p \cdot F}{|F^2|}$$

for observed velocities and forces and average over overdensity shells. In linear theory this ratio is unity while in the non-linear regime it is a function of overdensity. In order to test how well the deviations from spherical theory obey the extension of linear theory we can subtract the model from both the force field and velocity field. We calculate cosine of the angle between the force and velocity vectors.

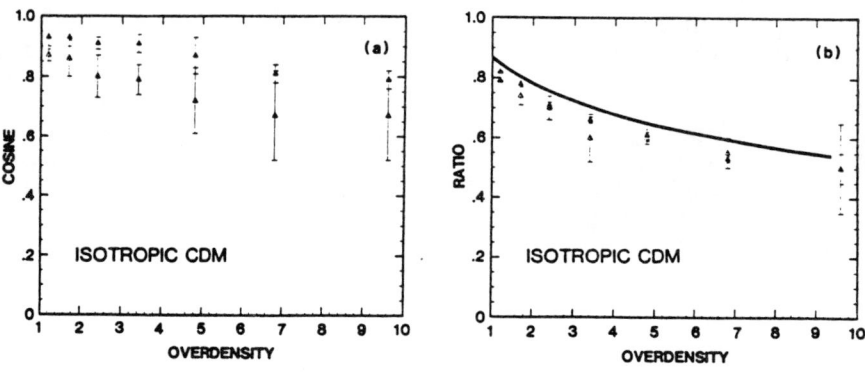

Figure 3: $\cos(F, V_p)$ Figure 4: Ratio α

Figure 3 shows the alignment of F and V_p with/without the spherical model subtracted. All quantitites are volume averaged and to minimize the effects of local non-linearities we exclude local high density regions in computing the averages. Up to an overdensity of $\delta = 4$ the alignment is better than 25 degrees with little scatter. With the spherical model subtracted the alignment is not as good and the scatter is larger but it is still good. Figure 4 shows the ratio α with/without the spherical model subtracted. Note that $\alpha < 1$ in the non-linear regime and that there is an excellent correlation between this ratio and the mean interior overdensity. With the spherical model subtracted, the scatter is larger but we see the same trend with overdensity. The solid curve shows the ratio predicted by the spherical model. The data track the curve remarkably well though at lower amplitude.

From this we conclude that there is a simple replacement of the spherical model that is not hampered by asphericities in the flow field, namely the excellent correlation of peculiar velocity and gravity. To use this correlation we need a full sky sample of galaxies with redshifts and distances so that we can compute the gravitational force vector assuming that the galaxies trace the mass on the scales of interest. The IRAS galaxy sample at 60μ is the best available in terms of uniformity and sky coverage and it is deep enough to measure our total expected peculiar acceleration. It has been shown (Meiksin and Davis 1986, Yahil et al. 1986) to contain a dipole anisotropy that points within 30 degrees of the microwave dipole anisotropy. If the IRAS galaxies roughly trace the mass distribution then our peculiar gravity should be directed close to the IRAS anisotropy, and the 30 degree misalignment with our peculiar velocity is just the typical deviation seen in the N-body models.

4. CONCLUSIONS

There is a systematic bias in the fits to the spherical model. The mean infall velocity measured in the N-body models are somewhat below the model values and the fractional deviation increases with δ. Only the spatial gradient in the peculiar flow field is observable, so the amplitude of the shear field is the observed quantity used to estimate the infall velocity. This will then lead to an underestimate of the infall velocity and of Ω. The seriousness of the bias depends on the proximity to Virgo of the innermost galaxies used in the fit to the spherical model. Possible remedies are to use the infall curve found for our models or to delete those galaxies inside the $\delta \approx 3$ surface.

We are situated near the Super Galactic Plane. If the exterior force field is zero the local infall velocity will be higher than average, but if the SGP extends to larger radii our infall velocity will be below average. A 50% variation in the local infall velocity will lead to an uncertainty in Ω of a factor of 3.

The peculiar velocity field tracks the peculiar force field extremely well. Comparisons of our peculiar velocity and peculiar force from the IRAS galaxy sample can be used to measure the mass in the Virgo supercluster.

ACKNOWLEDGEMENTS. This research was supported in part by NSF Grants AST-8419910, PHY-8440263, and by DOE contract DE-FG03-84ER40161.

REFERENCES

Aaronson, M., Bothun, G., Mould, J., Huchra, J., Schommer, R.A., and Cornell, M.E., 1985, *Ap. J.*, in press.

Bushouse, H., Melott, A.L., Centrella, J., and Gallagher, J.S., 1985, *M.N.R.A.S*, **217**, 7p.

Davis, M., and Peebles, J., 1983, *Ann. Rev. Astron. Astrophys.*, **21**, 109.

Lilje, P., Yahil, A., and Jones, B., 1986, preprint.

Meiksin, A., and Davis, M., 1986, *A. J.*, in press.

Villumsen, J.V., and Davis, M., 1986, *Ap. J.*, in press.

Yahil, A., Walker, D., and Rowan-Robinson, M., 1986, preprint.

VIRGO INFALL AND THE MASS DENSITY OF THE UNIVERSE

Adrian L. Melott
Department of Astronomy and Astrophysics
Enrico Fermi Institute
5640 S. Ellis Ave.
Chicago, Illinois 60637 U.S.A.

KEYWORDS/ABSTRACT: cosmology/superclusters/Virgo infall/gravitational clustering/galaxy formation.

Numerical models of gravitational clustering are studied in order to assess the expected error in estimates of the mass density of the Universe from Virgo infall. I find variation of the mean from cluster to cluster and large dispersion in infall velocity within a given spherical shell in all models. An apparent low density model can be reproduced in a critical density model using a biasing scheme which also has produced the correct two- and three-point correlations in cold-particle dominated universe models.

1. INTRODUCTION

I will report on some loopholes in estimating $\Omega=\rho/\rho_c$ from Virgo infall, using numerical simulations which can investigate more complex and realistic situations than analytic methods.

Previous studies[1,2] considered models with poisson and neutrino-dominated initial conditions. The Ω_{est} was found to be lower than unity, and there was a large dispersion in infall velocity within shells. One paper[3] argued that these results were due to binning and the use of linear theory, but did not correctly reproduce our binning algorithm in the tests; they also excluded particles with peculiar velocities away from clusters; this is incorrect since these must be included to get the correct center of mass velocity of the phase space. This exclusion biases upward[2] the estimate. Our work used linear theory in the weakly nonlinear regime, and it would be valuable to use the spherically symmetric nonlinear solution (A. Yahil, personal communication).

$$V_P/V_H = (\delta/3)(1+\delta)^{-0.25}\Omega^{0.6} \quad (1)$$

to test the importance of nonlinearity. Although the CIC method[4] suppresses discreteness, I will check this to confirm that it does not contribute to the dispersion.

I will report results when a cold-particle power spectrum is used. All previous studies have been done in $\Omega=1$ models, so I will report on the uncertainties in $\Omega_0=0.2$ models. Lastly, I will report on the validity of this method when galaxies form at only at density peaks, called biased galaxy formation.

2. THE MODELS

Four types of initial conditions were used here. They are neutrino, $\Omega=1$, called N; cold-particle, $\Omega=1$ called C; and cold-particle, $\Omega_0=0.2$ called CL. Models N, C, and CL have the Zel'dovich power spectrum processed by the appropriate transfer function[5]. A model with white noise initial conditions was also done. Lack of space prevents showing its results, which are not much different from C and N. At least three realizations of each model were run with 32^3 and one with 64^3 clouds in the same number of cells.

A biased subset of models of type C was constructed by including only particles which lay within regions of 1 Mpc^3 volume which were one standard deviation or more overdense. This is sufficient to amplify the autocorrelation enough to agree with observation[6]. To suppress noise, the biased subsets were taken from three 64^3 particle models, called CB models. In all runs, the power spectrum was appropriate to a cube 32 Mpc h^{-2} on a side.

In each model, the potential minimum was chosen as the cluster center. Concentric spherical shells were constructed about this center, and the peculiar velocity directed toward the center is binned in these shells. I have one and two orders of magnitude in particle density over the clusters studied in Bushouse et.al.

3. RESULTS

Discreteness effects can be measured by comparing the 32^3 with the 64^3 runs; no difference was found. The effect of binning coarseness was checked by varying the thickness of the shells, and no effect on the mean peculiar infall velocity was found. In the limit of *very* thick shells, the dispersion increased -- some measured dispersions [2] are a bit too large due to this artifact. They soon reach a minimum when a smaller shell thickness is used, but are still large.

In Table 1, I show the density contrast and the mean (peculiar) infall velocity (in units of V_P/V_H) of shells with 1σ error bars for each of the three versions of each model, for the shell nearest $\delta=2$. The quoted error is one standard deviation for infall velocity of the particles in the shell. In model CB, δ is defined as the *apparent* density contrast, i.e. the contrast in the density of tagged "galaxies", and only "galaxy" velocities are included.

The wide range of velocities in a shell diminishes prospects for deducing supercluster masses without complete three-dimensional velocity information; furthermore, there is quite a bit of cluster-to-cluster variation in the mean. In Figures 1-4 I show the Ω_{est} deduced from (1) for each of three versions of the models. There is a tendency to underestimate in all models, including low-density models, which increases with density contrast. The observations give a value of $\Omega \sim 0.2$ at our density contrast in the local Supercluster $\delta \sim 2$, which is not consistent with the $\Omega = 1$ models, although it is probably an underestimate. Since we do not have the full three-dimensional velocity field available in the observations, more work with models is needed to assess the effect of limited velocity information and incompleteness.

It is known[7,8] that model CB will give an incorrect result, basically because δ deduced from observing galaxies is not the true δ. One biasing algorithm was found to bring the two-point and three-point correlations of cold-particle Universes (which otherwise were in disagreement) into agreement with observation, by assuming that galaxies form only at the peaks of the density field, and also enhance certain measures of the existence of voids in the direction of agreement with observational data[6]. No calculations have yet shown what Ω_{est} to expect. We find that set CB gives the "right wrong answer"...reasonable agreement with the data, suggesting that critical density Universe may be compatible with data which measure much lower values of Ω.

4. ACKNOWLEDGEMENTS

This work was supported by an Enrico Fermi Fellowship at the University of Chicago and NSF grant AST-8313128. Computations were made possible by a grant of Supercomputer time on the Cyber 205 at Purdue University through the NSF Office of Advanced Scientific Computing; I thank the staff at PUCC for technical assistance. I thank the Department of Astronomy at the University of Chicago for assistance in manuscript preparation. For useful comments and conversations I thank Y. Hoffman, A. Yahil, and, as usual, J. Peebles.

1. A. Melott, *M.N.R.A.S.* **205**, 637 (1983).
2. H. Bushouse, A. Melott, J. Centrella, and J. Gallagher, *M.N.R.A.S.* **217**, 7P (1985).
3. H. Lee, Y. Hoffman, and C. Ftaclas preprint, (1985).
4. R. Hockney and J. Eastwood, *Computer Simulation Using Particles*, McGraw-Hill: New York, 1981).
5. J. R. Bond and A. Szalay, *Ap.J.* **274**, 443 (1983).
6. A. Melott and J. N. Fry, *Ap.J.* **305**, in press (1986).
7. G. Hoffman and E. E. Salpeter, *Ap.J.* **263**, 485 (1982).
8. J. N. Fry, *Phys. Lett.* 158B, 211 (1985)

Table 1 - Density Contrast and Infall Velocity (V_P/V_H)			
Realization	1	2	3
Model			
N	2.00	1.94	1.95
	0.39±0.23	0.39±0.19	0.47±0.23
C	2.09	1.78	1.94
	0.42±0.15	0.42±0.17	0.43±0.16
CL	2.07	1.87	1.77
	0.13±0.16	0.18±0.08	0.13±0.12
CB	1.94	1.73	2.02
	0.31±0.17	0.19±0.13	0.26±0.15

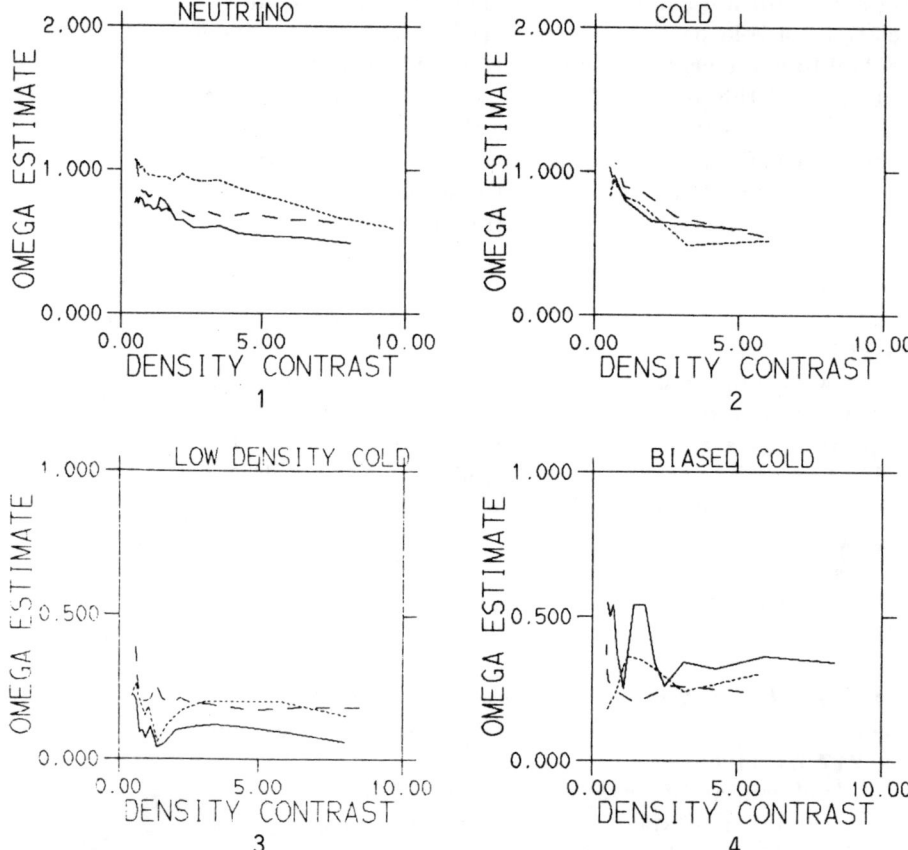

FIGURES 1 through 4 show the estimated Ω using equation (1) in models N, C, CL, and CB respectively, as a function of the density contrast interior to the shells, in three realizations of each model.

A Determination of Ω from the Dynamics of the Local Supercluster.

Ed Shaya
Caltech
Palomar Observatory
Pasadena, California 91125
U.S.A.

ABSTRACT. A method for determining the universal value for Ω is described that requires knowledge of the mass of one cluster and the velocity of infall for galaxies as they reach the periphery of the cluster. There is no dependence on the scale of the Universe or the age of the Universe, nor are there any assumptions on mass-to-light for this determination. The method is applied to the Virgo Cluster and a value of $\Omega \sim 0.1$ is found with an upper limit of $\Omega \sim 0.4$.

1. INTRODUCTION

I am going to discuss the quest for the *other* holy grail of cosmology, the determination of Ω. It is usually the case in searches for holy grails, that one seems forever to see it just on the horizon, but always just out of reach. According to legend, only a knight displaying true modesty and humility can find a holy grail (Saslaw 1985). Since I am not a knight, I will not be able to present the final value of Ω. Nevertheless, in the course of these quests, one covers a great deal of new territory and one learns much about the nature of things. In this adventure, we will examine galaxies falling into the Virgo Cluster and the motions of galaxies already within Virgo. Theoretically, the key to evaluating the cosmological arc-parameter and thus Ω rests here.

First, clear evidence will be presented for the skeptics that there is indeed an infall pattern into the Virgo Cluster. Next, the simplest cosmological model for the infall pattern is combined with the virial analysis of the Virgo Cluster to derive a method for obtaining Ω. Finally, the existing data for the velocities of galaxies in the Local Supercluster are used to crudely evaluate Ω.

2. THE 6° CONE TOWARD VIRGO

If galaxies are falling into the Virgo cluster from many different angles then it is plausible that the infall pattern can be directly observed by searching along the line-of-sight to Virgo for galaxies falling in. Fig. 1 shows galaxy distances derived using the Tully-Fisher relations versus recessional velocity for galaxies within 6° of the center of Virgo. The data come from the Catalogue of Nearby Galaxies (NBG) compiled by Tully (1985) and it is supplemented with HI data from Helou, Hoffman, and Salpeter (1984). The zero point for the Tully-Fisher relation is obtained by assuming $H_0=100$ km s^{-1} for galaxies in the NBG and removing all galaxies within 45° of Virgo from the sample. Any reasonable infall model has deviations from Hubble flow that drop to less than $\pm 25\%$ for galaxies more than 45° from Virgo and an average flow that is quite close to Hubble flow (Shaya, in preparation). Most of the distances were obtained using the relation between 21-cm line width and total blue magnitude. In the few cases where available, the H magnitudes were also used. Most of the galaxies pile up at the same distance, consistent with the usual value for the Virgo Cluster (here 13.2 h^{-1} Mpc), plus, there is a substantial number of galaxies at greater distances. Most of the time when diagrams like this are displayed, the obvious background galaxies are removed. This is another example of the saying "one person's garbage is another's treasure."

A dash-dotted straight line indicates the location galaxies would have if there were only constant Hubble flow in this region. Fig. 1 is the most direct and unambiguous evidence to date that there is an excess deacceleration of galaxies in the Local Supercluster centered on the Virgo Cluster. I have traced out two simple infall models by free-hand fitting through three positions known from the literature and the assumption of symmetry around the Virgo distance. The first (dotted lines) uses no motion of the Galaxy with respect to the local shell centered on Virgo, a velocity of 1004 km s^{-1} for the recessional velocity of Virgo, a turnaround radius at 28° or about halfway to the cluster (Hoffman, Olson & Salpeter, 1980), and an infall of 1000 km s^{-1} at 10° from Virgo (Tully & Shaya 1984). Spherical symmetry is assumed in the infall pattern. The second model (solid lines) is the same as the first except that an additional motion of 100 km s^{-1} toward Virgo for the Galaxy is included and this improves the fit marginally. For the second case the expansion velocity for the shell around Virgo at the Local Group position is $v_l = 1100$ km s^{-1}. An infall velocity relative to the 3K background for the Galaxy is seen to be 320 km s^{-1}, and for the local shell it is $u_l = 220$ km s^{-1}. The seven galaxies seen behind the Virgo Cluster at a recessional velocity of about 1000 km s^{-1} indicate the existence of a cloud of galaxies near the turnaround radius. There are also several galaxies at about a Virgo distance from the cluster but on the other side of Virgo from us. These background galaxies may be associated with either the Virgo W cloud (de Vaucouleurs, 1961) or the M cloud (Ftaclas et al. 1984) which was first identified by Shapley and Ames (1929).

Three galaxies in Fig. 1 lie far from the presumed infall pattern and there is no easy explanation for these. It is possible that they were shot out of Virgo during a three body encounter with much more massive galaxy partners. The discrepant galaxies are NGC 4569 in the foreground and NGC 4222 and NGC 4633 behind the cluster.

There is one galaxy in Fig. 1 in front of Virgo with a high velocity that is consistent with the infalling pattern. It is likely, since one galaxy is found in the foreground flow, that many more such galaxies will be found there. Only a small fraction of the UGC galaxies in the direction of the Virgo Cluster have had their distances determined. With the careful examination of many more galaxies in Virgo, the infall pattern will become increasingly clear. Perhaps there will always be large gaps at distances where there are real voids, but it is now clear that clouds of galaxies exist directly behind Virgo and possibly between us and Virgo. The infall pattern will become even clearer when we are able to determine relative distances to an accuracy of 10%, the accuracy with which the distances to Cepheids and RR Lyrae systems are now determined. With 10% accuracy, the front and back of the Virgo Cluster will be distinguishable from the middle and one can hope to discriminate the galaxies that are falling through the 6° radius and directly measure the velocity, v_6, at the cluster periphery. As it is v_6 must be inferred from less direct information. A measurement of v_6 is quite important; as I will now discuss, it can be used to infer the value of the universal Ω.

3. COSMOLOGICAL INFALL MODEL AND Ω

The Local Supercluster is modeled by a set of concentric shells each of which can be described by the equations of a Friedmann universe responding to the interior mass. For a start, spherical symmetry is assumed, and later the effects of non-sphericity and tidal stresses can be discussed. In the nomenclature of Schechter (1980), the motions of the shells can be described by an effective Hubble expansion h and an effective interior density ρ. Both of these are functions only of the constructed parameter η, the arc-parameter.

$$h(\eta) \equiv \frac{\dot{a}}{a} t_0 = \frac{\sin\eta(\eta - \sin\eta)}{(1 - \cos\eta)^2} \quad (1)$$

$$\rho(\eta) \equiv \frac{GM_a}{a^3} t_0^2 = \frac{(\eta - \sin\eta)^2}{(1 - \cos\eta)^3} \quad (2)$$

With no knowledge of the age of the Universe, t_0, or an absolute measure of scale, one can construct from measured velocities and just relative distances a run of the ratio of $h(\eta)$ for each shell to some arbitrary shell. If one assumes mass is distributed according to the light then one could also derive ratios of $\rho(\eta)$ for each shell to some arbitrary shell. One could then solve for the η at the reference shell that results in the best agreement between these ratios. One simple example would be to use the fact that the turnaround radius occurs at about half the Virgo distance (Hoffman, Olson & Salpeter 1980). This would imply that ρ is $\frac{\pi^2}{8}$ at this position. Thus the scale for the run of ρ can be set. To determine if the Universe is open it is necessary only that ρ drop in value from the turnaround radius by more than a factor of 5.5 ($\rho_{crit} = \frac{2}{9}$) at large distances. However, major discrepancies have arisen in the past when the assumption of constant mass-to-light have been made and here that assumption will not be made. A different method is sought to obtain η for at least one shell.

There is one shell for which η can be found, at least in principle. And, fortunately, η can be found without making assumptions for the age of the Universe or distance scale. The shell at the periphery of the Virgo Cluster is the special shell because the mass interior can be found from the virial theorem and therefore, the mass is independent of η or t_0. One can form the distance independent parameter λ by dividing the Virgo virial mass by the distance of the Local Group from Virgo;

$$\lambda \equiv \frac{M_a}{a_l} = c\frac{\sigma_v^2 \theta_\Omega}{G} \qquad (3)$$

The parameter λ depends only on the velocity dispersion σ_v and the angular harmonic mean separation of the galaxies in Virgo, both of which are distance independent.

A second determination of λ comes from the cosmological equation relating the velocity of infall to the mass interior to the given shell:

$$\frac{M_a}{a_6} = \frac{v_6^2}{G}(1 + \cos\eta_6)^{-1}. \qquad (4)$$

Multiplying this equation by $\frac{a_6}{a_l}$ results in λ. Solving for η_6 gives,

$$\eta_6 = 2\pi - \cos^{-1}\left(\frac{v_6^2 \sin 6°}{\lambda G} - 1\right). \qquad (5)$$

One now has η for one shell and h for this shell comes right from Eq. (1). The run of the ratio of h to h_6 for the different shells can now be used directly to determine $\rho(\eta)$ and η at every shell.

For the universe in general, one needs to know only the peculiar velocity relative to the 3 K background of the Local Group shell and the actual motion of the Local Group shell with respect to Virgo. For the effective h at infinity one has:

$$h_\infty = h_6 \frac{v_l + u_l}{v_6} \sin 6°. \qquad (6)$$

The universal value of η is found from Eq.(1) and Ω is calculated from the universal η via:

$$\Omega = \frac{2}{1 + \cos\eta_\infty}. \qquad (7)$$

4. THE λ FOR VIRGO

Table 1 gives the velocity dispersions, the angular mean harmonic separations, and λ calculated separately for several morphological types.

Table 1.

Virial Analyses for the Virgo Cluster

Type	No.	θ_Ω (degrees)	σ_v (km s^{-1})	λ (grams/cm)
E	30	1.4	678	1.7 x 10^{22}
S + S0	144	2.4	684	3.0 x 10^{22}
S0/a − Im	99	2.4	761	3.6 x 10^{22}
S0	45	2.6	515	1.8 x 10^{22}

The ellipticals sample a small extent of the entire Virgo Cluster and have a velocity dispersion slightly smaller than the spiral galaxies for classes S0/a through Im. The velocity distribution of Virgo ellipticals is close to gaussian (Tanaka 1986). Therefore, the elliptical system is probably mostly relaxed. Since it does not fully sample the cluster, the mass based on ellipticals must provide a lower limit to the total cluster mass. The inferred λ from the ellipticals is 1.7×10^{22} g/cm or about half of that inferred from the spirals. The spirals may contain a significant number of galaxies passing through, having fallen in from somewhat greater distances than the 6° perimeter. The value for λ based on the spirals, therefore, provides an upper limit to the mass. For now, a best estimate is a compromise between the two limits: $\lambda = 3.0 \times 10^{22}$ g/cm.

Parenthetically, it is perhaps not just a coincidence that the mass determined using the S0's is the same as the mass determined using the ellipticals even though the mean separations and velocity dispersions are quite different. It could be explained if both were virialized and seeing the same mass. This argument would imply that most of the mass is centralized within 3 degrees of the center. The low velocity dispersion in the lenticulars would be explained by the fact that their orbits take them farther from the center. While this argues for the lower limit value for λ, I feel this reasoning is still too speculative. The result of using a lower λ would be to lower the value determined for Ω.

The solid lines in Fig. 2 is a result of calculating Ω for a range of infall velocities. Two values for the motion of the Local Group shell with respect to the microwave background are used which span the range of values found in the literature; 220 km s^{-1} and 440 km s^{-1}. The velocity taken for the local shell, v_l, is 1100 km s^{-1}.

The infall velocity at 10° is known to be ~ -1000 km s^{-1} (Tully and Shaya, 1984) from a study of the velocity spread of galaxies in Virgo II, a cloud quite close to the Virgo Cluster. The error associated with the infall velocity is difficult to assess because the procedure to obtain it, namely fitting the upper envelope in velocity for galaxies as a function of angular distance from Virgo, is somewhat subjective. An error of 100 km s^{-1} is reasonably conservative. Since the velocity at 6° is not well known, at present, the same calculations for infall velocity have been done at 10° for two estimates of λ at this shell. In one case no additional mass above our nominal value for Virgo is assumed; the argument for this case is that there are almost no galaxies between 6° and the Virgo II cloud at 10°. The infall at 10° implies that $\Omega \sim .1$, but the lower limit on the infall velocity of 900 km s^{-1} does allow for Ω as high as 0.3. The second case assumes mass grows linearly with distance out to 10°. The expected infall velocity is now too great unless Ω is > 2. One can go the other way around; assume $\Omega = 1$ and $v_{10} = -1000$ km s^{-1}, then one requires more mass than the virial theorem can allow; so the invisible mass must extend beyond 6° but must grow more gradually than linearly with distance.

At 6 degrees from Virgo the velocity for infall was determined to be 1700 km s^{-1}, but this is very tentative since only 2 galaxies define the envelope of the velocity dispersion near these lines-of-sight. In fact, this velocity is higher than that permitted by the model for any value of Ω, but an error of 200 km s^{-1} would allow for $\Omega < 0.3$.

5. THE DILEMMA OF THE VELOCITY DISPERSION

There is a dilemma to deal with regarding the velocity dispersion of the spiral galaxies in Virgo. The infall velocity at 6° is around 1600 km s^{-1} as determined from the model and from Tully & Shaya. As the galaxies fall still closer to the center they fall down the potential well and their velocity climbs higher. If we distribute the mass in Virgo isothermally ($\rho \propto a^{-2}$) out to 6°, then the infall velocity is given by;

$$v_a = \sqrt{v_6^2 + 2\Delta U} = \sqrt{v_6^2 + \frac{2G\lambda}{\sin 6°}\ln\frac{a_6}{a}} \qquad (8)$$

At the 3° radius, the velocity is approximately 50% greater than at the 6° radius or ≈ 2400 km s^{-1}. If spirals are falling into Virgo now, then why do we not see any galaxies near this velocity?

The envelope of the velocity histogram of spirals in Virgo only spans a range of ± 1600 km s^{-1}. One possibility is that there are no infalling galaxies within the 3° radius with infall closely aimed along our line-of-sight. Another possibility is that the galaxies infalling during the present epoch have sufficient orbital angular momentum to avoid a pericenter much within 6°.

6. SUMMARY

1) The infall pattern can be observed in several of the galaxies along the line-of-sight to the Virgo Cluster by applying the Tully-Fisher method for determining distances. A cloud is found at about the turnaround radius behind Virgo and another cloud is found at a Virgo distance behind Virgo. From an examination of galaxies within 6° of Virgo, a peculiar motion of the shell at our distance from Virgo is crudely determined to be $\sim$200 km s^{-1}.

2) The virial theorem together with the infall velocity to any cluster can be used to determine the arc-parameter at the periphery of a cluster and this allows the scale to be set for the run of the effective expansion rate as a function of distance from the center of the Local Supercluster. If the sum of $v_l + u_l$ is known then setting the scale on the expansion rate amounts to determining Ω.

3) An infall velocity at 10° from the center of Virgo of -1000 km s^{-1} and a sum of $v_l + u_l = 1300$ km s^{-1} imply $\Omega \sim 0.1$, provided little mass exists between 6° and 10° of Virgo.

4) There is a lack of high velocity galaxies in the Virgo cluster if many galaxies are falling into Virgo in the present epoch.

REFERENCES

de Vaucouleurs, G. 1961, *Ap.J. Supp.*, **6**, 213.
Ftaclas, C, Fanelli, M., and Struble, M. 1984, *Ap.J.*, **282**, 19.
Helou, G., Hoffman, G., and Salpeter, E. 1984, *Ap.J. Supp.*, **55**, 433.
Hoffman, G., Olson, D. W., and Salpeter, E. E. 1980, *Ap.J.(Letters)*, **228**, L1.
Saslaw W. C. 1985. *Gravitational Physics of Stellar and Galactic Systems* (Cambridge UP) p. 141..
Schechter, P. L. 1976, *Ap.J.*, **203**, 297.
Shapley, H., and Ames, A. 1929 *Harvard College Obs. Bull.*, No. 865.
Shaya, E. 1986 in preparation.
Tanaka 1986, *Publ. Astron. Soc. Japan*, **37**, 427.
Tully, R. B. and Shaya, E. 1984, *Ap.J.*, **281**, 31.

Fig. 1 - Recessional velocity versus distance obtained from the Tully-Fisher relations. Zero point on the Tully-Fisher relation is chosen to give $H_0=100$ km s^{-1} Mpc^{-1}. The dotted-dashed line represents pure Hubble flow extending from our position. The existence of galaxies below this line behind Virgo is strong confirmation of a general infall pattern in the Local Supercluster. The dashed line is a "free-hand" model made to pass through the origin, 1000 km s^{-1} at half a Virgo distance, and 2000 km s^{-1} at the 10° shell. Symmetry around the mean Virgo distance of 13.2 h^{-1} is invoked. The solid line is similar to the dashed line except a motion of 100 km s^{-1} toward Virgo is given to the Galaxy.

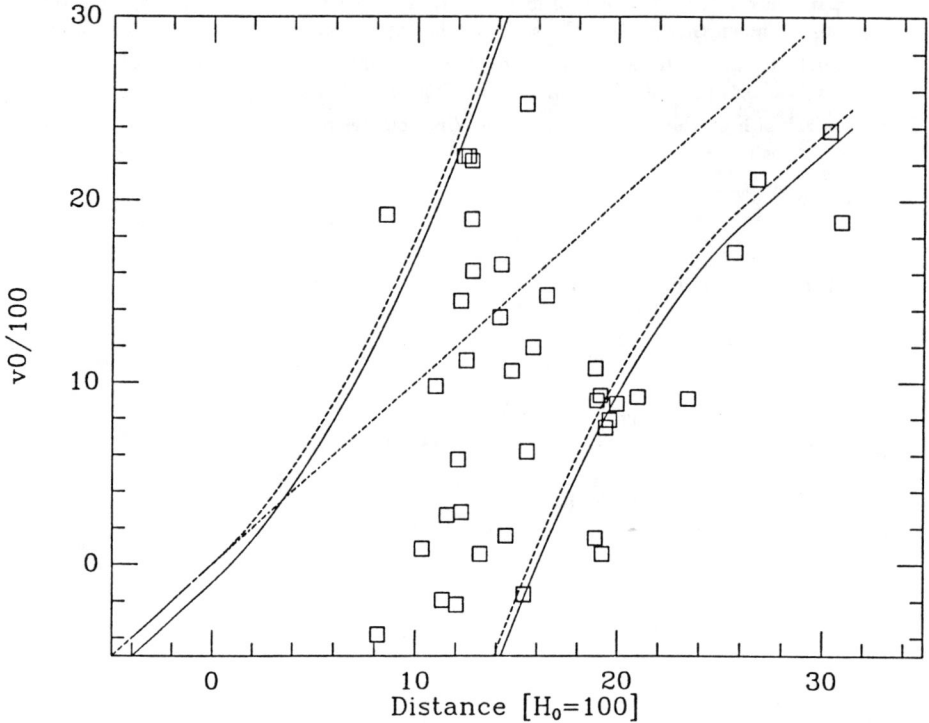

Fig. 2 - Infall velocity versus universal Ω. The pairs of thick solid lines are model calculations, the top pair are calculated at 6° and the other two pair are calculated at 10°. The top and bottom pair use $\lambda = 3 \times 10^{22}$ g/cm, the middle pair use $\lambda = 5 \times 10^{22}$ g/cm. Each pair consists of calculations with u_l of 220 and 440 km s^{-1}. The horizontal thin solid line at 1000 km s^{-1} represents the measurement of Tully and Shaya at 10° and the line at 1700 km s^{-1} is the more uncertain measure at 6°. The dotted lines represent errors around the measurement.

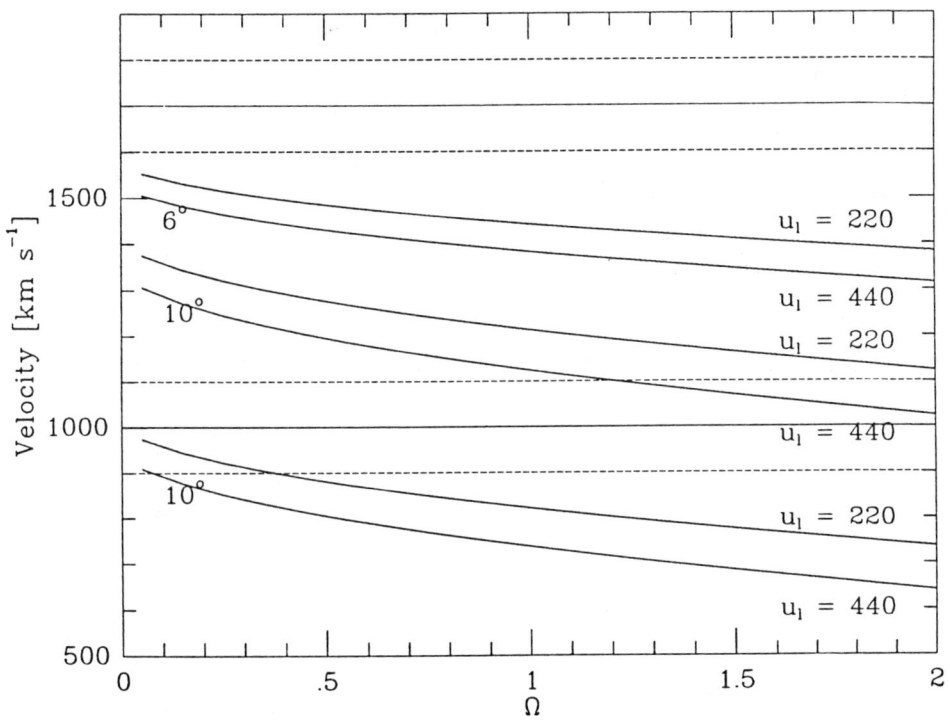

THE DYNAMICS OF SUPERCLUSTERS AND Ω_0

Yehuda Hoffman
Department of Physics
University of Pennsylvania
Philadelphia, PA 19104, USA

ABSTRACT. The dynamics of Local Supercluster-like objects is studied analytically by the Zeldovich quasi-linear formalism and by means of N-body simulations. Ensembles of such objects have been constructed by these two independent studies and an apparent value of Ω_0, Ω_{app}, is calculated for each object. In the case of an actual flat universe the distribution of Ω_{app} is found to have a dispersion of $\approx(0.2-0.3)$ around a mean value of ≈ 0.9. The value of Ω_{app} increases with the magnitude of the shear in the velocity field and non-shearing objects underestimate Ω_0. In addition, the formalism is applied to the decaying-particles scenario which is found to be compatible with the observed LSC.

1. INTRODUCTION

The dynamics of the Local Supercluster (LSC) has been studied intensively in the last decade and the prime target of that effort has been the determination of Ω_0. The common approach has been to model the LSC by the non-linear model of spherical infall (hereafter, the standard model) and to use it to derive Ω_0 (Silk, 1974). To study the applicability of the standard model we have used quasi-linear (QL) analytical calculations and N-body simulations to model LSC-like superclusters (SC) and to derive an apparent value of Ω_0, Ω_{app}. The analytical calculations are then used to study the Virgocentric infall problem within the framework of of the decaying particle scenario (DPS) of Turner et al (1984). To further study the applicability of the standard model we have analysed N-body simulations of a flat universe, identified LSC-like objects and calculated their Ω_{app}.

2. QUASI-LINEAR DYNAMICS

The growth of density perturbations to a QL order, and in particular its dependence on the anisotropy of the primordial velocity field, has been studied by Hoffman (1985, Paper I), who modeled the evolution of SCs by employing the Zeldovich (1970) QL formalism. Essentially, this formalism assumes the velocity field to be given by the linear theory and uses it to calculate the evolution in the non-linear regime. The comoving Eulerian and Lagrangian coordinates of a mass element, r and q respectively, are related by:

$$r(q,t) = q + b(t) \, v(q) \qquad (1)$$

where $b(t)$ is the linear growth function of δ and $v(q)$ is the (linear) peculiar velocity. Expanding the curl-free velocity field to first order in q around the center of a proto-SC one finds the peculiar and Hubble velocities relative to the center to be:

$$v(q,t)_{p,\alpha} = a(\dot{b}/b)\sigma_0(-1/3 \, \nu_0 \, \delta_{\alpha\beta} + \Sigma_{\alpha\beta})q_\beta = a(\dot{b}/b)\Lambda_{\alpha\beta} \, q_\beta \qquad (2)$$

$$v(q,t)_{H,\alpha} = a[\delta_{\alpha\beta} + (-1/3 \, \nu_0 \, \delta_{\alpha\beta} + \Sigma_{\alpha\beta})\sigma_0]q_\beta = aT_{\alpha\beta} \, q_\beta \qquad (3)$$

here ν_0 is the amplitude of the $\delta(0)$ in units of the r.m.s. value of δ, $\Sigma_{\alpha\beta}$ is the (normalized) traceless shear tensor, a is the universe scale factor and $\sigma_0 = \langle \delta^2 \rangle^{1/2} b(t)$. The time evolution of δ is given by $\delta(t) = \det(T^{-1}{}_{\alpha\beta}) -1$ and the ratio of the isotropic component of the infall to the Hubble velocities is:

$$v_p{}^{iso}/v_H = 1/3 \, f \, \sigma_0 \, B_{\alpha\alpha} \qquad (4)$$

where $f = (\dot{\delta}/\delta)/(\dot{a}/a)$ and $B_{\alpha\beta} = \Lambda_{\alpha\mu}T^{-1}{}_{\mu\beta}$. In terms of the eigenvalues of the shear tensor, $\Sigma_{\alpha\beta} = \lambda_\alpha \delta_{\alpha\beta}$, $B_{\alpha\beta}$ reads as:

$$B_{\alpha\beta} = [-\lambda_0 + \lambda_\alpha] / [1 + (-\lambda_0 + \lambda_\alpha) \, \sigma_0] \, \delta_{\alpha\beta} \qquad (5)$$

where $\lambda_0 = \nu_0/3$. In the QL regime the evolution of $\delta(t)$ and $v_p{}^{iso}$ is coupeled to the anisotropic shear term and this increases the collapse rate and, consequently, also Ω_{app}. Upon understanding the dynamics of a single SC we turn to study the statistics of such objects, assuming that structure has evolved from a Gaussian perturbation field. The three independent variables that characterize a proto-SC are ν_0 and the two independent eigenvalues of $\Sigma_{\alpha\beta}$, λ_x and λ_z say. Note, although the distribution of the elements of $\Sigma_{\alpha\beta}$ is Gaussian, its eigenvalues distribution is not, because of the non-linear nature of the rotation operation which does not preserve the original distribution. The assumed statistical homogeneity and isotropy of the δ-field imply that $\delta(0)$ and $\Sigma_{\alpha\beta}$ are uncorrelated, and therefore the distribution function of ν_0 and λ_x and λ_z can be written as $P(\nu_0, \lambda_x, \lambda_z) = P_G(\nu_0) P_{NG}(\lambda_x, \lambda_z)$. Here $P_G(\nu_0)$ is the Gaussian distribution and P_{NG} is found to be:

$$P_{NG}(\lambda_x, \lambda_z) = (27/4\sqrt{2}\pi) \, \exp[-15/2 \, (\lambda_x{}^2 + \lambda_z{}^2 + \lambda_x\lambda_z)]$$
$$(\lambda_x - \lambda_z)(2\lambda_x + \lambda_z)(2\lambda_z + \lambda_x) \qquad (6)$$

assuming $-\infty < \lambda_z < -(\lambda_x + \lambda_z) < \lambda_x < +\infty$. The distribution function $P(\nu_0, \lambda_x, \lambda_z)$ is then used to construct an ensemble of LSC-like SCs. The values of δ and v_p^{iso}/v_H are calculated for each object and this is used to calculate Ω_{app}, by the standard model. The overall normalization of the amplitude of δ is given by σ_0, which plays a crucial role in the determination of the mean value of Ω_{app} in the case where the ensemble of SCs is constrained to δs within a given range. Keeping that range fixed while increasing σ_0 implies that lower ν_0 perturbations can evolved to have δs at that range, and the lower is ν_0 the more dominant is the shear of that proto-object and therefore the larger is Ω_{app}. The case of a flat universe was studied in Paper I and the results are shown in Fig.1, which shows the dependence of the (ensemble) mean value of Ω_{app} and its r.m.s. dispersion, σ_Ω, on σ_0. To compare the predictions of the model with the LSC we here adopt σ_0 to be 0.42 (*cf* Paper I and Hoffman, 1986, hereafter Paper II). This yields Ω_{app}=0.92 and σ_Ω=0.25, for an ensemble of SC that are constrained to (present day) δ in the range of 2.0 to 4.0. These results depend on the range of δ and one finds that <Ω_{app}> increases as one considers smaller δ perturbations. To summarize, we find that the standard model does recover the actual value of Ω_0 only in the sense of a mean value. The dispersion of Ω_{app} is not random but is correlated with the magnitude of the shear.

2. DECAYING PARTICLES DYNAMICS

The DPS was suggested by Turner *et al* (1984) as a possible solution to the so-called 'Ω_0 problem',i.e. the discrepancy between the theoretical prejudice that the universe is flat and the actual 'observed' value of Ω_0. The DPS argues that the universe is flat but is radiation dominated at the present epoch. However, in order to enable the growth of structure via gravitational instability the universe should have undergone a phase of matter dominance, and therefor it is suggested that it was dominated by some massive, unstable, weakly interacting, particles that decayed at some redshift Z_D into relativistic remnants. Consequently, dynamical studies of Ω_0 which are sensitive only to the clustered, non-relativistic, component (baryons, say) would yield an apparent Ω_0 which is smaller than 1.0 . The growth of LSC-like objects within the framework of the DPS has been studied in Paper II, using the QL formalism of Paper I. Essentially, the QL evolution is still given by Eqs. 1-4 but with modified growth rates of a(t) and b(t) which now depend on Ω_B and Z_D (here, Ω_B is the density parameter of the non-relativistic component). A Monte Carlo ensemble of SCs of overdensities in the range of 2.0< δ <4.0 has been constructed and the dependence of <Ω_{app}> and σ_Ω on Z_D and Ω_B has then been studied. The results are shown in Fig.2. In all of the cases considered here σ_0 is set to be 0.42 and σ_Ω is found to be $\approx$0.3<Ω_{app}>. One finds that for an Ω_B=0.1 a typical object (i.e. $\Omega_{app}\approx$<Ω_{app}>) would yield Ω_{app}=0.26 for a Z_D which is as low as 6.0. Even an 'observed' Ω_0 of 0.14, which is lower than most of the published estimates of Ω_0, is only one standard deviation smaller than <Ω_{app}> for a Z_D=10.0. To summarize, the DPS is compatible with most of the current observation of the LSC for an Ω_B=0.1 and Z_D<10.0 .

4. N-BODY SIMULATIONS

The applicability of the standard model has been further studied by Lee, Hoffman and Ftaclas (1985, Paper III), who analysed N-body simulations of a flat universe starting with a 'white noise' of Centrella (1985). The analysis consists of identifying SCs which are similar to the LSC, calculating the (mean spherical) density profile and then fitting the infall velocity field of each SC by the standard model. The fitted velocity field is based on the 'observed' density profile and it uses the standard model to calculate the infall velocity at a given radius. Considering Ω_0 as a free parameter, Ω_{app} is defined as the one which minimizes the deviations from the fitted velocity field. This procedure has been applied to all the 23 SCs. Most of the SCs show a huge dispersion in their infall velocities at a given radius; SCs C and D of Figs.3 and 4 are typical cases. On the other hand, a few objects are found to show a much smaller dispersion; SCs A and B are the two extreme cases. The (v_p,r) distribution of each SC (see Fig.3) is then transformed into the $(v_p/v_H,\delta)$ plain (see Fig.4). An immediate result is that any determination of Ω_0 cannot be done based on the dynamics of a single particle because of the huge scatter. Even the best candidates for the application of a spherical infall model show a huge dispersion, in terms of the Ω_0's determined by individual particles. A surprising result is that SCs which show the least scatter are the ones that systematically underestimate Ω_0. One finds Ω_{app}=0.73 and 0.68 for SCs A and B, respectively, as opposed to Ω_{app}=1.07 and 1.05 for C and D. This trend is manifested by the distribution of (Ω_{app},σ) of all the SCs (shown in Fig.5). Here σ is the intrinsic standard deviation of Ω_{app} of each SC. The better a velocity field of a given SC is fitted by the standard model, the smaller is the value of Ω_{app}. The mean value of Ω_{app} is found to be 0.92 with a standard deviation of 0.14, but this depends somewhat on the range of δs over which the fit is done and the above results are obtained for $1.0<\delta<4.0$. For a range of (1.0 - 2.0) one finds Ω_{app}=0.96±0.20 and 0.89±0.19 for a range of (2.0 - 4.0). This is in a qualitative agreement with the QL calculations that predict that higher δ perturbations are (statistically) dominated by the isotropic component of the velocity field which leads to lower values of Ω_{app}. The N-body analysis shows that the standard model recovers the actual value of Ω_0 (to within $\approx$10%) only in the sense of a mean value taken over an ensemble of SCs. The scatter in Ω_{app} is found to be correlated with the structure of the velocity field, again this is in a qualitative agreement with the analytic model.

REFERENCES

Centrella, J., (1985), (unpublished)
Hoffman, Y., (1985), *Ap.J.* (Submitted)
          (1986), *Ap.J. Letter* (Submitted)
Lee, H., Hoffman,Y.and Ftaclas, C., (1985), *Ap.J. Letter* (in press)
Silk, J., (1974), *Ap.J.*, **193**, 525
Turner, M.S., Steigman, G. and Krauss, L.M.,(1984),*Phys. Rev.Lett.*, **52**, 2090

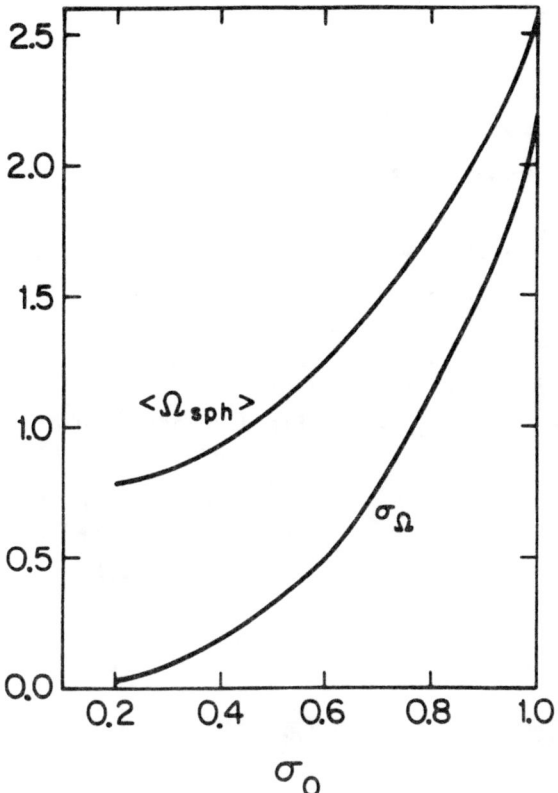

Fig.1 QL evolution of SCs in a flat universe. The dependence of $<\Omega_{app}>$ (upper curve) and σ_Ω on σ_0.

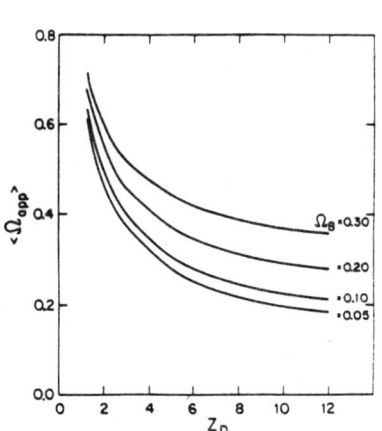

Fig.2 The dependence of $\langle\Omega_{app}\rangle$ on Z_D is shown here for a few values of Ω_B.

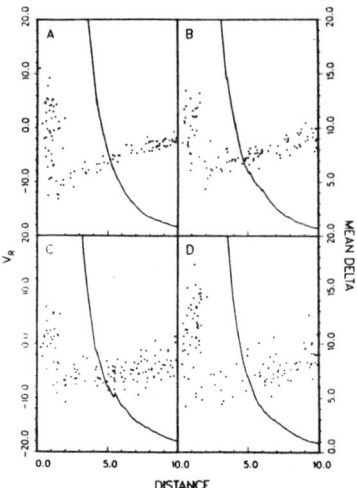

Fig.3 N-body simulations: The distribution of particles peculiar (radial) velocities as a function of radius is shown here for four representative cases.

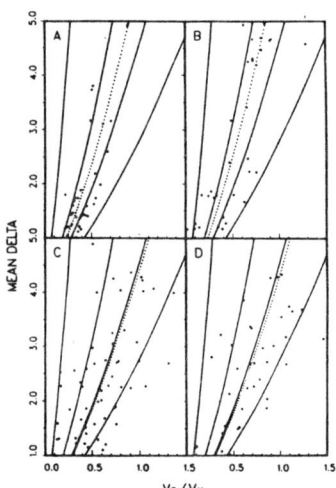

Fig.4 The distribution of Fig.3 is transformed here into δ and v_p^{iso}/v_H. The four solid lines correspond to the prediction of the standard model for $\Omega_0=0.1$, 0.5,1.0 and 2.0 (from left to right), and the dashed line corresponds to the best fit.

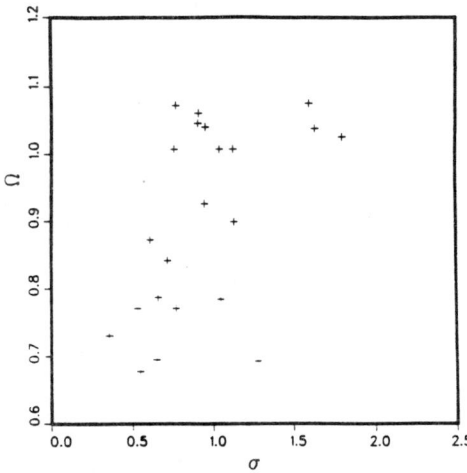

Fig.5 The distribution of the values of $\langle\Omega_{app}\rangle$ vs. its internal dispersion, σ_Ω, of each SC.

THE DISTRIBUTION OF GALAXIES VERSUS DARK MATTER

Avishai Dekel
Department of Physics
The Weizmann Institute of Science
Rehovot 76100, Israel

I. MOTIVATION

The assumption that galaxies trace mass enables us to estimate Ω and to interpret the large-scale structure based on observations of galaxies. However, it would be astonishing if galaxy formation was not affected by environmental effects (e.g. Dressler 1980), resulting in *biased galaxy formation*. I will summarize the evidence for this suspicion, argue that a segregation of one sort or another is a natural outcome of every cosmogonic scenario (although it is not always of the desired sort), and point at key observational tests before expressing my biased conclusions (see Dekel 1986 for a detailed discussion).

We are dealing with two dark matter (DM) problems (e.g. Dekel, Einasto and Rees 1986): while the mass associated with luminous matter amounts only to $\Omega \sim 0.01$, the dynamically inferred mass in galactic halos and in clusters of galaxies indicate $\Omega \sim 0.1 - 0.3$ (if M/L is similar everywhere), still short by a factor ~ 5 from 'closing' the universe. If the matter were all baryonic, $\Omega = 0.1$ was compatible with standard big-bang nucleosynthesis if $h \leq 0.65$.

There has been, however, a development in theorists' attitudes, partly because we are much more willing now to consider non-baryonic DM, but mostly because of 'inflation'. Being so appealing in resolving certain cosmological problems, inflation gives high credibility to the prejudice in favour of $\Omega=1$, which introduces a *second* DM problem. For the presently observable universe to be in causal contact at some earlier time (required by background isotropy), the universe must have gone through an inflation by a factor $\geq 10^{25}$. The exponential growth, once started, rapidly continues for many expansion timescales: it is likely to overshoot, hence stretching any small chaotic hypersurface to become essentially flat over our present horizon scale, yielding $\Omega = 1$ to a very high precision. For inflation to yield the dynamically inferred value $\Omega \sim 0.2$, the inflation factor would have to be 'just' 10^{25}, making the present radius of curvature 'just' of order the Hubble radius. This coincidence demands a *fine-tuning* which one would prefer to avoid. In addition, our presently observable universe would have to grow from an initial hypersurface whose curvature was uniform to one part in 10^5, otherwise, the curvature fluctuations would produce too large quadrupole effects in the microwave background. This, again, requires special initial conditions.

How could we explain the low value for Ω? The mass estimates on scales of $\sim 10 \ h^{-1} Mpc$ are based either on modeling our infall into the *Local Superclus-*

ter (LSC), or on *cosmic virial analysis* of pairs of galaxies. In a linear, spherical model for the LSC, for a given infall velocity of the shell containing the Local Group towards Virgo, Ω is related to the density enhancement δ within it by $\Omega \propto \delta^{-1.5}$, and for given mean pair velocities the cosmic virial theorem gives $\Omega \propto \xi^{-1}$, where $\xi(r)$ is the two-point correlation function. The results suggest again $\Omega \simeq 0.1 - 0.3$, but they are obtained using the quantities corresponding to *galaxies*: their number overdensity δ_g or the galaxy correlation function $\xi_g(r)$. If, however, the galaxies cluster more than the matter, such that $\delta_g = f\delta$ and accordingly $\xi_g(r) = f^2\xi(r)$, the obtained value of Ω is larger and compatible with $\Omega = 1$ if $f \simeq 2 - 3$.

The emerging picture of the distribution of galaxies on large scales is that '*voids*' $\sim 50h^{-1}Mpc$ in diameter are quite common (Kirshner et al 1981; Einasto, Joevier and Saar 1986; de Lapparnet, Geller and Huchra 1986; Giovanelli and Haynes, this volume; Tully, this volume; Oemler, private comm.). The number density of galaxies in these 'voids' is typically less than 10% of the mean. Based on spherical models (e.g. Hoffman, Salpeter and Wasserman 1982), a similar underdensity in the mass corresponds at recombination to $|\delta| \geq 10^{-2}$ if $\Omega \simeq 1$, and $\geq 5 \times 10^{-2}$ if $\Omega \simeq 0.1$. This is incompatible with the microwave isotropy on $10' - 1°$ in any of the cosmogonic scenarios, unless reionization has washed out the fluctuations. The large-scale N-body simulations demonstrate the difficulty: even in pancake scenarios (e.g. White, Frenk and Davis 1983; Centrella and Melott 1983; Dekel and Aarseth 1984) such large regions are not found with a density below 25% of the mean; they cannot be substantually evacuated dynamically by the present epoch, which is defined by matching the correlation functions of the simulated mass and the observed galaxies. The situation is worse if $\Omega < 1$, where the 'voids' are evcuated even less efficiently.

What is the real mass density in 'voids'? Consider a 'toy' universe which consists of 'top-hat' superclusters (sc) and uniformly underdensed 'voids', where

$$(\delta_g/\delta)_{void} = (\delta_g/\delta)_{sc} = f \qquad (1)$$

results trivially from the definition of the δ's. Assuming that the LSC and the Bootes 'void' are typical, we can adopt the corresponding observed values for the *galaxies*: $\delta_g \simeq 2.5$ and -0.9 respectively. If $\Omega = 1$, the real mass overdensity in the LSC must be $\delta \simeq 0.85$ ($f \sim 3$), so using (1) we get for the *mass* in the 'voids' $\delta \simeq -0.32$. (The fractional volume in the 'voids' is then 73% and the mass fraction is $\sim 50\%$.) These mass densities in superclusters and in 'voids' are both compatible with $|\delta| \simeq 9 \times 10^{-4}$ at recombination. If most of the mass is non-baryonic, this corresponds to $\delta T/T \simeq 3.5 \times 10^{-5}(\Omega h^2)^{-1}$, which is compatible with the isotropy constraints if Ωh^2 is not much smaller than unity. An open universe with $\Omega \simeq 0.2$ would be in trouble: it would require no bias in the LSC, and therefore a real deep mass underdensity of 10% in the 'voids'. The corresponding $\delta T/T \sim 5 \times 10^{-3}$ would be hard to reconsile with observations. From another angle, the relative 'efficiency' of galaxy formation in an environment 'x' can be quantified by $\epsilon = (n_x/\rho_x)/(n/\rho)$, where n corresponds to galaxy number density and ρ to matter density. In 'voids', using the above values, the efficiency is only $\epsilon \simeq 0.15$ while in superclusters $\epsilon \simeq 1.9$.

Finally, studies of *scenarios* dominated by either 'cold' DM (CDM) or massive neutrinos have shown that neither can reproduce the observed distribution of galaxies unless it is *biased*. In both cases the matter correlation function steepens in time, and the stage of the simulation to be regarded as the present epoch is determined by matching its logarithmic slope to the observed $\gamma = 1.8$ of galaxies. The CDM correlation length at this time turns out to be only $r_0 \simeq 1 \; (\Omega h^2)^{-1}$ (Davis et al. 1985). Hence, for the galaxies to match the observed $r_0 \simeq 5 \; h^{-1} Mpc$, with $\Omega = 1$, the galaxies must be biased by $\xi_g(r) = (5 - 20) \; \xi(r)$ ($h = 0.5 - 1$), in agreement with the f deduced above. In the case of $\sim 30 \; eV$ neutrinos, at the time when the slope is 1.8 the structure is still young (e.g. Dekel and Aarseth 1984); collapse to pancakes must have occured at $z \sim 1$. This poses a timing difficulty for any 'dissipative pancake' scenario which assumes that galaxies are 'daughters' of pancakes, if one considers the evidence for galaxy formation at $z \geq 3$ (e.g. based on high-redshift quasars). The difficulty is also one of scaling: the neutrino correlation length, by the time when its slope is right, has already grown to be $r_0 \simeq 8 \; (\Omega h^2)^{-1}$, which is too large in comparison with galaxies unless $\Omega h > 1$. If the galaxies form only in the collapsed regions the constraints become even tighter. The bias required here is therefore of an opposite sense: the galaxies should somehow be less clustered than the neutrinos.

II. BIAS MECHANISMS

The bias mechanisms are of three general types: (i) the 'voids' may be filled uniformly with DM different from the clustered DM, or (ii) there is only one kind of DM but the baryons are segregated from the non-baryonic DM even on scales $\sim 30 \; h^{-1} Mpc$ (e.g. from 'isothermal' fluctuations, from dissipation relative to a collisionless DM, or from explosions that have pushed the gas), or (iii) the baryon distribution does trace the DM on scales larger than a few Mpc but the efficiency with which baryons turn into luminous galaxies is sensitive to other environmental effects. This possibility is less extravagant in energy. The bias could be determined in each protogalaxy individually, or be a result of feedback from other galaxies. This influence may propagate by gas transport to limited distances or by radiation or fast particles to larger distances. The result might be either destructive, suppressing galaxy formation locally ('under-clustering') or far away ('over-clustering'), or constructive, enhancing galaxy formation in the neighborhood of other galaxies.

II.1. A Uniform Component

One possibility is that the universe is dynamically dominated by very 'hot' weakly interacting particles which do not cluster because they have velocities larger than $10^3 \; km \; s^{-1}$. The universe could in principle contain 'hot' enough non-baryonic DM while the clumped DM is all baryonic, but the problem is that the mass of the 'hot' particles would have always been dynamically dominant over the baryons, and would have inhibited gravitational clustering altogether. This would also yield an unacceptably fast expansion timescale during nucleosynthesis. A way around this difficulty involves supposing that these particles arise from non-radiative *decay* of heavy particles with lifetimes only slightly shorter than the age of the universe (Turner, Steigman and Krauss 1984). Assuming that these decay products are substantially lighter than their unstable parents, they would be very 'hot'. Galaxies and clusters formed during the era of matter domination by the unstable particles, but they then expanded or even became unbound as a result of the mass-loss during the decay. An elaboration of this idea suggests that the universe finally becomes dominated by a stable primordial CDM species (Olive, Seckel and Vishniac 1985) which helps explain the survival of structure on both small and large scales, but it requires certain *ad-hoc* fine-tuning among various DM components. Observational constraints on such models arise from the isotropy of the microwave background (Silk and Vittorio 1986) and from the effects of the decay on the dynamics of structures (e.g. the LSC, Efstathiou 1985). For example, Flores, Blumenthal, Dekel and Primack (1986) have found that galactic rotation curves would not have remained flat if the universe were dominated by relativistic decay products.

The universe could be flat ($k = 0$) with $\Omega < 1$ if a *cosmological constant* contributed to the curvature such that $\Omega + \Lambda/(3H_0)^2 = 1$. In some respects this idea resembles the alternative discussed above, but contrariwise, the Λ-term is unimportant at early epochs and so it would not have had such a serious inhibiting effect on galaxy formation. Actually, the large-scale structure in a flat CDM scenario with $\Lambda \neq 0$ is quite successful in reproducing the two and three point galaxy correlation functions and their peculiar velocities (Davis *et al.* 1985) and it is also compatible with the isotropy of the microwave background (Vittorio and Silk 1986). However, for the Λ contribution today to be comparable to that of the ordinary matter, the required fine-tuning is as *ad-hoc* as the one we intended to avoid adopting $\Omega = 1$.

II.2. Bias In Hierarchical Clustering (e.g. Cold Dark Matter)

An enhanced clustering of galaxies over the background matter arises if the galaxies formed only from *high peaks* of the density distribution smoothed on galactic scales; peaks with an overdensity δ above a threshold $\nu\sigma$, where $\sigma^2 \equiv < \delta^2 >$. If the local distribution function of δ has a steeply decreasing tail, like a *Gaussian*, and the power spectrum is not a white-noise, high peaks occur with enhanced probability in the crests rather than the troughs of a large-scale small-amplitude fluctuation mode, so they display enhanced clustering (Kaiser 1984; Jensen and Szalay 1986). The crucial question is what astrophysical mechanism prevents lower-amplitude peaks from also turning into galaxies, thereby neutralising the effect. This should be a fairly sharp cutoff in the efficiency of galaxy formation at $\nu \sim 2.5$, the number density of such peaks in the case of CDM being comparable to that of bright galaxies (Bardeen 1985).

The high-ν peaks would collapse earlier, and have higher density at turnaround, than more typical fluctuations on a given mass scale. This could, in principle, in itself account for the biasing if star formation were highly sensitive to (for instance) Compton cooling on the microwave background (Rees 1985) – an effect that depends on time like $t^{-8/3}$.

The bias may result from intrinsic processes detemined by the local background density. For example, Dekel and Silk (1986) have argued that the 'normal' galaxies *must* originate from high density peaks ($2\sigma - 3\sigma$) in the initial fluctuation field, while typical ($\sim 1\sigma$) peaks either cannot make a luminous galaxy at all because the gas is too hot and too dilute to cool in time, or, if only their virial velocity is less than $\sim 100\ km\ s^{-1}$, they make diffuse dwarf galaxies by losing a substantial fraction of their mass in supernova-driven winds out of the first burst of star formation. This would lead to a *selective bias*, in which the *bright* galaxies are biased towards the clusters and superclusters, while the *dwarf* galaxies do trace the mass, and should provide an observational clue for the real distribution of the DM. The evidence for such a segregation between the high and low surface-brightness galaxies is still inconclusive (Giovanelli, Haynes and Chincarini 1986) and in particular there are searches going on for dwarfs in nearby 'voids' of bright galaxies (Dekel, Eder, Oemler and Schombert, in preparation).

There are several ways whereby the first galaxies ($> \nu$) could have influenced their environment so as to modify the formation of later galaxies ($< \nu$). Various physical processes have been suggested (e.g. Bardeen 1985; Rees 1985; Silk 1985; Peebles 1986), but most of them would not seem to do the job very convincingly. In order to unbind a protogalaxy one has to heat the intergalactic gas to $> 100\ eV$, the potential well of a typical galaxy. Unfortunately, photoionization by available sources is capable of heating the gas only to a few eV, the binding energy of hydrogen. Furtheremore, in order to be relevant, any influence must propagate sufficiently fast over large distances, from proto-clusters to proto-voids, and maintain a continuous suppression of galaxy formation for a long time. UV radiation is capable of affecting the IMF in second generation galaxies by dissociating H_2, the cooling agents assumed to be responsible for efficient fragmentation in first generation galaxies (Silk 1985). The resulting massive stars may be highly disruptive via supernovae, leaving behind only 'failed' galaxies from the second generation. Alternatively, 'cosmic ray' particles may raise the Jeans mass by heating the gas (if $< 0.1c$) or raising its pressure (if relativistic), provided that they can diffuse appropriately (Rees 1985). A constant pressure gradient may produce a constant drift of the baryons, which, if larger than the escape velocity from the DM potential wells, would be sufficient to prevent further galaxy formation. It would be hard to expect, though, that any mechanical heat source, such as explosive winds, would be capable of suppressing galaxy formation at a distance.

II.3. Bias in Pancake Scenarios (e.g. Neutrinos)

A bias is generated automatically in any 'top-down' scenario where the perturbations below $\sim 30\ h^{-1} Mpc$ have damped out (neutrinos or adiabatic baryonic fluctuations). First, there are large-scale motions: collapse into flat pancakes accompanied by streaming toward their lines of intersections ('filaments'), and toward the knots where rich clusters form. The gas then contracts dissipatively into the high density regions, within which the conditions become ripe for cooling and galaxy formation; galaxies are thus limited to very specific regions. However, if the efficiency of galaxy formation is similar in all the collapsed regions, this natural bias just makes the timing-scaling difficulty described above more severe. One has to invoke a mechanism which would rather suppress galaxy formation in the high density regions and/or enhance their formation in the low density regions. If, for example, galaxies form exclusively in the flat pancakes but not in the denser filaments and clusters, the constraints are somewhat loosened (Braun, Dekel and Shapiro, in progress), but a physical mechanism that would produce such an anti-bias is still lacking and although it is not impossible *a priori*, it is not a natural outcome of the dissipative pancake scenario.

What may help is the alternative non-dissipative pancake scenario (Dekel 1983) arising from a *hybrid scenario* (Dekel 1981; Wasserman 1981; Dekel and Aarseth 1984; Dekel 1984; Valdarnini and Bonometto 1985; Umemura and Ikeuchi 1985). If galaxies form independently of pancakes, from density perturbations of a different origin ('isothermal'?) or in a separate ('cold'?) DM component, the timing constraint becomes irrelevant: galaxies could have formed at $z \geq 3$ and large-scale pancakes at $z \sim 1$. Galaxies would not be limited to pancakes but rather be present everywhere, subject to the biasing mechanisms that are relevant in 'bottom-up' scenarios. The hybrid scenarios are successful where the single-dark-matter models fail, e.g., in reproducing the observed structure both on galactic scales and on supergalactic scales, and in smearing the anisotropies in the microwave background. The large-scale structure and large velocities reported in this meeting indeed confirm the inavitable need for such a hybrid scenario.

II.4. Bias in the Explosion Scenario

Biasing also results if galaxies *enhace* the formation of other galaxies close to them. In the 'explosion' scenario (Ostriker and Cowie 1981; Ikeuchi 1981) nuclear energy from a first generation of objects helps gravity in forming further galaxies nearby via spherical blast waves that push the gas out of their interiors, expand, cool and fragment into a new generation of galaxies. The last generation forms at $z \sim 5 - 10$, after which the shells cannot cool efficiently. The galaxies hence form in spherical shells while cool gas still fills the inter-shell volume, and the non-dissipative DM is still distributed everywhere. This introduces an automatic bias in the galaxy positions similar to that in pancake scenarios. The life time of this bias is limited: in addition to the accretion of DM onto galaxies, the evacuation of the baryons from the shell interiors creates under-densed regions which eventually tend to develop gravitationally into 'voids' empty of DM. N-body simulations (Saarinen, Dekel and Carr 1985; Weinberg, Dekel and Ostriker, in progress) show that a bias enough to yield $\Omega = 1$ persists until a universal expansion by ~ 8 after the formation of the last generation of galaxies, which is compatible with the favoured epoch for galaxy formation in this scenario, $z \sim 7$.

II.5. Non-Random Phases

It seems that the structure on *very large scales* requires a formation (and bias) mechanism over and above the type of biasing process that could plausibly account for the galaxy correlations on scales $< 10\ h^{-1}Mpc$. One crucial observation in addition to the presence of big 'voids' is the enhanced superclustering of rich Abell clusters over the galaxies (and the matter), characterized by a correlation function which is nonlinear over a few tens of Mpc and stays positive at least out to $100\ h^{-1}Mpc$ (Bahcall and Soneira 1983; Klypin and Kopylov 1983). This effect is hard to reproduce in any of the standard scenarios that assume initial fluctuations with a scale-invariant spectrum and random-phases (Barnes, Dekel, Efstathiou and Frenk 1985). In particular, the CDM scenario predicts that the matter correlation function should go negative beyond $\simeq 20\ h^{-1}Mpc$, which, if the clusters are biased as peaks in a Gaussian noise, implies that the cluster correlation function should also become negative there, in conflict with the observations. The large-scale velocities reported by Bahcall in this meeting seem to point at the same problem from another angle.

If the density fluctuations originated in the inflation era, they are indeed expected to be Gaussian (Bardeen et al. 1986), but it is also possible that the fluctuations arose from a different mechanism, in which they would not in general be Gaussian, and would not have random phases. A specific model that incorporates this feature is the scenario in which the density fluctuations were induced by *cosmic strings* (Vilenkin 1984). The strings are curvature singularities with a random-walk topology generated in a phase-transition in the early universe, which turn into closed smoother 'parent' loops on entering the horizon and then chop themselves into stable 'daughter' loops, all in a scale-free self-similar fasion (Albrecht and Turok 1985). The resultant loop-loop correlation function has the general dependence (Turok 1986) $\xi(r) \propto r^{-2}$, somewhat resembling that of galaxies and of clusters, as expected from 'beads' along locally-linear 'strings'. The string-induced fluctuations in the DM (by accretion onto the loops), which would have a scale-invariant Harrison-Zeldovich spectrum, are the seeds for the formation of galaxies and clusters, which are therefore expected to be aligned in space along the same 'parent' linear structures. The effect of such *phase-correlations* on the correlation functions of the galaxies and the clusters relative to the matter was found to be very pronounced indeed in a string model that incorporated DM gravity by Blumenthal, Dekel and Primack (this volume). It provides a natural galaxy biasing mechanism, as well as an explanation for the excess of cluster clustering on very large scales.

III. CONCLUSIONS

The key observations, which may distinguish between the possible bias mechanisms, focus on the 'voids'. (i) How big and empty are the 'voids'? We need to quantify the data using a meaningful statistic. (ii) Are voids empty of galaxies of all types, or only those that are most conspicuous? Any evidence that galaxies of different types, and real *dwarfs* in particular, display unequal degrees of large-scale clustering would be relevant. (iii) How much gas is there in the voids? Absorption systems along the lines of sight to quasars passing through 'voids' may be detectable or, if the gas is at $10^5\ ^\circ K$, some features characteristic of neutral He may reveal its presence. Also of relevance is the relationship between 'parent' DM halos and 'daughter' galaxies, asking whether there are any 'barren' galactic-mass dark halos with no luminous galaxy within them (gravitational lenses?), or any 'orphan' galaxies which lack dark halos.

I have tried to argue that the idea of biasing is not just an *ad-hoc* idea introduced by theorists to save the philosophically attractive $\Omega=1$ model when confronted with apparently conflicting evidence. A bias is essential in order to understand even the gross features of the large-scale structure, and to reconcile any of the cosmogonic scenarios with the observed universe. Therefore, the search for an appropriate bias mechanism, and for confirming observational evidence, are of great importance. I believe that what might have looked at first as a frustrating idea for astronomers, who can only observe the luminous 'tips' of the 'icebergs', should become an exciting area of observational search for the relevant evidence, which would require, though, inteligent interpretation.

On the theoretical side, finding the 'correct' bias mechanism is intimately related to finding the 'correct' cosmogonic scenario, or the 'correct' nature of the DM. Although some of the proposed bias mechanisms seem somewhat contrived, there are a few which are very plausible, and they desrve to be worked out in more detail. My current personal bias would be towards the possibility that the baryons still trace the mass and the 'bias' arises from high peaks in a Gaussian fluctuation field rather than from a feedback mechanism. My favorite in this category is the selective bias of dwarfs versus bright galaxies in the CDM scenario. Nevertheless, other possibilities, such as the bias arising in the explosion scenario, are also quite plausible. In particular, I would take seriously the evidence for non-Gaussian fluctuations on very large scales, and look into the bias arising in the cosmic-string theory; this seems to be the most promising explanation for the very-large-scale structure, although the origin of this scenario may be regarded as somewhat speculative.

I gratefully thank Martin Rees for his contribution to this paper.

REFERENCES

Albrect, A. and Turok, N. 1985, Phys. Rev. Lett. **54**, 1868.
Bahcall, N. and Soneira, R. 1983, Ap. J. **270**, 20.
Bardeen, J., Bond, J.R., Kaiser, N. and Szalay, A. 1986, Ap. J., in press.
Bardeen, J. 1985, in Inner Space/Outer Space, ed. Kolb, E.W. and Turner, M.S. (University of Chicago Press).
Barnes, J., Dekel, A., Efstathiou, G. and Frenk, C. 1985, Ap. J., **295**, 368.
Centrella, J. and Melott, A. 1983, Nature **305**, 196.
Davis, M., Efstathiou, G., Frenk, C.S. and White, S.D.M. 1985, Ap. J. **292**, 371.
de Lapparnet, V., Geller, M.J. and Huchra, J.P. 1986, Ap. J. (Lett.), in press.
Dekel, A. 1981, Astron. Ap. **101**, 79.
Dekel, A. 1983, Ap. J. **264**, 373.
Dekel, A. 1984, in the Eighth Johns Hopkins Workshop on Particle Theory, eds. Domokos, G. and Koveski-Domokos, S. (World Publishing Co.: Singapur), p. 191.
Dekel, A. 1986, Comments on Astrophysics, in press.
Dekel, A. and Aarseth, S.J. 1984, Ap. J. **283**, 1.
Dekel, A., Einasto, J. and Rees, J. 1986, Rev. Mod. Phys., in preparation.
Dekel, A. and Silk, J. 1986, Ap. J., **303**, April 1.
Dressler, A. 1980, Ap. J., **236**, 351.
Efstathiou, G. 1985, M.N.R.A.S. **213**, 29.
Einasto, J., Joeveer, M. and Saar, E. 1986, in Dark Matter in the Universe, eds. J. Kormendy and G. Knapp (Dordrecht: Reidel), in press.
Flores, R., Blumenthal, G.R., Dekel, A. and Primack, J.R. 1986, Ap. J., submitted.
Giovanelli, R., Haynes, M.P. and Chincarini, G. 1986, preprint.
Hoffman, G.L., Salpeter, E.E. and Wasserman, I. 1982, Ap. J. **263**, 485.
Ikeuchi, S. 1981, Pub. Astron. Soc. Japan **33**, 211.
Jensen, L.G. and Szalay, A.S. 1986, preprint.
Kaiser, N. 1984, Ap. J. (Lett.) **284**, L9.
Kirshner, R.F., Oemler, A. Schechter, P.L. and Shectman, S.A. 1981, Ap. J. **248**, L57.
Klypin, A.A. and Kopylov, A.A. 1983, Soviet Astr. Lett. **9**, 41.
Olive, K., Seckel, D. and Vishniac, E. 1985, Astrophys J., in press.
Ostriker, J.P. and Cowie, L. 1981, Ap. J.(Lett.) **243**, L127.
Peebles, P.J.E. 1986, Nature, in press.
Rees, M.J., M.N.R.A.S. 1985, **213**, 75P.
Saarinen, S., Dekel, A. and Carr, B. 1986, in IAU Symp. no. 117 on Dark Matter in the Universe, eds. Kormendy, J. and Knapp, G. (Dordrecht: Reidel).), in press.
Silk, J. and Vittorio, N. 1985, preprint.
Silk, J. 1985, Ap. J. **297**, 1.
Turner, M.S., Steigman, G. and Krauss, L.M. 1984, Phys. Rev. Lett. **52**, 2090.
Turok, N. 1985, preprint.
Umemura, M. and Ikeuchi, S. 1985, preprint (Tokyo Observatory).
Valdarnini, R. and Bonometto, S. 1985, Astron. Ap. **146**, 235.
Vilenkin, A. 1984, Phys. Reports **121**, 264.
Vittorio, N. and Silk, J. 1985, preprint (Berkely).
Wasserman, I. 1981, Ap. J. **248**, 1.
White, S.D.M., Frenk, C.S. and Davis, M. 1983, Ap. J. Lett. **274**, L1.

PECULIAR VELOCITIES AND GALAXY FORMATION

Nicola Vittorio
Department of Astronomy, University of California,
Berkeley, CA 94720

I.INTRODUCTION

The cosmic microwave background (CMB) dipole anisotropy (DA) (Lubin and Villela, this volume) implies a Local Group (LG) velocity $v_D = 640 \pm 50 Km\,s^{-1}$ relative to the comoving reference frame, in a direction which is 45° away from the Virgo cluster. The CMBDA seems to be consistent both in amplitude and in direction with the DA of the Hubble flow ($780 \pm 188 km\,s^{-1}$) measured by Aaronson et al. (1985) in a set of 10 clusters at a distances $40 \to 100\,h^{-1}Mpc$. Also, the IRAS point source catalogue exhibit a DA in the galaxy distribution in reasonable agreement in direction with the CMBDA (Yahil et al., 1985; Meiksin and Davis, 1985).

It is very important to determine if the CMBDA is primarely due to local dynamics or is mainly the coherent effect of large scale density inhomogeneties. An ingenious way for clarifying the locality of the DA consists in selecting galaxies in shell of radius r big enough not to be strongly affected by local nonlinearity. If these galaxies are unperturbed tracers of the Hubble flow on large scales the peculiar velocity of the LG relative to this shell should be exactly equal to the velocity of the LG relative to the CMB. If the two velocities are different, this implies a coherent motion relative to the CMB of all the matter inside the shell.

de Vacouleurs and Peters(1984) have measured the velocity of the LG relative to shells of galaxies of different radius. They concluded that a shell of $25\,h^{-1}Mpc$ has a peculiar velocity of $350\,km\,s^{-1}$ relative to CMB and comment on the fact that on scales $> 40\,h^{-1}Mpc$ matter and radiation are at rest.

On the other hand, Staveley-Smith and Davies (1985) have concluded that the body of the DA is determined on scale >40 Mpc. More recently Collins et al. (this volume) report a peculiar velocity of $v = 970 \pm 300 km\,s^{-1}$ relitive to CMB for a shell of radius $\sim 50h^{-1}Mpc$.

We want to discuss the theoretical implications of peculiar velocities of shells of galaxies relatively to the CMB. We want to report preliminary results of a larger project (Turner and Vittorio, 1986) for predicting peculiar velocities in different galaxy formation scenario, in particular low density cold dark matter (CDM) dominated universe, open or flat because of a non zero cosmological constant $\Lambda = 1 - \Omega_0$ (Λ is expressed in units of $3 H_0^2/c^2$).

II. METHOD

It is generally assumed that galaxies and clusters of galaxies form through the gravitational amplification of initially small density fluctuations. Such fluctuations naturally generate peculiar velocities (Clutton-Brock and Peebles, 1981). Since in the linear regime the velocity and the density fields are proportional, observations of large scale peculiar velocities can be used to constrain the density fluctuation amplitude on large scales.

In the linear regime, the amplitude of the Fourier component of the density contrast is $\delta(k,t) = \delta(k,t_{in}) T(k) D(t)$. $D(t)$ describes the time evolution of the perturbation after the matter-radiation decoupling, while $T(k)$ describes any modification of the initial density fluctuation spectrum occurred before and during recombination. We will consider at the initial time t_{in} a Harrison-Zeldovich density fluctuation spectrum: $|\delta(k,t_{in})|^2 = A k$. For a CDM dominated universe $T(k) = (1 + \alpha k + \beta k^{1.5} + \gamma k^2)^{-1}$: $\alpha = 1.7 h^{-2} Mpc$, $\beta = 9 h^{-3} Mpc^{1.5}$, $\gamma = h^{-4} Mpc^2$, and h is the Hubble constant in units of $100\, km\, s^{-1} Mpc$ (Davis et al., 1985). The introduction of a cosmological constant Λ changes only the perturbation growth rate.

We normalize the spectrum requiring that the r.m.s. density fluctuation averaged inside a randomly placed sphere of radius $r = 8 h^{-1} Mpc$ is unity. This is equivalent to fit the variance in the counts of bright galaxies (Davis and Peebles, 1983) and to assume that galaxies are a good tracer of the overall mass distribution.

The peculiar velocity induced by a density fluctuation can be calculated in the linear theory, e.g., from the continuity equation. After expanding in the Fourier serie, it is found (see,e.g., Peebles,1980): $v_k = \frac{a}{k}\delta(k,t) H(t) \frac{d\log D(t)}{d\log a}$, valid for an arbitrary cosmological background. Here a is the scale factor normalized to one at the present time and $H(t) = \dot{a}/a$. In an Einstein-de Sitter universe $D(t) \propto t^{\frac{2}{3}} \propto a$ and the previous equation simplifies in $v_k = H_0 \frac{\delta(k,t_0)}{k}$.

The residual velocity of a gaussian sphere of radius r relative to the CMB is evaluated (Clutton-Brock and Peebles, 1981 ; Kaiser, 1983 ; Vittorio and Silk, 1985a) by convolving v_k with a spherical symmetric gaussian window function:

$$v_{rms}^2(r) = \frac{1}{2\pi^2} \int_0^\infty k^2 dk |v_k|^2 exp(-k^2 r^2) \qquad (1)$$

It is clear that only perturbations of wavenumber $k < r^{-1}$ or wavelength $\lambda > r$ contribute to this integral.

Since the density field is assumed to be a random gaussian field, the statistics of the velocity field is completely determined: the modulus of the velocity has a χ^2-like distribution with three degrees of freedom. Eq.(1) predicts the r.m.s. peculiar velocity as observed by a random placed observer. For $r \to \infty$, $v_{rms}^2(r) = \frac{A^{\frac{1}{2}}}{2\pi} \frac{H_0}{r} \frac{d\log D(t)}{d\log a}$.

The probability of measuring a peculiar velocity in the interval $v_1 \to v_2$ is given by $P = \sqrt{\frac{2}{\pi}} \int_{v_1}^{v_2} \left(\frac{v}{v_{rms}}\right)^2 e^{-\frac{1}{2}(\frac{v}{v_{rms}})^2} \frac{dv}{v_{rms}}$: then, there is a probability of 90% of measuring a velocity $\frac{1}{1.6} < \frac{v}{v_{rms}} < 3$.

The contribution to the CMBDA from the linear portion of the spectrum (hereafter v_D) is calculated from eq.(1) setting $r = 10h^{-1}Mpc$: in this way the small non linear scales are filtered out. We also will refer to v_{25} and v_{50} as the rms velocity, relative to the CMB, of a sphere of radius 25 and $50\,h^{-1}Mpc$, respectively.

III. RESULTS

If the universe is CDM dominated the expected contribution to the DA from the large scale density fluctuations is:

$$v_D = 322 kms^{-1} \quad \begin{array}{ll} \Omega_0^{+0.03} h^{-0.57}; & \Lambda = 0 \\ \\ \Omega_0^{-0.01} h^{-0.57}; & \Lambda = 1 - \Omega_0 \end{array} \qquad (2)$$

The expected peculiar velocities of a sphere of radius 25 and 50 $h^{-1}Mpc$ are:

$$v_{25} = 156\,kms^{-1} \quad \begin{array}{ll} \Omega_0^{-0.18} h^{-0.78}; & \Lambda = 0 \\ \\ \Omega_0^{-0.22} h^{-0.78}; & \Lambda = 1 - \Omega_0 \end{array} \qquad (3)$$

$$v_{50} = 83\,kms^{-1} \quad \begin{array}{ll} \Omega_0^{-0.33} h^{-0.92}; & \Lambda = 0 \\ \\ \Omega_0^{-0.38} h^{-0.92}; & \Lambda = 1 - \Omega_0 \end{array} \qquad (4)$$

These are fits to the numerical results obtained by integrating eq.(1) and are accurate to $< 10\%$. The prediction for the CMBDA is quite insensitive to Ω_0, while the predictions both for v_{25} and v_{50} depend inversely on Ω_0, despite the well known dependence $\frac{d\log D(t)}{d\log a} = \Omega_0^{0.6}$. This is due to the fact that T(k) has its own dependence on Ω_0 such that the amplitude of the density fluctuation on large scale is, for a fixed normalization, higher for low values of Ω_0.

The introduction of a cosmological constant $\Lambda = 1 - \Omega_0$ does not change substantially the peculiar velocity prediction. Infact, although the growth of fluctuations in the $\Lambda = 1 - \Omega_0$ case is more efficient than in the $\Lambda = 0$ case, the factor $\frac{d\log D(t)}{d\log a}$ looks very similar ($= \Omega_0^{0.57}$). However, peculiar velocities on large scales tend to be higher in the Λ case.

One way of reconciling the theoretical prejudice for an $\Omega_0 = 1$ universe with the observed low values of Ω_0 is to require that galaxies are a biased tracer of the overall mass distribution (see, e.g., Bardeen et al., 1986). In this scenario galaxies form only in the highest peaks ($\nu \equiv \frac{\delta}{\delta_{rms}} \gg 1$) of the density field: this would imply that the overall mass distribution is more uniform and that the amplitude of the density fluctuations is, for a fixed normalization criterium and at the 1σ level, $\propto \nu^{-1}$.

It has been argued (Vittorio and Silk, 1985a) that in an $\Omega_0 = 1$ CDM dominated universe the biased galaxy formation scenario may predict too low peculiar velocities. Infact, at the 90% confidence level one has $200 < \frac{v_D h^{0.57} \nu}{km\ s^{-1}} < 970$ and $80 < \frac{v_{25} h^{0.78} \nu}{km\ s^{-1}} < 470$ and $50 < \frac{v_{50} h^{0.92} \nu}{km\ s^{-1}} < 250$. Definetly for $\nu \sim 3$ the predictions are too low for taking into account peculiar motions of $970 \pm 300\ km\ s^{-1}$ on scale of $50 h^{-1} Mpc$.

The theoretical predictions for a flat massive neutrino dominated universe are $v_D = 400 km\ s^{-1}$ and $v_{50} = 100 h^{-2} kms^{-1}$. It seems then unlikely to fit the high peculiar velocities on large scales (Collins et al.) without having an excessive contribution to the DA from the linear region of the spectrum.

The present upper limit on the CMB small scale anisotropy (Uson and Wilkinson, 1984) implies $\Omega_0 h > 0.2$ if $\Lambda = 0$ (Vittorio and Silk, 1984; Bond and Efstathiou, 1984) and $\Omega_0 h > 0.05$ if $\Lambda = 1 - \Omega_0$ (Vittorio and Silk, 1985b).

Since we need low values of Ω_0 and h in order to have higher peculiar velocities, it seems possible to reconcile the present upper limit on the CMB small scale anisotropy with large values of the peculiar velocity field only if the universe is low density, CDM dominated universe, with $\Lambda = 1 - \Omega_0$.

References

Aaronson,M., Bothun,G., Mould, J., Huchra, J., Schomer, R.A., Conell, M.E., 1986, *Ap.J.*, in press.
Bardeen, J., Bond, J. R., Kaiser, N. ,and Szalay, A. 1986, *Ap.J.Lett.*, in press.
Clutton–Brock, and Peebles, P. J. E. 1981, *Astron.J.*, **86**, 115.
Davis, M., Efstathiou, G., Frenk, C., and White, S. D. M. 1985, *Ap. J.*, **292**, 371.
Davis, M., and Peebles, P.J.E. 1983, *Ap.J.*, **267**, 465.
De Vaucouleurs, G. H. and Peters, W.L. 1984, *Ap. J.*, **287**, 1.
Kaiser, N. 1983, *Ap. J.*, **273**, L17.
Meiksin, A., and Davis, M., 1986, *Ap.J.*, in press.
Peebles, P. J. E. 1980, *The Large Scale Structure of the Universe*, Princeton press.
Staveley–Smith, A., and Davies, R., 1985, *preprint*, .
Turner, M., and Vittorio, N. 1986, *in preparation*, .
Uson, J. M., and Wilkinson, D. T. 1984, *Nature*, **312**, 1119.
Vittorio, N., and Silk, J. 1984, *Ap. J.*, **285**, L39.
Vittorio, N., and Silk, J. 1985a, *Ap. J.*, **293**, L1.
Vittorio, N., and Silk, J. 1985b, *Ap. J.*, **297**, L1.
Yahil,A., Walker,D., Rowan-Robinson, M. 1985, *Ap.J.*, in press.

LARGE SCALE STRUCTURE IN UNIVERSES DOMINATED BY COLD DARK MATTER

J. Richard Bond
Canadian Institute for Theoretical Astrophysics
Toronto, Canada

ABSTRACT. The theory of the peaks of Gaussian random density fields is applied to a numerological study of the large scale structure from adiabatic fluctuations expected in models of biased galaxy formation in $\Omega = 1$, $h = 0.5$ universes dominated by cold dark matter (CDM). The number density of rich clusters is sensitive to model assumptions but can plausibly bracket the observed number. The amplitude of the rich cluster correlation function is too small for $r > 20\ h^{-1}Mpc$ and generally has the wrong shape. The angular anisotropy of the cross-correlation function $\xi_{cl,g}$ demonstrates that the far field regions of cluster-scale peaks are asymmetric as recent observations indicate. These regions will generate pancakes or filaments upon collapse. One-dimensional singularities in the large-scale bulk flow should arise in these CDM models, appearing as pancakes in position space. They are too rare to explain the CfA bubble walls, but pancakes that are just turning around now *are* sufficiently abundant and would appear to be thin walls normal to the line of sight in redshift space. Large-scale streaming velocities on $\sim 30\ h^{-1}Mpc$ scales are significantly smaller than recent observations indicate. To explain the reported $\sim 700\ km\ s^{-1}$ coherent motions, mass must be significantly *more* clustered than galaxies with a biasing factor $b < 0.4$ and a nonlinear redshift z_{nl} at cluster scales > 1 for both massive neutrino and cold models. With biasing and CDM, $b \sim 2$ and clusters are rare events.

1. Adiabatic CDM Universes

In inflationary cosmologies, the global density parameter Ω is almost exactly one and the density fluctuations are scale-invariant and Gaussian. There are two possible modes for the fluctuations: adiabatic and isocurvature. The case for an isocurvature mode is not as strong as for the adiabatic one and will not be considered in this paper. It can be ruled out very convincingly on the basis of producing large-angle microwave background anisotropies in excess of those observed (Efstathiou and Bond 1986) and having a redshift of galaxy formation too close to the present (Bardeen, Bond, Kaiser and Szalay 1986, hereafter BBKS).

Since primordial nucleosynthesis abundances seem to constrain the baryon density parameter $\Omega_B < 0.1$, to satisfy $\Omega = 1$ the dark matter must be non-baryonic, a massive relic of the early universe. The relic could be cold, warm or hot (Bond and Szalay 1983). A substantial fraction of the dark matter could also be vacuum energy associated with a nonzero cosmological constant, with, for example, $\Omega_{vac} \equiv \Lambda/(3H_0^2) \sim 0.8$, $\Omega_X \sim 0.2$, where Ω_X is the density parameter of the CDM. Though without much motivation from particle physics, such a model has a number of attractive features as Peebles has emphasized; however, the redshift of galaxy formation may be too low (§3.5). In this paper, we use as our standard example the CDM model with $\Omega_X \approx 1$ and $h = 0.5$, where $h \equiv H_0/(100\ km\ Mpc^{-1})$.

The shape of the density and velocity linear fluctuation spectra for adiabatic perturbations with cold dark matter depends only upon the combination $\Omega_X h^2$, provided $\Omega_B \ll \Omega_X$. The power spectrum for the density, $|\delta_X(k)|^2$, flattens from the natural initial Zeldovich-Harrison $n = 1$ slope, which remains on very large wavelengths, to $n \approx -1$ on cluster scales, to $n \approx -2$ on galactic scales, and to $n \approx -3$ on smaller scales. N-body research of the late seventies on the power law required to reproduce the $r^{-1.8}$ form of galaxy correlation function if light traces mass gave the phenomenological value $n \sim -2$. The natural physical way this arises in the adiabatic CDM model should not be forgotten in assessing the merits of this theory as it confronts the large scale structure problem. Nonetheless, for $h = 0.5$ the power spectrum falls off rapidly with increasing wavelegth beyond cluster scales, making still larger scale structure quite rare. Whether the amount and clustering properties are sufficient will depend on how reliable the observables turn out to be, as there are significant discrepencies between the theory and the data.

There remains one overall *normalization amplitude*, which we parameterize by b, the *biasing factor*, defined so that mass traces light for $b = 1$. The technical definition of b relates a statistical average of mass density fluctuations to a volume average of fluctuations in the density of bright galaxies

$$\langle \frac{\Delta M}{M}(< r = 10\ h^{-1}Mpc)\frac{\delta\rho}{\rho}(0)\rangle \equiv b^{\mp 2} \langle \frac{\Delta N_g}{N_g}(< r)\frac{\delta n_g}{n_g}(0)\rangle \equiv \frac{3J_{3g}(r)}{b^2 r^3} \approx \frac{0.81}{b^2}. \quad (1.1)$$

The numerical value for J_{3g} is the result from the CfA redshift survey. If $b > 1$, mass is *less* clustered than light, the preferred possibility in the biased galaxy formation theories considered here, and if $b < 1$ mass is *more* clustered than light. $b < 1$ might arise naturally in neutrino-dominated models (Szalay, Bond and Silk 1983).

2. Biasing Galaxy and Cluster Formation

To discuss the formation and clustering of objects of mass M, we smooth the mass density perturbation field $\delta(\vec{r}) = \delta\rho(\vec{r})/\rho$ on mass scale M; that is, we filter the density fluctuation spectrum on (comoving) length scale R_f. The relation between the mass scale and (Gaussian) filtering length is

$$M = 1.1 \times 10^{12} \, (R_f/Mpc)^3 \, h_{50}^2 \, M_\odot. \qquad (2.1)$$

Thus, a galactic scale smoothing would be $R_g \sim 1 \, Mpc$, $M \sim 10^{12} \, M_\odot$, and a rich cluster scale smoothing would be $R_{cl} \sim 10 \, Mpc$, $M \sim 10^{15} \, M_\odot$.

We introduce the *rms* level of the fluctuations in the density and velocity at time t, $\sigma_0(R_f,t)$ and $\sigma_v(R_f,t)$, respectively. The linear growth law is used to extrapolate these *rms* levels into the nonlinear regime; for the $\Omega_X = 1$ universes considered here, the redshift at which the *rms* fluctuations of scale R_f reach nonlinearity is $1 + z_{nl}(R_f) = \sigma_0(R_f, \, now)$.

We assume the cosmic structures form at the smoothed peaks of the the density field. The height of a peak is characterized by $\nu \equiv (\delta\rho/\rho)/\sigma_0$. A selection function must be chosen for peaks which are to form a given class of cosmic objects. Kaiser (1984) adopted the natural selection criterion for *rich clusters* as $\sim 10^{15} \, M_\odot$ peaks which have collapsed by the present:

$$\nu\sigma_0(R_{cl}, \, now) > f_c = 1.69. \qquad (2.2)$$

The value of the collapse factor f_c is somewhat arbitrary. $f_c = 1.69$ corresponds to a uniform spherical mass of cold particles in an $\Omega = 1$ universe having completely collapsed to a point in the center. Turn-around of the spherical perturbation occurs at 1.06, hence a value in between may be more appropriate. Further, one-dimensional collapse treated using the Zeldovich (1970) approximation has $f_c = 1$, and turnaround would correspond to the value 0.5. In the simplest version of the biased galaxy formation theory, *'bright' galaxies* - those used in the determination of the galaxy correlation function - are assumed to form only above some global threshold value $\nu_t \sim 3$. The immediate question - what are the $\nu \sim 1$ *rms* objects which are so numerous? - has yet to be satisfactorily answered.

3. BBKS Semi-analytic Treatment of Galaxy Clustering

The mathematical framework used here is the theory of the statistics of peaks of Gaussian random fields. The accretion rate into the neighbourhood of a point is maximized at density field peaks, a result which remains valid in the pre-merging weakly nonlinear regime. The relevant tools for working with such maxima are developed in BBKS. Here, we outline the main results.

3.1 **Fixing the Galaxy Scale** R_g is set by the galaxy mass. We take $R_g = 0.7 \, Mpc$, though values in the range $0.4 - 1 \, Mpc$ might be entertained.

3.2 **The Bright Galaxy Threshold Height** The *global* threshold ν_t for biased galaxy formation is then *determined* by requiring that the number density of peaks above threshold match the number of bright galaxies, the objects we wish the peaks to become: $n_{pk}(\nu_t; R_g) = n_{bright\ gals} = (9\ Mpc)^{-3}$. For $R_g = 0.7\ Mpc$, we obtain $\nu_t = 2.8$, using a selection function which falls off rapidly below ν_t (the $q = 8$ model of BBKS). An infinitely sharp threshold gives $\nu_t = 2.6$. The spacing between peaks of this mass scale of arbitrary height is $4\ Mpc$, the average height being $\sim 1 - 1.5$. Irrespective of biasing, there are a factor of 13 more peaks than bright galaxies, a mystery that any theory of galaxy formation must address.

3.3 The Galactic Peak Correlation Function The correlation function of the peaks consists of two parts. Within the initial conditions themselves the peaks are *statistically clustered*: $\xi^{stat}_{pk,pk}(r) = A\psi(r)$, where $\psi(r) \equiv \xi_{\rho,\rho}(r)/\sigma_0^2$ is the normalized density-density correlation function ($\psi(0) = 1$). The amplification factor A depends only upon ψ and ν_t. For the fuzzy threshold case, A ranges upward from an asymptotic small-ψ value $A_\infty = 3$, much smaller than the extremely high threshold result $A_\infty = \nu_t^2$. The BBKS approximations break down for large ψ. It is quite remarkable that, for the adiabatic CDM model, the combination of the rise of A with ψ together with the gradual flattening of ψ towards 1 with decreasing radius leads to $\xi^{stat}_{pk,pk}(r) = (r/r_{0,stat})^{-1.8}$ over at least the range from $1.5 - 7\ h^{-1}Mpc$, with $r_{0,stat} \approx 2\ h^{-1}Mpc$.

In addition, as the mass density field grows in amplitude, the peaks are carried along with the mass, giving an extra *dynamical* contribution to the correlation function. This term is difficult to treat generally except by N-body methods. A simple analytic model adopted by BBKS for the total peak-peak correlation function,

$$\xi^{stat+dyn}_{pk,pk}(r) \approx (A^{1/2}/\sigma_0 + 1)^2\ \xi_{\rho,\rho}(r), \tag{3.1}$$

seems to accord well with N-body studies (Davis etal 1985, DEFW) provided the Universe is not too dynamically 'old' on galactic scales (to avoid the correlation function steepening precipitously with age). The biasing factor b defined by eq.(1.1) is only slightly greater than the asymptotic value $b_\infty = A_\infty^{1/2}/\sigma_0 + 1$.

3.4 Spectrum Normalization Using the Galaxy Correlation Function We require that the galactic peak-peak correlation function (3.1) agree with the observed ξ_{gg}. In particular, we determine $\sigma_0(R_g) = 2.9$ by requiring that the J_3-normalization, eq. (1.1), be satisfied with $b = 1.7$. (Hence $b_\infty = 1.6$.) The correlation function reaches unity at $r_0 = 5.6\ h^{-1}Mpc$, to be compared with the value $5.4 \pm 0.3\ h^{-1}Mpc$ of the CfA redshift survey (Davis and Peebles 1983). (In BBKS, we required that $\xi_{pk,pk}$ be unity at $5\ h^{-1}Mpc$, which gave $\sigma_0 = 2.4$, reflecting the degree of uncertainty in normalization.)

Since the density fluctuations are lowered by a factor of b from the value they would have if mass traces light, all quantities dependent upon this normalization are also lowered. For example, angular anisotropies in the microwave background scale as $\Delta T/T \propto b^{-1}$, lowering the prediction (Bond and Efstathiou 1984, Vittorio and Silk 1984) for this $h = 0.5$ CDM model for the Uson and Wilkinson (1984)

4.5' experiment from the mass traces light value to $\Delta T/T = 2.5 \times 10^{-6}$, well below their upper limit of 3×10^{-5}.

3.5 Redshift of Galaxy Formation The redshift when peaks with $\nu = \nu_t$ collapse is given by $1 + z_g = \nu_t \sigma_0(R_g)/f_c$. Hence, $z_g = 3.7$ for $f_c = 1.69$. Through the relation $1 + z \propto \nu$, the BBKS expression for the number of peaks within ν and $\nu + d\nu$, and the selection function, the galactic-scale peak collapse rate can be determined. For the slightly fuzzy selection function used here for the bright galaxies, collapses would be spread out from $z \sim 6 - 3$. Such relatively low redshifts for galaxy formation could prove to be an embarassment for the theory if a great burst of energy accompanies collapse. Since smaller scale clouds within the protogalaxy would collapse first in this hierarchical model, the period of maximum energy release could be earlier than z_g. The rate of peak collapse (without a selection function being folded in) rises with decreasing z until its maximum around $z \sim 0.5$. Again the fate of these peaks enters as an unresolved issue.

The $\Omega_{vac} = 0.8$ CDM model with biasing already has substantial statistical clustering for the peaks ($A_\infty = 3.8$ for a sharp threshold $\nu_t = 2.4$), with little room left for dynamical evolution ($\xi^{stat}_{pk,pk}(5\ h^{-1}Mpc) \approx 0.7$). The result is an unacceptably small redshift of galaxy formation ($z_g = 0.2$ with $b_\infty = 3.6$). This is true even if light traces mass, for $\sigma_0(R_g) = 2.7/b$, hence, with the appropriate fluctuation growth law for vacuum-dominated universes, $z_{nl,g} \approx 3$, with complete collapse of the *rms* structures not occurring until $z \approx 1.1$.

4. Large Scale Structure

4.1 Rich Cluster Abundances Once $\sigma_0(R_g)$ is determined, σ_0 is determined on all other scales; in particular, $\sigma_0(R_{cl}) = 0.42$ for our standard choice $R_{cl} = 10\ Mpc$. Through eq.(2.2) this determines the threshold above which cluster-scale peaks would have collapsed by now, $\nu_t(f_c)$, and their average spacing, $d(f_c) \equiv n_{pk}^{-1/3}(> \nu_t)$. As shown in Table 1, d is very sensitive to the choice of collapse factor. These numbers should be compared with the observed average spacing between clusters of richness class > 1, $d_{cl} = 110\ Mpc$. $f_c = 1.2$ ($\nu_t = 2.8$) is required to reproduce this spacing. White has recently suggested that the choice $R_{cl} = 7\ Mpc$ may more nearly correspond to Abell's cluster selection criterion. In this case, there are clearly no abundance problems for clusters even with $f_c = 1.69$.

The spacing between richness class > 2 is $160\ Mpc$ (Bahcall and Soneira 1983); if the filtering mass just scales with the number of galaxies in an Abell radius, then the associated filtering scale is 1.2 times that for richness > 1, not a large enough difference to present a significant problem. However, for richness class 4, R_f would rise by a factor of 1.6, making the collapsed clusters *extremely* rare. However, I doubt that one should take this simple one-to-one correspondence with galaxy count in an Abell radius and R_f very seriously.

Table 1: Peak and Pancake Thresholds $\nu_t(f_c)$ and Spacings $d(f_c)$

Type	$R_f(Mpc)$	$\sigma_0(R_f)$	$\nu_t(1.69)$	$d(1.69)$	$\nu_t(1)$	$d(1)$	$\nu_t(0.5)$	$d(0.5)$
peak	7	0.62	2.7	74	1.6	42	0.8	36
peak	10	0.42	4.0	335	2.4	83	1.2	52
panc	10	0.42	-	-	5.3	706	2.7	100
peak	16	0.25	6.8	5(4)	4.0	496	2.0	101
panc	16	0.25	-	-	8.9	3(6)	4.5	385
peak	25	0.14	12	6(11)	7.1	2(5)	3.6	478

4.2 Rich Cluster Correlation Function Table 2 compares $\xi_{pk,pk}$ with the observed ξ_{cl} of Bahcall and Soneira (1983). Though the amplitude is not wildly off at $r = 17\ h^{-1}Mpc$, serious difficulties begin beyond this point in shape and amplitude. The approximation methods used to make the $\xi_{pk,pk}$ calculation tractable involved neglect of derivatives of ψ; estimation of the error by comparing $\xi_{pk,\rho}$ with and without gradients indicate this neglect will only slightly lower $\xi_{pk,pk}$ over the range of Table 2, though, for example, the position of the zero does move. For $r < 5\ h^{-1}Mpc$, the peak-peak correlation function will plunge toward zero, reflecting the fact that peaks cannot lie on top of each other. More detailed calculations are in progress (Bardeen, Bond, Jensen and Szalay 1986).

Table 2: Rich Cluster Correlation Function Estimate ($R_{cl} = 10\ Mpc$, $\nu_t = 3$)

$r\ (h^{-1}Mpc)$	$\xi_{pk,pk}$	ξ_{cl}	ψ	$\xi_{\rho,\rho}$
16	1.7	2.2	0.20	0.03
20	0.84	1.5	0.08	0.015
25	0.24	1.0	0.03	0.005
36	≈ 0	0.52	0	0

4.3 Anisotropy of the Cluster Neighbourhood For Gaussian random fields, clusters are predicted to have triaxial tails extending to $\sim 20\ h^{-1}Mpc$. Evidence for this comes from the cluster-peak mass density correlation function, $\xi_{pk,\rho}(r,\theta) = \xi_{S\rho}(r) - \xi_{A\rho}(r)(1 + 3\cos(2\theta))/6$, where θ is the angle between the principal axis of the largest eigenvalue of the moment of inertia tensor of the cluster and the mass point. The results in Table 3 indicate a strong effect (much larger than on galaxy scales), which would be even stronger if one includes nonlinear dynamics in the outer regions which enhances anisotropy and if one considers $\xi_{pk,g}$ since biasing also increases asymmetry. The simple approximation $\xi_{pk,g} = exp[b_\infty \xi_{pk,\rho}] - 1$, with $b_\infty = 1.6$, includes these effects to first order. The last 2 columns in Table 3 are respectively $exp[b_\infty \xi_{S\rho}] - 1$ and $\xi_{Ag} = \xi_{pk,g}(90°) - \xi_{pk,g}(0°)$. Argyres etal. (1985) have estimated $\xi_{cl,g}$ from the Shane-Wirtanen catalogue and find angular dependences and anisotropy amplitudes similar to the theoretical ones found here.

Table 3: Far Field Anisotropy of Clusters $(R_{cl} = 10\ Mpc,\ \nu = 3)$

$r\ (h^{-1}Mpc)$	$\xi_{S\rho}(r)$	$\xi_{A\rho}(r)$	$\xi_{cl,g}$	ξ_{Ag}
5	1.0	0.16	4.0	1.2
12.5	0.32	0.37	0.7	0.9
20	0.05	0.16	0.1	0.3

4.4 Rare Large Scale Pancakes and Peaks Superclusters, filaments, pancakes and voids must all appear in any successful theory of galaxy formation (Oort 1983, de Lapparent etal. 1985). Here, we consider supercluster and pancake features that should naturally arise on large scales in the CDM model. Bahcall (1986) has proposed on phenomenological grounds that clusters should have asymmetric tails similar to those found in §4.3. When such regions collapse, they will form either far field pancakes or filaments; which depends upon the triaxial parameters. A convenient way to consider such collapses is to use density peaks smoothed on scales larger than clusters, but to use a one-dimensional collapse factor $f_c = 1$. Table 1 indicates these structures would be spaced very far apart, even for $R_f = 16\ h^{-1}Mpc$. On the other hand, the factor $f_c = 0.5$ corresponds to structures that would just be turning around and these would appear to be collapsed structures along the line of sight in redshift space, as Kaiser emphasized at this meeting. Such structures *are* sufficiently abundant. One is tempted to view superclusters as rare events consisting of rich clusters with aligned tails forming bridges between the clusters. They should be much more numerous perpendicular to the line of sight than parallel to it.

Another way of viewing large scale structure formation utilizes the beautiful work of Arnold, Shandarin and Zeldovich (1982) on catastrophes in the velocity field, appropriately smoothed (see §4.5), determined by the behaviour of the shear tensor. Pancakes begin at the maxima of the largest eigenvalue λ_1 of the shear tensor of the peculiar velocity field, spreading out from that point to cover a two-dimensional curved surface. In approximately a Hubble time, the matter flows transversely away from the pancake maximum point, accumulating in the deepest potential wells, the peaks of the field: pancakes are transient phenomena on any given scale. The analogue of the height ν of peaks for pancakes is $\nu_p \equiv \lambda_1\ \sqrt{5}/\sigma_0$. The pancaking criterion is $\lambda_1 = f_c$, with $f_c = 1$ for position space collapse and $f_c = 0.5$ for redshift space collapse along the line of sight. The spacing between pancakes assumed to exist only for a Hubble time is given in Table 1 using an abundance formula of Doroshkevich and Shandarin (1978). Note that $2\pi R_f$ corresponds to the wavelength scale usually quoted in pancake discussions (e.g. Bond etal. 1984).

Biasing enhances the density contrasts, but the dynamical phenomenon of pancaking either in real space or in redshift space is required to generate thin bubble walls. Recent unpublished results from the DEFW collaboration also indicate that pancake-like structures *are* sufficiently abundant in the biased galaxy formation theory to reproduce the CfA redshift survey pattern.

4.5 Large Scale Streaming Velocities The velocity field on scale R_f is a superposition of a bulk and a thermal contribution, the thermal contribution arising from collapse on smaller scales. When filtered over R_f-scales, only the bulk contribution should remain. Note that the maximum in the linear velocity fluctuation spectrum occurs at cluster scales for $h = 0.5$, $R_f \sim 10\ Mpc$. The probability of finding a given speed (smoothed over R_f and assuming the mass distribution is still linear on this scale) is distributed like χ^2 with 3 degrees of freedom. If $\sigma_v(R_f)$ denotes the 3-dimensional *rms* velocity dispersion constructed from the linear fluctuation spectrum, then there is only a 5% probability of finding a bulk speed above $1.6\sigma_v$, and a 5% probability of finding one below $\sigma_v/2.97$. In Table 4, these values are compared with the streaming motions reported by three observational groups at this meeting. The corresponding top hat scale is $R_{TH} \approx 1.56 R_f$. The Aaronson etal. number is corrected for their points (clusters) being peaks of the density field smoothed on cluster scales. The biasing factor b is treated as arbitrary.

Table 4: Large Scale Streaming Velocities and the Biasing Factor

$R_f\ (h^{-1}Mpc)$	$b\sigma_v/3$	$b1.6\sigma_v$	$\bar{v}_{obs}$	obs
5	200	970	~ 500	Aaronson etal. (1986)
25	90	420	~ 700	Burstein etal. (1986)
40	60	300	$\sim 1000 \pm 300$	Collins etal. (1986)

$\Omega_X = 1$, $h = 0.5\ CDM$ model : $b\sigma_0(R_g) = 4.9$, $\quad b\sigma_0(R_{cl}) = 0.71$.
$\Omega_\nu = 1$, $h = 0.5$ massive neutrino model : $\sigma_0(R_{f\nu} = 10.6\ Mpc) = 0.7/b$.
$\Omega_{vac} = 0.8$, $\Omega_X = 0.2$, $h = 0.5\ CDM$ model : $b\sigma_0(R_g) = 2.7$, $\quad b\sigma_0(R_{cl}) = 0.75$.

Thus $b \sim 0.4$ would be required to be consistent with the data, implying the mass is *more* clustered than the galaxies. Somehow, galaxies would have to avoid forming in and accreting onto the large number of massive clusters full of dark matter that would exist if $b < 1$. This is precisely the dilemna that the massive neutrino model (with damping scale $R_{f\nu}$ and velocities as given in Table 4) needs to circumvent to be viable (Szalay, Bond and Silk 1983, Doroshkevich 1983, White, Frenk and Davis 1984, Bond etal. 1984). The $\Omega_{vac} = 0.8$ CDM model does not quite have large enough streaming velocities if mass traces light ($1.6b\sigma_v = 550$ and $450\ km\ s^{-1}$ at 25 and 40 $h^{-1}Mpc$), and is very far off with biasing ($b = 3.6$).

Whether the case against the CDM model with biasing is considered strong enough to rule it out depends upon the reader's assessment of the reliability of the data. We can certainly expect solidification of the evidence over the next few years. The stimulating interplay with my fellow statisticians, B,J,K, and S, and the cosmic simulators, D,E,F and W, is gratefully acknowledged. This work was supported by a Sloan Fellowship, NASA grant NAGW-299 at Stanford, the Canadian Institute for Advanced Research and the NSERC at CITA.

REFERENCES

Aaronson, M., Huchra, J. and Mould, J. 1986, this proceedings.

Argyres, P.C., Groth, E.J., Peebles, P.J.E. and Struble, M.F. 1985, preprint.

Arnold, V.I., Shandarin, S.F. and Zeldovich, Ya.B. 1982, Geophys. Ap. Fluid Dynamics 20, 111.

Bahcall, N. and Soneira, R. 1983, Ap. J. 270, 70.

Bahcall, N. 1986, preprint.

Bardeen, J.M., Bond, J.R., Kaiser, N. and Szalay, A.S. 1986, Ap. J., in press. (BBKS)

Bond, J.R. and Szalay, A.S. 1983, Ap. J. 277, 443.

Bond, J.R., Centrella, J., Szalay, A.S. and Wilson, J.R. 1984, M.N.R.A.S. 210, 515.

Bond, J.R. and Efstathiou, G. 1984, Ap. J. Lett. 285, L45.

Burstein, D., Davies, R., Dressler, A., Faber, S., Lynden Bell, D., Terlevich, R., and Wegner, G. 1986, this proceedings.

Collins, C.A., Josephs, R.D. and Robertson, N.A. 1986, this proceedings.

Davis, M. and Peebles, P.J.E. 1983, Ap. J. 267, 465.

Davis, M., Efstathiou, G., Frenk, C. and White, S.D.M. 1984 Ap. J. 292, 371. (DEFW)

de Lapparent, V., Geller, M. and Huchra, J. 1985, CfA Preprint No. 2231.

Doroshkevich, A.G. and Shandarin, S.F. 1978, Sov. Astron. 22, 653.

Doroshkevich, A.G. 1983, Sov. Astron. Lett. 9, 271.

Efstathiou, G. and Bond, J.R. 1986, M.N.R.A.S. 218, 103.

Kaiser, N. 1984, Ap. J. Letters 284, L49.

Oort, J. 1983, Ann. Rev. Astron. Ap. 21, 373, and references therein.

Szalay, A.S., Bond, J.R. and Silk, J. 1983, Proc. 3rd Rencontre de Moriond, ed. Audouze, J. (Dordrecht: Reidel).

Uson, J.M. and Wilkinson, D.T. 1984, Ap. J. 283, 471.

Vittorio, N. and Silk, J. 1984, Ap. J. 285, L41.

White, S.D.M., Frenk, C. and Davis, M. 1984, M.N.R.A.S. 209, 27P.

Zeldovich, Ya.B. 1970, Astron. Ap. 5, 84.

DARK MATTER, COSMIC STRINGS, AND LARGE SCALE STRUCTURE

Joel R. Primack and George R. Blumenthal
Boards of Studies in Physics and Astronomy and Astrophysics
University of California
Santa Cruz, California 95064

and

Avishai Dekel
Department of Physics
The Weizmann Institute of Science
Rehovot 76100, Israel

ABSTRACT. Although the hypothesis of cold dark matter with a Zeldovich spectrum of primordial Gaussian fluctuations appears to give a picture of galaxy and cluster formation that is in reasonably good agreement with the available observations, there are indications that this model leads to less structure on very large scales than is observed. This paper gives a progress report on our efforts to study the formation of large scale structure in models with either (a) an additional feature in the fluctuation spectrum on large scales, such as can arise in a hybrid model with comparable amounts of baryonic and cold dark matter, or (b) non-Gaussian fluctuations, for example those that arise from cosmic strings. Although we cannot yet tell whether either approach will ultimately be sucessful, our preliminary results confirm that cosmic strings can lead to the sorts of rich cluster correlations that are observed.

I. INTRODUCTION

Given for one instant an intelligence which could comprehend all the forces by which nature is animated and the respective situation of the beings who compose it — an intelligence sufficiently vast to submit these data to analysis — it would embrace in the same formula the movements of the greatest bodies of the universe and those of the lightest atom; for it, nothing would be uncertain and the future, as the past, would be present to its eyes. The human mind offers, in the perfection which it has been able to give to astronomy, a feeble idea of this intelligence. (Laplace)

We have all been excited by the evidence presented at this conference for the existence of large-scale velocity fields and by the prospect of a rapid increase in the amount and quality of independent galaxy distance and redshift data. It might almost seem that Laplace's dream could soon come true for cosmology, as he himself helped to realize it for the solar system. But, just as the uncertainty principle frustrates such a deterministic goal on atomic and sub-atomic scales, our inability to measure velocities perpendicular to the line of sight and our ignorance about the position as well as the velocity of most of the matter in the universe — the dark matter — means that we will still need much observational and theoretical cleverness as well as much more data before the future and the past of the universe will be present before our eyes!

The cold dark matter (CDM) hypothesis appears to be able to provide a coherent picture of the origin and evolution of galaxies and clusters, as we review briefly in the next section, but CDM alone does not appear to lead to as much structure as observed on supercluster and larger scales. The question we consider here is how to build a theory which keeps the nice features of the CDM picture on the kpc to Mpc scales, and also gives an accurate account of very large scale structure.

II. COLD DARK MATTER: SUCCESSES AND PROBLEMS

There is persuasive observational evidence that the greater part of the mass in the universe is "dark" — that is, invisible. The most conservative possibility for the composition of the dark matter is that it is entirely baryonic: made of protons, neutrons, and electrons, like the ordinary matter of which all visible astronomical objects are made. But it is now known that this is inconsistent with the observational upper limits on the small scale anisotropy of the cosmic microwave background radiation, assuming that the initial fluctuations were adiabatic (as predicted in most current theories for the origin of the primordial fluctuations) and that reionization does not wash out the $\Delta T/T$ fluctuations. Cosmologists have therefore been led to contemplate the possibility that at least some of the dark matter in the universe is nonbaryonic, perhaps some sort of elementary particle. Popular candidates have been grouped into two categories for astrophysical purposes: hot DM, including light stable neutrinos (with $m_\nu \sim$ 30 eV) and majorons, for which free streaming erases fluctuations smaller than superclusters, and cold DM, such as axions or photinos, for which free streaming is cosmologically irrelevant (Bond and Szalay 1984, Primack and Blumenthal 1983).

Recent research has shown that the hypothesis that the universe is dominated gravitationally by cold DM leads to a picture of galaxy and cluster formation which is in fairly good agreement with the observational data. (Primack 1985 is a recent review with much more complete references to the original literature than can be included here; a revised version is in preparation in collaboration with George Blumenthal for publication in *Reviews of Modern Physics*.) Two theoretical possibilities fit this data about equally well (Blumenthal, Faber, Primack, and Rees 1984 [BFPR]): an open universe with $\Omega \approx 0.2$ and $h \approx 1.0$, and an Einstein-de Sitter universe with $\Omega = 1$, $h \approx 0.5$, and biased galaxy formation — i.e., the assumption that bright galaxies form only at the higher peaks of the underlying density distribution (Davis, Efstathiou, Frenk, and White 1985 [DEFW]). (Here $\Omega \equiv \bar{\rho}/\rho_c$ is the cosmological density parameter and $h \equiv H/(100 \text{ km s}^{-1}\text{Mpc}^{-1})$ is the Hubble parameter.) However, the large value of the Hubble parameter in the open universe case leads to a relatively short age for the universe $t_0 \approx 7$ Gy in serious conflict with globular cluster age determinations. Also, the $\Omega \approx 0.2$ model is in conflict with $\Delta T/T$ limits (Bond and Efstathiou 1984, Vittorio and Silk 1984). Moreover, various theoretical arguments, especially the hypothesis of cosmic inflation, favor $\Omega = 1$ — or at least a flat universe with $\Lambda/3H^2 + \Omega = 1$, although this requires an as yet unexplained fine-tuning of the cosmological constant Λ (Peebles 1984). The age in the flat $\Omega \approx 0.2$ model with a cosmological constant is increased and the conflict with $\Delta T/T$ is removed (Vittorio and Silk 1985).

Simple explanations have been given in the context of the CDM picture for key observational regularities of galaxies and clusters such as the variation of total mass with rotational or virial velocity, including the Tully-Fisher and Faber-Jackson relations $M \propto v^4$ for bright galaxies (BFPR), for the flat rotation curves of spiral galaxies (Blumenthal, Faber, Flores, and Primack 1986), for the existence and properties of globular clusters and dwarf galaxies, and possibly for the origin of biasing in galaxy formation (Rees 1985, Fall and Rees 1985, Dekel and Silk 1986).

The main potential problem for the CDM model is in accounting for the observed very large scale structure in the universe, in particular
1. the enhanced clustering of Abell clusters compared to galaxies, and the existence of very large superclusters of Abell clusters (Bahcall and Soneira 1983, Bahcall, these proceedings);
2. the existence of filamentary structure on scales larger than 100 Mpc (Tully, these proceedings), for example the Perseus-Pisces supercluster (Giovanelli and Haynes 1985, Haynes, these proceedings);
3. the existence of large voids in the distribution of bright galaxies (Kirshner, Oemler, Schechter, and Shectman 1981, private communications from Koo and Oemler 1985); and
4. the existence of large-scale correlations in peculiar velocities of galaxies and clusters (Burstein et al. and Collins et al., these proceedings).

Some of this data, such as the voids in the distribution of rich clusters, may be compatible (Politzer and Preskill 1986) with the cold dark matter fluctuation spectrum. But the weight of the data on large scale structure suggests that CDM is not enough. In particular, it is hard to explain the amplitude of the cluster-cluster correlation function ξ_{cc} with CDM and Gaussian fluctuations with a primordial Zeldovich spectrum. As Kaiser (1984) pointed out, on statistical grounds one expects the cluster-cluster correlation function to be proportional to the galaxy-galaxy correlation function, i.e. $\xi_{cc}(r) = A\xi_{gg}(r)$, where A is an amplification factor which is larger for richer clusters. An argument in favor of this picture is the fact that ξ_{cc} and ξ_{gg} both fall with distance $\propto r^{-1.8}$ (although the relevant data in the two cases is in different regions of r with relatively small overlap). It is however difficult to understand the large value of A, and the fact that ξ_{cc} apparently remains positive at least to $r \sim 100$ Mpc while ξ_{gg} is very small and may even be negative beyond $r \sim 25$ Mpc. Moreover, the CDM N-body simulations just do not seem to have as much large-scale structure — superclusters and voids — as the observations indicate, nor would one expect the large-scale coherent peculiar velocity fields.

III. HYBRID SCENARIOS

It may be possible to generate the observed large scale structure and cluster-cluster clustering in models with adiabatic Gaussian primordial fluctuations if the fluctuation spectrum has a feature at large scales such as can arise in hybrid models with two (or more) sorts of dark matter. For example, in a model universe with $\Omega \approx 0.2$, with comparable amounts of cold DM and baryons, there is a bump in the fluctuation spectrum after recombination at about $M_J \sim 10^{17} M_\odot$. It arises from the fact that fluctuations on these scales can grow as soon as they enter the horizon (that is, they are larger than the Jeans mass M_J) while fluctuations on smaller scales cannot grow until after recombination (Gott 1977). In numerically integrating the fluctuation equations, one sees that the fluctuations in baryons and cold DM are in phase on large sclaes, but oscillate in and out of phase below M_J (Bond and Efstathiou 1985; Blumenthal and Primack, in prep.).

Earlier work (Dekel 1984, Barnes, Dekel, Efstathiou, and Frenk 1985) has indicated that such a feature in the fluctuation spectrum could give rise to pancake-type collapse on very large scales — "super-superclusters" — and perhaps explain the enhanced clustering of Abell clusters and the apparent existence of very large voids and filamentary structures stretching several hundred Mpc across the universe. We are presently investigating this in more detail, starting from the sorts of fluctuation spectra that arise in physically realizable models such as the one just mentioned and investigating the resulting large scale structure via both analytic techniques (e.g., the Zeldovich approximation) and N-body simulations.

Why consider $\Omega_{CDM} \approx \Omega_b \approx 0.1$? Comparison of the standard nucleosynthesis calculations with the abundances of the light nuclides requires $\Omega_b \leq 0.035 h^{-2}$ (Boesgaard and Steigman 1985). The role of the cold DM in this scheme is to preserve galaxy-size fluctuations, which in the baryon component are wiped out by Silk damping. Thus if Ω_b were much greater than Ω_{CDM}, galaxy formation would be hindered, and if the opposite were to hold, the cold DM would dominate and the fact that the baryonic fluctuations are out of phase would not lead to a significant effect in the fluctuation spectrum.

It must be confessed that, even with the addition of a cosmological constant $\Lambda/3H^2 = 1 - \Omega$ to make the model compatible with inflation, increase t_o, and remove the discrepancy with $\Delta T/T$, this model seems rather unattractive. Besides being rather contrived, the model must confront the question what becomes of the majority of the baryons which must be a component of the dark matter. By the usual arguments (Hegyi and Olive 1986), they probably must be in jupiter-size gas balls — but we do not know how such objects form nor why they have a distribution so different from that of the visible baryons.

IV. PHASE CORRELATIONS FROM COSMIC STRINGS

If the primordial fluctuations that grow gravitationally and eventually collapse to form galaxies, clusters, and large scale structure, all began as thermal/quantum fluctuations of an essentially free scalar field during the era of cosmic inflation, then the fluctuations are indeed Gaussian. But it is also possible that the fluctuations responsible for galaxies and larger-scale structures arose from a different mechanism, in which case they would not in general be Gaussian. As Zeldovich (1980) and Vilenkin (1981) pointed out, cosmic strings provide an alternative mechanism for generating fluctuations that, like inflation, also leads to a Harrison-Zeldovich ($|\delta_k|^2 \propto k$) primordial spectrum.

Vilenkin (1985) has reviewed the rapidly growing literature on cosmic strings. Recent work by Turok and collaborators has shown that in this model the dominant mechanism for generating fluctuations is probably the creation of long-lived "daughter loops" as horizon-size "parent loops" self-intersect; the daughter loops then begin to accrete dark matter and (after recombination) baryons, and form the seeds for galaxies and clusters. Turok (1985) has argued that the fact that the daughter loops form along essentially linear structures guarantees that their correlation function $\xi(r) \propto r^{-2}$, which may be responsible for the corresponding behavior of the cluster-cluster correlation function. Moreover, the fact that the formation of parent and daughter loops is a self-similar process suggests that one should compare $\xi(r/d)$, where d is the average spacing of the objects in question (galaxies and clusters) (Schramm and Szalay 1985); then the

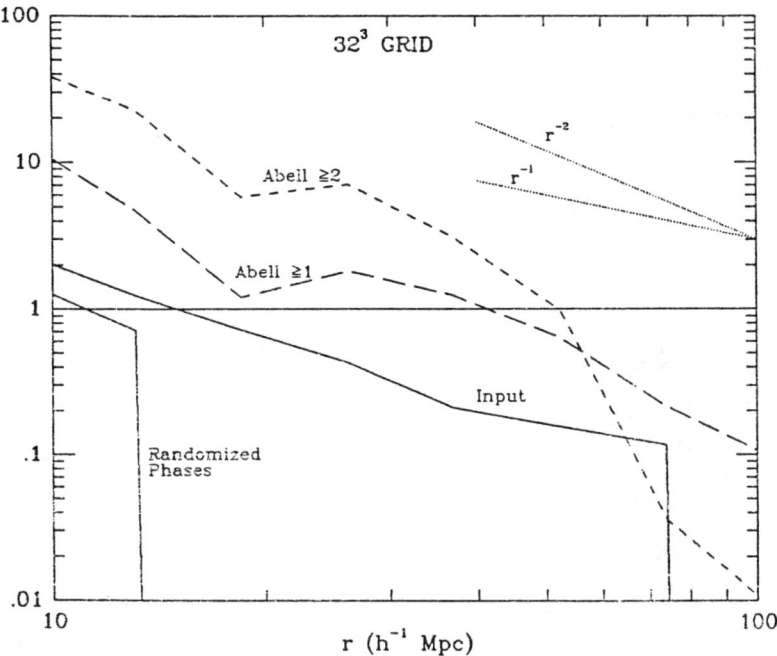

Figure 1. Cluster correlation functions from cosmic string model.

fact that $\xi_{gg}(r/d_g) > \xi_{cc}(r/d_c)$ is understood as reflecting the fact that galaxies form earlier and their clustering is gravitationally enhanced.

We are in the process of examining this class of models in detail, adding dark matter and gravity to existing calculations and focusing on the effects of phase correlations. We have thus far studied a toy model for parent loops based on the Albrecht and Turok (1985) numerical simulation of loop evolution, with the daughters spaced randomly along these parent loops and treated thereafter as static point masses. The calculational procedure that we have used in the work we are reporting here has been to calculate the dark matter and baryon perturbations induced by these daughter loops by Fourier transforming to k-space (on 32^3 or 64^3 grids), multiplying by k^2 to simulate the growth of fluctuations ($\delta \propto R$) after entering the horizon in the matter-dominated era, Fourier transforming back to r- space, using a cluster-finding algorithm with thresholds appropriate for the number densities of Abell clusters of richness ≥ 1 and ≥ 2, and finally calculating the corresponding correlation functions, which are displayed in Fig. 1. This linear calculation is obviously crude, but we hope that it is not misleading. Our tentative conclusions are that the resulting correlation functions are compatible with the observed rich cluster correlation functions. We have also verified that the phase correlations are crucial: models with the same power spectrum but random phases have much smaller correlation functions for their rich clusters.

We are presently improving these linear calculations, and then plan to study the nonlinear clustering process in detail using the Zeldovich approximation and N-body simulations. The questions that we expect to be able to address with these calculations include the following: Can string-type models account both for galaxy and cluster formation? Are the ranges of masses and densities predicted for such objects reasonable? Can such a model account for observed features of superclusters and voids, as well as the correlation function of rich clusters? And more generally, how can observations be used to distinguish between Gaussian and non-Gaussian models for the formation of large scale structure?

ACKNOWLEDGMENT. This work was supported in part by National Science Foundation grant PHY-8415444.

REFERENCES

Albrecht, A., and Turok, N., 1985 *Phys. Rev. Lett.* **54**, 1868.
Bahcall, N., and Soneira, R., 1983 *Ap. J.* **270**, 20.
Barnes, J., Dekel, A., Efstathiou, G., and Frenk, C., 1985 *Ap. J.* **295**, 368.
Blumenthal, G. R., Faber, S. M., Primack, J. R., and Rees, M. J., 1984 *Nature* **311**, 517. [BFPR]
Blumenthal, G. R., Faber, S.M., Flores, R., and Primack, J. R., 1986 *Ap. J.*, Jan. 15.
Bond, J. R., and Efstathiou, G., 1984 *Ap. J. Lett.* **285**, L45.
Bond, J. R., and Efstathiou, G., 1985 *Inner Space/ Outer Space Proceedings* (in press).
Bond, J. R., and Szalay, A., 1984 *Ap. J.* **274**, 443.
Boesgaard, A. M., and Steigman, G., 1985 *Ann. Rev. Astron. and Astrophys.* **23**, 319.
Davis, M., Efstathiou, G., Frenk, C., and White, S. D. M., 1985 *Ap. J.* **292**, 371. [DEFW]
Dekel, A., 1984 in *Eighth Johns Hopkins Workshop on Current Problems in Particle Theory: Particles and Gravity*, edited by G. Domokos and S. Kovesi-Domokos (Singapore: World Scientific), p. 191; *Ap. J.* **284**, 445.
Dekel, A., and Silk, J., 1986 *Ap. J.*, Feb. 15.
Fall, S. M., and Rees, M. J., 1985 *Ap. J.*, **298**, 18.
Giovanelli, R., and Haynes, M. P., 1985, preprint.
Gott, J. R., 1977 *Ann. Rev. Astron. Astrophys.* **15**, 235.
Hegyi, D., and Olive, K., 1986 *Ap. J.* in press.
Kaiser, N., 1984 *Ap. J. Lett.* **284**, L9.
Kirshner, R. F., Oemler, A., Schechter, P. L., and Shectman, S. A., 1981 *Ap. J. Lett.* **248**, L57.
P.S. de Laplace, *A Philosophical Essay on Probabilities*, translated by F.W. Truscott and F.L. Emory (New York, Dover Publications, 1951) p. 4.
Peebles, P. J. E., 1984 *Ap. J.* **284**, 439.
Politzer, H. D., and Preskill, J. P., 1986 *Phys. Rev. Lett.* **56**, 99.
Primack, J. R., and Blumenthal, G. R., 1983 in *Formation and Evolution of Galaxies and Large Structures in the Universe*, edited by J. Audouze and J. Tran Thanh Van (Reidel).
Primack, J. R., 1985, preprint SLAC-PUB-3387, to appear in Proceedings of the International School of Physics "Enrico Fermi", Varenna, Italy, 1984.
Rees, M. J. 1985, preprint.
Schramm, D. N., and Szalay, A. S., 1985, preprint.
Turok, N., 1985, preprints.
Vilenkin, A., 1981 *Phys. Rev.* **D24**, 2082.
Vilenkin, A., 1985 *Phys. Rept.* **121**, 263.
Vittorio, N., and Silk, J., 1984 *Ap. J. Lett.* **285**, L39.
Zeldovich, Ya. B., 1980 *M.N.R.A.S.* **192**, 663.

CLUSTERING IN REAL SPACE vs REDSHIFT SPACE

Nick Kaiser
Institute of Astronomy
Cambridge

Summary

Four aspects of the mapping from true spatial coordinates to redshift space coordinates are considered:

a) Estimation of the net acceleration vector from a complete redshift survey. For small values of Ω, the net acceleration vector can be reliably estimated as the sum of the inverse squared redshifts of the galaxies. This acceleration can then be compared with our peculiar velocity to obtain an estimate of Ω. For Ω close to unity the distortion of the mapping from real to redshift space introduces correction terms to the calculated quantity which are typically larger than the quantity one would like to measure. Estimating a large acceleration vector does not then count as evidence against a closed universe.

b) An alternative method for estimating Ω is provided by the anisotropy of the clustering pattern at large-scales. This shows up as an Ω-dependent flattening of the clustering pattern in the line of sight direction. This anisotropy is most simply expressed in terms of the power spectrum of galaxy clustering. Consideration is given to the practicality of detecting this anisotropy with the next generation of redshift surveys.

c) The revised Shapley-Ames catalogue has the best sky coverage for the purpose of estimating the density enhancement of the local supercluster. This survey has been analysed by Yahil, Sandage and Tamman, who found a large density contrast and therefore inferred a small value for Ω. We have constructed a simple but dynamically consistent model of the density and velocity field in the LSC, under the assumption that the universe has closure density. The density field was populated with galaxies drawn from a realistic luminosity function, and a simulation of a magnitude limited survey was then constructed. We find that, for reasonable values of the redshift of the Virgo cluster and our peculiar infall velocity, the resulting 'density' distribution measured in redshift space is very similar to that observed. We conclude that the data are quite compatible with closure density, though a value of Ω smaller or larger by about a factor of 3 cannot firmly be ruled out.

d) Using the simple, spherical infall model as a guide, we argue that, in addition to the well known elongation of the dense relaxed inner regions as seen in redshift space, rich clusters should be accompanied by large pronounced transverse features in redshift space. These artefacts can be very large, extending to 15 h^{-1} Mpc or more from the centre of a massive cluster. For the cold, spherical infall model many of the galaxies lie on a caustic surface which looks rather like two trumpet horns glued face to face. If a redshift survey is made for a slice through such a structure the result is that low density regions (voids) appear to be bounded by sharp convex shells. We suggest that this type of distortion can account for the features seen in some recent redshift surveys without recourse to powerful cosmological explosions.

COSMOLOGICAL SHELLS AND BLAST WAVES

Jeremiah P. Ostriker
Princeton University Observatory
Princeton, NJ 08544

1. INTRODUCTION

The microwave background is extremely smooth, showing possible deviations from isotropy at only two scales. The dipole anisotropy (cf. Lubin, this conference) is attributable to the motion of the Local Group with respect to the background at approximately 600 km/s. At a much smaller scale (1') Partridge et al. (1986) may have found fluctuations attributable to the era of galaxy formation (Ostriker, Vishniac and Sunyaev, 1986). But fluctuations in the background at intermediate scales due to initial density perturbations have not been detected, implying (cf. DeZotti, this conference) embarrassment for some of the standard models for the development of large-scale structure.

The simplest explanation for this would be that there is, in fact, very little large-scale structure; small fluctuations in the background are to be expected. But there is a great deal of evidence, much of it newly presented at this conference, that large-scale perturbations in density and velocity do exist and have been found. Burstein (this conference) and Bahcall (this conference) find large-scale velocity variations in the 500-1000 km/s range (quite consistent with the sun's motion), and bubbly large-scale density variations were reported by Haynes at this conference and deLapparent et al.(1986) elsewhere.

Where does the structure come from? Could it have been produced after decoupling in ways that would not produce significant fluctuations in the microwave background? I would like to argue here for the possibility that cosmic explosions, proposed by Ostriker and Cowie (1981, ("OC") and Ikeuchi (1981) as influential in promoting, or amplifying galaxy formation, can also also produce significant amounts of large-scale structure.

2. DIRECT IMPLICATIONS OF OBSERVATIONS

The CfA optical survey and the Arecibo radio survey of galaxy positions accord well with pencil beam redshift surveys reported by Koo and Szalay (this conference) and others. They are consistent (but not uniquely so) with the majority of galaxies lying on the surfaces of bubbles with radii of (5-20) Mpc/h [where $h \equiv H_0/(100 \text{ kms/Mpc})$]. To move all the particles from the interior of a filled sphere (radius R) to its surface, the average particle (half mass point) will move a distance of 0.21 R. If the time available is the Hubble time, $2/(3H_0)$, then the mean proper velocity required to evacuate the sphere is

$$\overline{\Delta v} = \frac{R}{5} \frac{2}{3H_0} = 0.3 R\, H_0 \,, \quad \text{or} \quad \left(\frac{\overline{\Delta v}}{v_H}\right) = 0.3 \,,$$

where v_H is the radius of the bubble in Hubble space. This rough estimate, based on the simplest kinematic considerations, shows that large spatial voids necessitate large velocities.

To better calculate the current velocities expected, two approximately balancing effects must be considered. First, the voids are growing not stationary in co-moving coordinates, so the mean velocity will be larger than required to evacuate a fixed co-moving sphere. But current velocities will always be smaller than the time averaged mean velocities due to both gravitational and purely kinematic decelerations (the latter due simply to choice of the co-moving reference frame). For self-similar growing voids or bubbles both effects are easily calculated. If, in an inertial frame, the bubble radius expands as power law

$$R_b = R_0(t/t_0)^\eta \,, \quad \text{then} \quad v_b = \eta(R/t) \,,$$

but

$$v_H = b(R/t) \,, \quad \text{where} \quad b = \frac{2}{\Omega + 2} \,,$$

so

$$\Delta v = (v_b - v_H) = (\eta - b)(R/t), \text{ and } \left(\frac{\Delta v}{v_H}\right) = \frac{\eta - b}{b} = \eta\left(1 + \frac{1}{2}\Omega\right) - 1 \,.$$

For the case of an adiabatic cosmological bubble found by Schwarz, Ostriker and Yahil ("SOY" 1975) and displayed explicitly by Ikeuchi, Tomisaka and Ostriker ("ITO" 1983) and Bertschinger ("B1" 1983) ($\Omega = 1$, $\eta = 4/5$), then $(\Delta v/v_H) = 1/5$. For the uncompensated, adiabatic void found by Fillmore and Goldreich (1984) ($\Omega = 1$, $\eta = 8/9$) then $(\Delta v/v_H) = 1/3$.

Thus the initial rough estimate of $(\Delta v/v_H) = 0.3$ is seen to be reasonable. It implies that, if the typical bubbles have radii of ~ 10 Mpc/h or widths of 2,000 km/s, then we should <u>expect</u> velocities of order 600 km/s with respect to the microwave background. This is just the order of magnitude of the presently detected non-Hubble motions so a pleasing consistency exists between the observed density and velocity inhomogeneities. We should not take this as evidence for or against any particular dynamical theory that might have produced the perturbations, since it is simply a consequence of mass conservation on the cosmic scale. But it is an argument against the existence of some forms of "biased galaxy formation", in which the apparent density inhomogeneities are not real. If the galaxies on the edges of voids were merely "lit up", then there would be no reason for them to have large proper velocities.

3. DEVELOPMENT OF NONLINEAR POSITIVE ENERGY COSMOLOGICAL PERTURBATIONS.

I would now like to address more specifically the dynamics of growing isolated voids. In addition to the work already referenced (SOY, ITO, B1), important papers have been written by Peebles (1982), Hausman et al. (1983) Hoffman et al. (1983), Bertschinger (1985, "B2") and Maeda and Sato (1985a,b,c). Here are some of the essential points to be concluded from the work done to date.

1) The nonlinear behavior is very different from that found for linear perturbations and displays an exceedingly robust soliton-like character.

2) Independent of the initial geometry, and whether the energy is initially in the form of kinetic, thermal or gravitational form, solutions tend to develop into thin spherical self-similarly expanding shells with approximately balancing gravitational and bulk kinetic energy. The "thermal" energy (random motion in the shell) is typically only a few percent of the total energy so the expanding bubble is in many respects simply a rearrangement of the Hubble flow.

3) The nature of the shells is almost independent of the microphysics assumed. Thus the overall structure and expansion rate of shells made up of collisionless particles (neutrinos, galaxies), collisional adiabatic gas (with different assumed values of Γ) or even radiating, momentum conserving gas, all expand at roughly the rate $R = R_0(t/t_0)^{0.8}$ in an $\Omega = 1$ universe.

4) If one allows cooling to occur, then under some circumstances (cf. OC1; Vishniac, Ostriker and Bertschinger "VOB", 1985) the shells will be gravitationally unstable to fragmentation and the formation of new galaxies. The details of this process are not yet well understood.

5) When seen in the frame moving with the shell, the phenomena variously called pancakes, voids, and explosively generated shells are variants of a common physical situation. After long times, the origins of the positive energy perturbations are forgotten and

all of these phenomena have, in terms of observable effects, similar astronomical manifestations.

The reader is referred to the literature for details of the various solutions. I will list here some of the properties of the prototypical adiabatic ($\Gamma = 5/3$) cosmological solution found by SOY, ITO and Bl.

$$R_b = 1.887\ldots(EGt^4)^{1/5} \quad ; \quad \left(\frac{\Delta r}{R_b}\right)_{shell} = 0.032$$

$$(\Delta v/v_H) = 1/6 \quad , \quad (\Delta v/v_H) = 1/5$$

$$(E_{kin}/E) = 1.58 \quad ; \quad (E_{th}/E) = 0.02 \quad ; \quad (E_{gr}/E) = -0.60$$

4. CONSEQUENCES OF COSMIC EXPLOSIONS

The basic theory has been outlined in (OC, Ikeuchi (1981), ITO and VOB). If there are seed explosions with energy of order 10^{61} ergs originating in young galaxy starburst events, quasars, or even interactions involving cosmic strings, then blast waves propagate out from the sources into the surrounding space. At first these are of the classic Sedov-Taylor type. They eventually develop a thin shell structure when their age $(t-t_0)$ exceeds the Hubble time at their origin t_0 and then, for certain epochs and ranges of energy, the shells will cool and become gravitationally unstable. This typically occurs at (co-moving) shell radius of $R_b = 5$ Mpc/h and should result in a new generation of galaxies formed in groups having random velocities of order 100 km/s located on shells expanding with several hundred km/s radial velocities.

Such shells of baryonic material will gradually sweep up any dark matter that may exist and develop into the self-similar solutions described in the previous section. Since this is a gradual process, there is at any given time a certain degree of dynamically induced "bias", that is, the baryonic material is more clustered than the dark matter and thus measurements of its properties will lead to an underestimate of the cosmological density parameter Ω. Preliminary investigation by Dekel and Saarinen (unpublished) indicates that a factor of ~ 2-5 underestimate is quite possible.

In the later development these shells of galaxies and accumulating dark matter will intersect one another. We now turn to that question.

5. QUESTIONS OF SHELL OVERLAP AND MERGER

The issue of the filling factor of regions shock heated by cosmic explosions was first addressed by SOY in 1975. They found that the formal filling factor is likely to exceed unity unless Ω_b is close to unity. As a byproduct of such overlapping explosions, it is likley that the intergalactic medium would have been effectively heated by shocks in the same way as is the largest part of the interstellar medium. But what happens when spherical shocks overlap? In the classical case of two isolated adiabatic blast waves in a uniform medium, the result is clear. The result of a merger of two blast waves will be, after a transient period, the emergence of a new blast wave having an energy equal to the sum of the two merged components. We do not know what will happen for cosmological shocks (or voids), but it seems likely on general principles that the same result will hold. Numerical simulations now being performed will test this contention, but if it is assumed to be true, then one can make simple experiments to see what will be the final result of multiple overlapping spherical cavities. I have performed some trivial simulations on a microcomputer and come to certain highly tentative conclusions. Imagine a co-moving density of explosive centers n_0 each one of which would propagate, without interactions, in a given length of time to fill a volume V_0 ($\equiv \frac{4}{3} \pi R_0^3$). Thus we may define a formal filling factor $\eta_0 \equiv n_0 V_0$, such that the probability that a given point is inside one of the cavities is $f_0 = [1 - \exp(-\eta_0)]$. Then the results of my simulations indicate that a hierarchical set of merging events will occur that is bounded. What I mean by this is that, even when $\eta_0 > 1$, merging does not tend to produce infinitely large bubbles. On the contrary, a distribution of bubble sizes is produced from the unit size R_0 to much larger bubbles with the rms average being

$$R_{rms} = 1.25 \, \eta_0^{1/2} \, R_0 \ .$$

Examples may make the meaning clearer. If $\Omega_b = 0.2$ and we consider blasts with energy 2×10^{61}, then an isolated explosion would reach to 5.0 Mpc at the current epoch. If we then imagine a co-moving density of explosive centers $n_0 = 6 \times 10^{-3}$ Mpc^{-3} (comparable to the space density of giant elliptical galaxies), we find that, including all mergers the typical bubble's size is $R_{rms} = 9.0 \pm 0.3$ Mpc. For the slightly less conservative case where [E = 4×10^{61} ergs, $R_0 = 5.83$ Mpc, $n_0 = 5 \times 10^{-3}$ Mpc, $\eta_0 = 4.1$) then $R_{rms} = 17.4 + 0.6$ Mpc. Thus relatively large bubbles will be produced from a plausible merging scheme. The distribution around R_{rms} seems to be roughly log-normal, but further work must be done to test this.

6. PRELIMINARY CONCLUSIONS

   a) Observations appear to indicate the presence of bubble-like structures with radii in the range (5-20)/h Mpc. This may be indicating that positive energy perturbations, whatever their origin, were more prevalent than negative energy perturbations, since the latter would produce a prevalence of cluster-like (not bubble-like) irregularities.

b) Energy input from processes occurring during galaxy formation can release $10^{61}-10^{62}$ ergs per event. These can produce isolated bubbles with radii in the range $(5-6)/h$ Mpc. Such energy input would not overly disturb the microwave background (Vishniac and Ostriker 1986). Such bubbles of galaxies are likely to overlap, merging to produce larger bubbles. Very rough simulations indicate that typical (final bubble radii in the range $R_{rms} = (10-20)/h \times$ Mpc would result. Since this merging process occurs among dissipationless particles (galaxies) the growth of such large bubbles would have no further effect on the microwave background.

c) If much larger bubbles of radii 50-100 Mpc are found, then another energy input is likely to be the cause with processes involving cosmic strings a possibility (cf. also Primack, this conference).

I would like to express thanks to NSF (grant #AST83-17118) and NASA (grant #NAGW-765) for support of the Princeton work.

REFERENCES

Bertschinger, E. W. 1983, Ap. J., 268, 17 ("B1").
Bertschinger, E. W. 1985, Ap. J. Suppl., 58, 1 ("B2").
de Lapparent, V., Geller, M. J. and Huchra, J. P. 1986, preprint.
Fillmore, J. A. and Goldreich, P. 1984, Ap. J., 281, 9.
Hausman, M. A., Olson, D. W. and Roth, B. D. 1983, Ap. J., 270, 351.
Hoffman, G. L., Salpeter, E. E. and Wasserman, I. 1983, Ap. J., 268, 527.
Ikeuchi, S. 1981, Pub. Astron. Soc. Japan, 33, 211.
Ikeuchi, S., Tomisaka, K. and Ostriker, J. P. 1983, Ap. J., 265, 538 ("ITO").
Maeda, K. and Sato, H. 1983a, Prog. Theor. Phy., 119, 70.
Maeda, K. and Sato, H. 1983b, Prog. Theor. Phy., 119, 772.
Maeda, K. ans Sato, H. 1983c, Prog. Theor. Phy., 119, 1276.
Ostriker, J. P. and Cowie, L. L. 1981, Ap. J. (Lett.), 243, L127 ("OC").
Ostriker, J. P., Vishniac, E. and Sunyaev, R. 1986, Ap. J. (Lett.), submitted.
Partridge, R. B. 1986, IAU General Assembly "Highlights in Astronomy'.
Peebles, P. J. E. 1982, Ap. J., 257, 439.
Schwarz, J., Ostriker, J. P. and Yahil, A. 1975, Ap. J., 202, 1 ("SOY").
Vishniac, E. and Ostriker, J. P. 1985, Societa Italiana di Fisica, 1, 157.
Vishniac, E., Ostriker, J. P. and Bertschinger, E. W. 1985, Ap. J., 291, 399 ("VOB").

MORE ON STRUCTURE INVOLVING 10^{18} $M_\odot$

R. Brent Tully
University of Hawaii
Institute for Astronomy
2680 Woodlawn Drive
Honolulu, Hawaii USA 96822

The claim has been made (Tully 1986a,b) that we reside in an overdense region that involves 10^{18} $M_\odot$ and extends across a dimension corresponding to 10% of the distance to the event horizon. An important part of the basis for the claim is the observation that there is a significant enhancement of rich clusters of galaxies coincident with the supergalactic equator; that is, coincident with the plane of the Local Supercluster as defined by Shapley-Ames galaxies.

The limited purpose of this communication is to report that the putative adherence of nearby rich clusters to a plane is still evident in a third iteration sample of clusters and to illustrate this fact with a map.

The discovery of the phenomenon (Tully 1986a) was founded on a study of an all-sky sample of 214 rich clusters, with corroborative evidence provided by the distribution of less rich clusters studied by Shectman (1985). With the enlargement of the all-sky sample to 309 clusters, the effect was considerably more evident (Tully 1986b). Recently, Huchra (private communication) has made more richness classifications and cluster redshifts available to us. The new redshifts provide us with 75 new clusters, while the richness information causes us to reject 13 clusters with a richness <0 from our old sample, with the result that we now have a sample of 371 clusters.

The rejection of low richness clusters that tend to be nearby and the addition of clusters with newly measured redshifts that tend to be farther away means that the new sample deemphasizes the local region and suffers less from incompletion as a function of distance compared with the earlier samples. Moreover, the new sample incorporates more data from the southern celestial hemisphere, though coverage there is still quite incomplete.

The enhancement of clusters along the supergalactic equator is equally significant in the less biased sample of 371 clusters to that with the sample of 309 clusters. There is a 4.6 σ excess of rich clusters in an 80 h_{75}^{-1} Mpc interval perpendicular to the supergalactic equator centered 10 h_{75}^{-1} Mpc below the plane of the Local Supercluster. The prominent components that constitute this excess stretch across a long dimension of 500 h_{75}^{-1} Mpc.

The maps represent an attempt to display this concentration toward the supergalactic equator with a minimum of manipulation. There is the problem, though, that the effect is somewhat diluted by line-of-sight confusion in the most straightforward edge-on projections of the sample. Specifically, there are clusters in two regions on the periphery of our sample that lie at intermediate latitudes and broaden the apparent plane in edge-on projections.

We will show what the edge-on projection looks like with these two regions excised. Figure 1 is a polar view of the surface density of <u>all</u> the clusters in the new sample (Gaussian smoothing with σ = 14 Mpc; contour levels 0.05, 0.1, and 0.2 clusters/Mpc). Figure 2 is an edge-on view of the cluster surface density in the region contained within the rectangle imposed on Figure 1. The abscissa in Figure 2 is the long axis of the rectangle, which is rotated 20° to the SGY axis of Figure 1. The incorporated depth of the projection in Figure 2 is 245 h_{75}^{-1} Mpc (the width of the rectangle). There are 241 of the 371 clusters in the volume displayed.

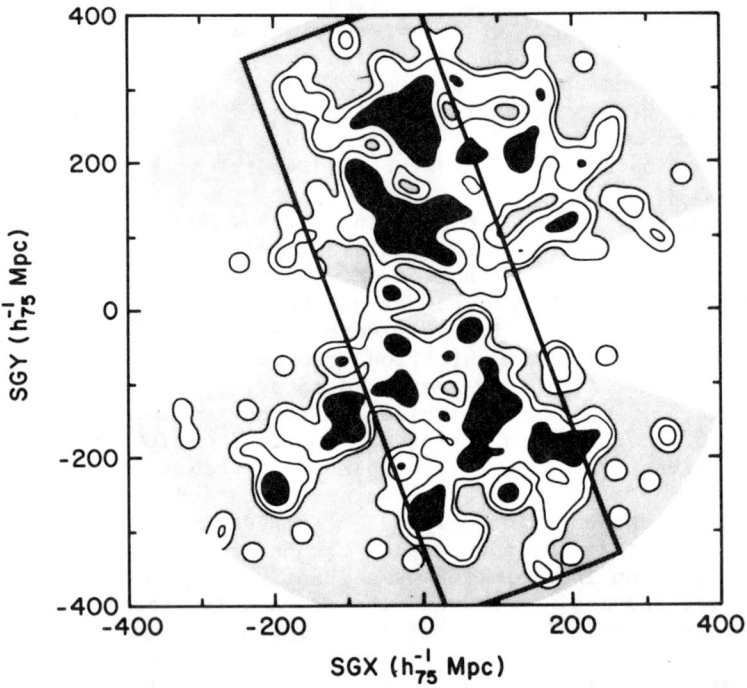

Figure 1

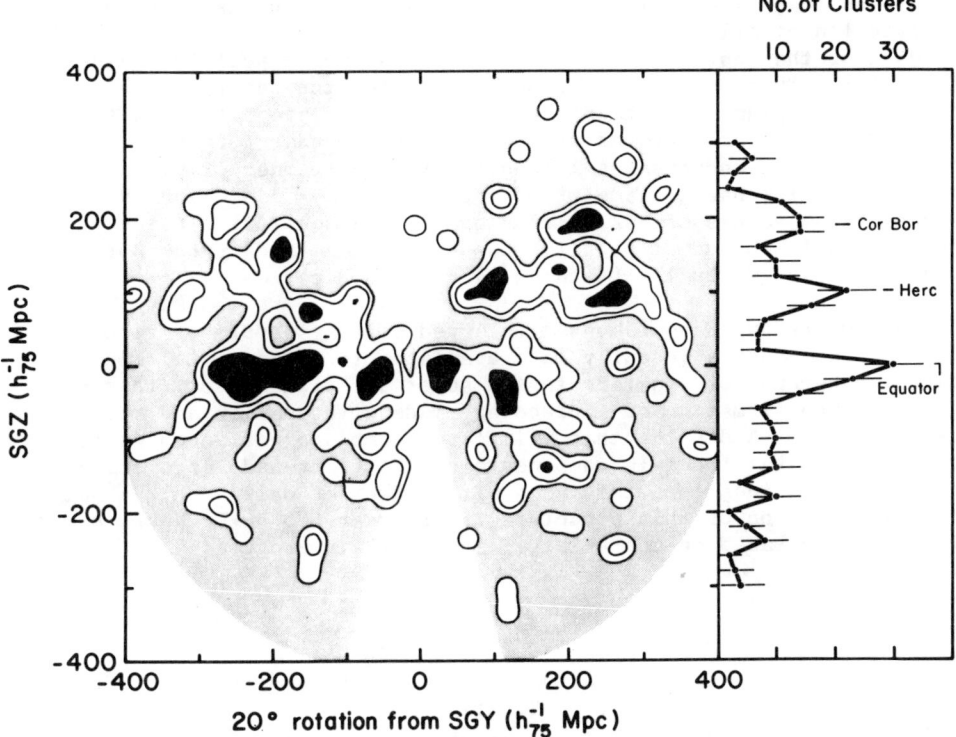

Figure 2

The map is collapsed to a one-dimensional representation of cluster counts as a function of supergalactic latitude on the right border of Figure 2. These counts have been corrected for obscuration and geometric effects in the manner described in the discovery paper.

The enhancement of rich clusters in the plane of the Local Supercluster is confined to a narrow latitude range in this restricted sample. There is a 4 σ excess of clusters in only three 20 h_{75}^{-1} Mpc bins coincident with the equator. There is a clean separation between this plane and the Hercules Supercluster, and beyond it the Corona Borealis Supercluster, at positive values along the abscissa of Figure 2 and at positive supergalactic latitudes.

A brief word will be said about the excised material. It is seen in Figure 1 that there is a prominent feature outside the rectangle at negative values of SGX and SGY. This region lies in the south celestial hemisphere and has not been completely explored. However, the principal feature that is known can be described as an elongated structure that is inclined at roughly 10° with respect to the supergalactic equator and that rises out of the equatorial plane but is thin in the SGZ direction at all places.

Most of the rest of the excised clusters lie at positive values of SGX and SGY. There are features here that blur the separation between the equatorial plane and the Hercules region in a projection of the whole sample. These features are part of the boundary of the vast empty region that encompasses the Boötes Void (Kirshner et al. 1981). This empty region has a diameter of roughly 160 h_{75}^{-1} Mpc and is bounded at low values of SGZ by clusters in the equatorial plane, at high values of SGZ and at negative SGX by Hercules and Corona Borealis, and at positive SGX by the features alluded to here.

In summary, we have taken an enlarged all-sky sample of rich clusters that John Huchra recently made available to us and still find an apparent adherence of clusters to the supergalactic equator, thus providing some confirmation of a discovery made with smaller and more biased samples. A display that eliminates some projection confusion effects shows that the main components of the flattened large-scale feature are confined to a thickness with a FWHM of only 40 h_{75}^{-1} Mpc. The long dimension of this structure is of order 500 h_{75}^{-1} Mpc. A more detailed study based on the enlarged sample is in progress.

REFERENCES

Kirshner, R. P., Oemler, A., Jr. Schechter, P. L. and Shectman, S. A.: 1981, Astrophys. J. Lett. 248, L57.
Shectman, S. A.: 1985, Astrophys. J. Suppl. 57, 77.
Tully, R. B.: 1986a, Astrophys. J. 303 (April 1).
Tully, R. B.: 1986b, in G. R. Knapp and J. Kormendy (eds.), Dark Matter in the Universe, IAU Symposium 117, Reidel, Dordrecht, in press.

GALAXY DISTANCES AND DEVIATIONS FROM THE HUBBLE FLOW; SUMMARY REMARKS

Jeremiah P. Ostriker
Princeton University Observatory
Princeton, NJ 08544

1)   INTRODUCTION

Cosmology is an elusive subject. The discipline is somewhat like the mythical hero Proteus who had the ability to change his form, unpredictably altering himself from one type of beast to another. Whenever we get a bead on it, it changes. Whenever we feel we are about to answer the important questions, the subject wriggles out from under us and the questions change. It is not that our patient progress is not real. The scientific method does really work and we typically are able to close in on the well defined questions. It is rather that we wake up to find that the questions we have been asking are hopelessly naive.

An example from the past makes the point clearly. In the teens of this century the cosmological question was of course what it always was and will be; the nature, the structure, of the universe. But the universe then was what we would now call the Galaxy. The relatively tidy picture, which was summarized by the Kapteyn (sun centered) model, the careful and quantitative culmination of a research program begun over a century before by Herschel. The Kapteyn model was seriously upset by the results of Shapley's studies of the spatial distribution of globular clusters that ultimately led to the famous Shapley-Curtis debate. We now think of this event as concerned with the nature of the nebulae but a careful reading shows that not to be the case. The debate was about the scale of the universe, i.e. the galaxy, and the nature of the nebulae was a peripheral issue. But, while people were arguing about the Galaxy, the subject of cosmology was blown apart by Hubble's (and others) monumental discoveries. Resolving nearby nebulae into stars, finding the number-magnitude relation for nebulae, and then their overall expansion totally changed the discipline. Lively and important debate about the structure and nature of our Galaxy continued and continues today. But cosmology changed in an irrevocable way.

This has happened repeatedly. Sometimes the changes were due to observations such as Hubble's discoveries or those of the microwave background and sometimes due to theoretical analyses or conceptions. Often the terms of the discussion were established by theory. As an example, after the standard Friedmann cosmological models were developed, the "big questions" were reduced, it seems, to the determination of two numbers (H_0, q_0) or three if Λ_0 were included. Massive observational programs were initiated to determine those numbers. Some, like the efforts to determine H_0, have continued to the present day and are, as we shall see, bearing fruit in a way that would have delighted Hubble. But others, like the galaxy counts relation, $N(m)$, begun to determine q_0, have really run aground. There were observational problems, but these were not really the primary difficulty. What happened is that many scientists, theoreticians and observers, realized (in the 1970s) that the standard candles and standard meter sticks could not be trusted to be standard over the time periods needed to determine q_0. An evolutionary change of .03 magnitudes per billion years, if not modeled by the observer, is enough to produce an error of unity in q_0, from a closed to an open universe. In principle, all of the types of stellar and dynamical evolution proposed by various theorists can be detected and removed by appropriate calibration. But the project began to look immense and more like finding the way out of a dark swamp than the neat path to determination of model parameters.

Later on the gradual realization that dark matter is pervasive gave us another wrench. If we have been only looking at the 10% of foam on the hitherto invisible waves how can we proceed? Now we are only beginning to understand the issue of dark matter well enough to return to the determination of q_0 or Ω_0.

Simultaneously, the realization that our standard objects were changing made us take a hard look at them causing a change of focus and an alteration in the relevant questions. Aware that our "standard objects" had not always existed, we were forced to ask how did they develop and, in fact, how did any structure develop in the universe. Thus, before we were able to answer the last set of questions concerning H_0 and q_0, the issues had changed. We asked where did our markers come from? A whole new subdiscipline--galaxy formation was born.

This conference has shown that we are another juncture, our intellectual space ship seems to be making another stomach wrenching turn. As an example, it appears that, just as methodologies and the results from different techniques are converging with respect to the Hubble parameter, the darn thing is wriggling out from under us. It may be that an accuracy of 10% is unattainable or meaningless if H_0 is different in different directions and different if we average over different volumes. That has clearly happened with the galactic constants (A, B) where we now know that, though the galactic rotation field may not be a fractal, it is sufficiently messy to make the local definition of $(dV_c/dR)_0$ quite ambiguous.

GALAXY DISTANCES AND DEVIATIONS FROM THE HUBBLE FLOW

Thus I maintain that this conference will be seen as an historic turning point at which many scientists realized that there was an exciting and nasty set of new questions to be answered before the receding older ones could be addressed properly. I will divide up my discussion into the important but relatively standard parts and the quite new issues. I apologize in advance for the fact that my discussion will be partial in both senses of the word. Partial because I will omit to discuss important work which I didn't attend to properly and partial because I will be unfair to things I did not agree with, no doubt through incomprehension.

2) CLASSICAL COSMOLOGICAL STUDIES

2.1 Distances to Nearby Objects

Progress in this old and important field reported at this meeting was dramatic. Two of the more spectacular items were Prichet and van den Bergh's report of R-R Lyrae stars in M31 and Aaronson's report of Cepheids detected in M101. In both cases the results were preliminary, but showed the potential of ground-based telescopes using CCD detectors to make observations previously thought to be impossible. Other relatively new techniques such as Schommer's main sequence fitting for clusters, Freedman and Madore's IR Cepheid work and Brent Tully's use of A supergiants seemed to me very promising additions to the classical methods described by Sandage and de Vaucouleurs. As several speakers stressed, including Feast and Sandage, there is agreement on the distances to the local calibrating galaxies to better than or approximately ± 0.1 mag in general. Further improvements by the same methods of a factor of two seem quite possible, and the Hubble Space Telescope should certainly clear up most of the remaining local difficulties. In a related matter, there is a particular appeal to me in Craig Wheeler's work on supernova. The Baade-Wesselink method should be applicable now that we are beginning to understand the stellar atmospheres of Type I supernovae and this will, in principle, offer a completely independent method of obtaining the extragalactic distance scale. I would put a high priority on both the theoretical and observational work required to put this method on a sound footing.

2.2 Intermediate Scale Distance Determinations

Here almost all methods are new and improved versions of the color-magnitude or velocity magnitude relations found one or two decades ago by Sandage, Visvanathan, Tully, Faber and others. One distance independent quantity is measured such as color, velocity dispersion or surface brightness; then a luminosity (or physical size) is inferred, which, with a measured flux (or angular size), determines a distance. In some of the newer variants, more than one distance independent quantity is measured in order to improve the accuracy of the predicted luminosity or size. In principle this is simple, in practice it is of course terribly complicated. The intrinsic dispersion in the relationships is not small, ranging from about 0.5 magnitude for the elliptical color-magnitude relation described by Visvanathan to slightly better than this for the velocity dispersion D_Σ relation used by Burstein et al., the similar relation found by Djorgovski and the IR Tully-Fisher relation used by several investigators. My own guess is that more and more work to reduce the cosmic scatter in these relations will not be well rewarded. What is more important is to use, of these methods, those that are quick and accurate observationally, since it becomes increasingly clear that large samples will be needed. There are two reasons for needing a large sample; first to reduce the statistical errors in the mean distances of sampled objects by $\sigma/\sqrt{N}$, and second to obtain the redshift coverage and angular sky coverage which we are now finding is necessary to obtain reliable results.

The second set of problems is that of data handling. As an example we must allow for observational biases such as the famous Malmquist bias and the like. This is not a new problem in astronomy. Here good scientists all agree on how the work should be done. Complete data sets should be taken according to clearly defined restrictive criteria. Then the implicit biases can be modeled and removed. I have very little to add to the current debate and my guess is that the various groups shouting at one another will ultimately produce intelligently analyzed data sets. In fact, I am not convinced that this has not happened already and that much of the work we have seen at this conference, protestations to the contrary notwithstanding, is free from serious bias.

2.3 The Local Peculiar Velocity

This is something we now know. The direction and movement of the sun's motion with respect to the cosmic background radiation is now known, according to Lubin's report, to better than 10%. After it was first discovered many were surprised by the 600 km/s motion of our galaxy with respect to the background radiation. Meiksen and Yahil now find that the IRAS moderately distant galaxies (50-100 Mpc) define essentially the same frame. In other words, the force on us due to the mass distribution inferred from the IRAS flux distribution would produce a motion in the same direction as our observed motion with respect to the microwave background. Finally as Burstein summarized, several different galaxy samples (with distances and velocities known or estimated) are consistent with a solar motion in this same direction. The microwave result is most accurate, and most basic, so it would seem best, for the present moment, to simply adopt that frame as the rest frame and treat the cosmic solar motion as known to good accuracy. This is in fact the procedure that Mould et al. have used.

2.4 The Local Hubble Constant and Other Age Parameters

These subjects are so classic that there is bound to be serious disagreement among analysts. In new fields there is not always time enough for competing views to crystallize out. I am not really competent to judge the contentious issues, since they depend crucially on a myriad of small points of observational technique and analysis. It was reassuring that there is increasing agreement on the local calibrators (nearby galaxies) and on the best methods to use at intermediate distances. But it is also clear that different groups, normalizing their results to essentially the same calibrators, obtain quite different results for H_0, even if using variants of the same method. My own guess as to the reason for this is that the differences are primarily due to sampling different parts of the universe. It appears that the velocity field has significant gradients in it, so the nearby Hubble constant may be smaller than the distant one (cf. paper by Tully), and that, locally, H_0 may vary significantly with direction. Thus large data bases will be needed, which reach out to at least 10,000 km/s. These must be large for the statistical regions mentioned earlier (given the intrinsic $\Delta m = 0.4$ of the standard candles), and large to average over the large scale real fluctuations in H_0. For the present I will be happy enough to take the coward's approach of adopting $H_0 = 75$ and hoping that I will not make too big an error. The age determination summarized by Renzini and the nuclear chronometers reviewed by Thieleman are consistent with this result, but that, unfortunately, has always been the case even when H_0 was much larger.

2.5 Ω_0 from Virgo Infall

The computer simulations presented by Melott and Vilumsen indicated how difficult it will be to model this phenomenon well. They showed that previous simplified analyses apparently have underestimated Ω_0. Also Yahil pointed out that for a given model, and given infall velocity, since the overdensity of Virgo may be smaller than we thought, therefore Ω_0 may be still larger than previously calculated. Values of $\Omega_0 \approx 1$ by this method seem within the range of observational and theoretical uncertainty at this time.

3) NEW COSMOLOGICAL DEVELOPMENTS

3.1 Large-Scale Inhomogeneity

To me the most exciting new results concern large scale inhomogeneities and velocity variations. Martha Haynes' 21-galaxy survey clearly shows the same bubbly structure found in the recent Geller-Huchra CfA optical survey of a different part of the sky. The results are dramatic, and in my view indicate the likelihood of either dominantly explosive (positive energy) perturbations, or the hot dark matter (Zeldovich) picture revived by Vittorio, Szalay, Melott and others. Neta Bahcall also presented strong evidence of structure on the very large scales of superclusters which is difficult to accommodate with any of the standard scenarios. There are clearly going to be a number of big programs developed to map out the sky.

3.2 Large-Scale Deviations from the Hubble Flow

We still do not know what Δv between adjacent galaxies is. Is it less than 50 as Yahil, Toomre and others have said or more like 300 km/s? But, on larger scales, it is clear that peculiar motions of galaxies with respect to the microwave background of 600 km/s are typical rather than anomalous. Burstein's group found dramatic variations of that magnitude over the sky. Mould et al. did not find this, but their result is probably due to their more limited sky coverage and the relative smoothness of the flow. Bahcall found even larger velocities on the supercluster scale of 1400 km/s which, however, may be reduced somewhat with a more detailed analysis of the velocity errors.

3.3 Origin of the Velocity and Density Inhomogeneities

The new results concerning irregularities in density and velocity are mutually re-enforcing as I pointed out in my own talk. The continuity equation forces such a relation (in the absence of "bias"), the details of which are dependent on the process. But what caused these large-scale structures? The theoreticians confessed themselves to be a bit surprised, happily or unhappily depending on their previous work, but to me the developing picture looks most like the old Zeldovich pancake model. It seems to be really ideal for making such large-scale structures. Of course we now know that this scheme fails miserably for making galaxies, galaxy groups etc. but perhaps we can leave the origin of the "local structure" to other processes. The string picture presented by Primack is also very attractive for producing the large scale structure without violating constraints imposed by the microwave background. Clearly the last word is not in. There is always a tendency for the most recently proposed model to seem best.

4) PROSPECTS

Where will we be in 10 years? I believe that the nearby distance scale will be determined to better than 10%. I think that the "observed", dynamical values of Ω_0 will be creeping up to the point where the discrepancy with Ω_{baryon}, given by cosmic nucleosynthesis of the light elements will be serious, and the discrepancy with the theoreticians' favorite value of unity will seem manageable. There will also of course be new theories. But what would I most like to see? The most boring possible thing. Surveys. Large complete surveys to map our own galactic neighborhood in great detail out to 10,000 or 20,000 km/s would be immensely valuable.

Also there will be surprises. As we have gone to larger and larger scales using stars or gas as test particles to probe the universal gravitational potential, we have found that our previous understanding simply failed. We needed to invoke either "dark matter" or a new theory of gravity. More recently, using photons as probes and observing gravitational lenses produced by intervening matter between us and the quasars I think we again have come up against severe difficulties, even with currently acceptable amounts of dark matter. Thus old fashioned gravity is capable of presenting us with ever new surprises. I suspect that this situation will continue.

5) ACKNOWLEDGEMENT

I am happy to thank Brent Tully for organizing such an exciting meeting and NSF grant AST83-17118 and NASA grant NAGW-765 for supporting my own work in this area.

GALAXY SURFACE BRIGHTNESS AND CLUSTERING: A TEST FOR BIASED GALAXY FORMATION SCENARIOS?

S. Djorgovski[1] and Marc Davis[2]
[1]) Harvard-Smithsonian Center for Astrophysics,
   60 Garden St., Cambridge, MA 02138.
[2]) Departments of Astronomy and Physics,
   University of California, Berkeley, CA 94720.

ABSTRACT. Davis & Djorgovski (1985) described a correlation between the clustering properties and characteristic surface brightness of UGC galaxies. The existence of such an effect is of considerable importance for theories of galaxy formation. In this contribution, we are trying to clarify some questions and problems about this intriguing phenomenon.

Biased galaxy formation is a general and appealing idea, which promises to solve, or at least alleviate, many important problems in our understanding of large-scale structure and galaxy formation. A biasing scheme may operate in many different formation scenarios (cf. Davis et al. 1985; Silk 1985; Dekel, this volume). One important prediction of this family of theories is that some of the global properties of galaxies, such as their luminosities, densities, morphologies, etc., may be closely related to their large-scale environment.

Having this in mind, and following a suggestion of Bothun (1985, private communication), we investigated clustering properties of UGC galaxies as function of their surface brightness (Davis & Djorgovski 1985). Briefly, we find that the high surface brightness (SB) galaxies are more strongly clustered than the low SB galaxies, and that the contrast in clustering increases with the contrast in SB. For example, we find that the angular clustering amplitudes differ by a factor of 1.3 between the top and the bottom SB quartiles, and that the slopes of two-point correlation functions also differ significantly. This result also suggests that the luminous matter is not necessarly a good tracer of mass distribution. An example of the effect is illustrated in Figure 1. More details and quantitative results are given by Davis & Djorgovski (1985).

In order to compensate for the well-known morphology-clustering correlation (Davis & Geller 1976; Dressler 1980), we subdivided the UGC sample in morphological types, selected the top and the bottom SB quartiles within each type, and then mixed those subsamples together. Thus, we believe that the morphology-clustering (or environmental density) correlation may be a consequence of a more general, galaxy surface brightness (or luminosity density) relation with the large-scale structure.

Let us make one important clarification: we find that the *projected luminosity density* of galaxies – not the *luminosity* itself – is strongly related to their clustering properties. Sometimes the word "dwarf" is used to denote both kinds of objects, which is a cause of much confusion. There is little or no evidence at the present that galaxies of different luminosity cluster differently.

The SB-clustering effect is clearly present on the sky, and it is of interest to see how much of the effect there is in the three-dimensional space. In order for the effect to disappear, the low-SB UGC galaxies should be in average much further away than the high-SB ones; this seems intuitively inconceivable in a diameter-limited catalog, such as the UGC. Thus, the effect almost has to be present in the three dimensions as well.

It is not necessary to have the redshifts for all galaxies: for a simple comparison, only a "typical" redshift is needed, e.g., median redshift for the sample. We tried to estimate characteristic redshifts for both kinds of objects by using the available data from the Huchra's ZCAT. The problem is that the redshift completeness of UGC is different for the two groups, viz., most of the ZCAT low-SB galaxies are the members of a volume-limited, Local Supercluster sample, whereas most of the high-SB ZCAT galaxies are the members of a magnitude-limited CfA redshift survey.

Now here is an important point: redshift-incompleteness is not detrimental, as long as it can be estimated, and compensated for; and this is what we did. We tried to compensate for incompleteness by dividing the samples in SB bins, computing the redshift completeness in each bin, and evaluating redshift distributions with galaxy weights inversely proportional to the ZCAT completeness at their SB. The "weighted" median redshifts evaluated in this way were practically the same as the "straight" median redshifts from the available ZCAT data. This means that *we do have fair estimates of characteristic redshifts for the catalogs, even though the redshift information is incomplete.* The ratio of three-dimensional clustering amplitudes for the top and the bottom SB quartiles for the whole UGC is about 2.9 .

Bothun *et al.* (1985; 1986) presented results of their redshift survey of low-SB, late-type UGC galaxies, designed to overcome the known biases and incompleteness of the ZCAT. Their diameter cutoff removes most of the Local Supercluster galaxies from their sample, and consequently the global redshift distribution they obtain is quite different from the ZCAT redshift distribution of low-SB galaxies presented and used by us (Davis & Djorgovski 1985). Since there is no high-SB control sample with the similar selection criteria, it is not easy to compare the biases of the two studies, and their implications. For example, the southern Galactic hemisphere subsample, shown in Fig. 1, has practically the same redshift distribution as Bothun *et al.* sample, and it shows dramatic SB-clustering effects. Bothun *et al.* state that there is no evidence that low-*luminosity*, late-type galaxies in their sample fill in the voids in the distribution of high-SB, all-type galaxies; this is not in contradiction with our results. They also compare the counts of high-SB and low-SB galaxies in bins 4^h wide in RA, and 10° - 36° wide in DEC, and find no differences in such "coarse-grained" distribution. We found that the two-point correlation functions for our samples merge and become too noisy at angles greater than several degrees; again, there is no contradiction, since we are probing different angular scales. Bothun *et al.* are right in their conclusions – they interpret their data in a straightforward manner – but we do not think that their skepticism towards the SB-clustering effects in three dimensions is warantied; as we have shown here, our two studies probe different regimes, with samples selected in different ways, and use considerably different methods of analysis.

Clearly, we need much more complete redshift data base for the further investigations of SB-clustering effects. Perhaps even more important would be improvements of SB measurements for a large number of galaxies. We believe that the SB-clustering effects will stay with us, and may provide a powerful test for the theories of biased galaxy formation.

Another important question concerns the possible enviromental dependence of distance-indicator relations: Burstein *et al.*, Bothun, and Djorgovski & Davis (elsewhere in this volume) introduced surface brightness terms in Faber-Jackson and Tully-Fisher relations, with strong beneficial results. If such a fundamental quantity as surface brightness (or projected luminosity density) depends on the large-scale environment of galaxies of all morphological types, then it is possible that there are systematic differences in zero-points and/or slopes of distance-indicator relations. Such systematics, if they exist, would masquerade as large peculiar velocities of different clusters and groups, accounting, perhaps, for the interesting phenomena reported here by Burstein *et al.* The problems of galaxy formation and our tools for the mapping of the Local Supercluster and vicinity are deeply related.

We acknowledge stimulating discussions with G. Bothun and M. Geller. This work was supported in part by the NSF grant AST-8419910.

REFERENCES:

Bothun, G., Beers, T., Mould, J., and Huchra, J. 1985, *Astron. J.* **90**, 2487.
Bothun, G., *et al.* 1986, preprint.
Davis, M., Efstathiou, G., Frenk, C., and White, S. 1985, *Astrophys. J.* **292**, 371.
Davis, M., and Geller, M. 1976, *Astrophys. J.* **208**, 13.
Davis, M., and Djorgovski, S. 1985, *Astrophys. J.* **299**, 15.
Dressler, A. 1980, *Astrophys. J.* **236**, 351.
Silk, J. 1985, *Astrophys. J.* **297**, 1.

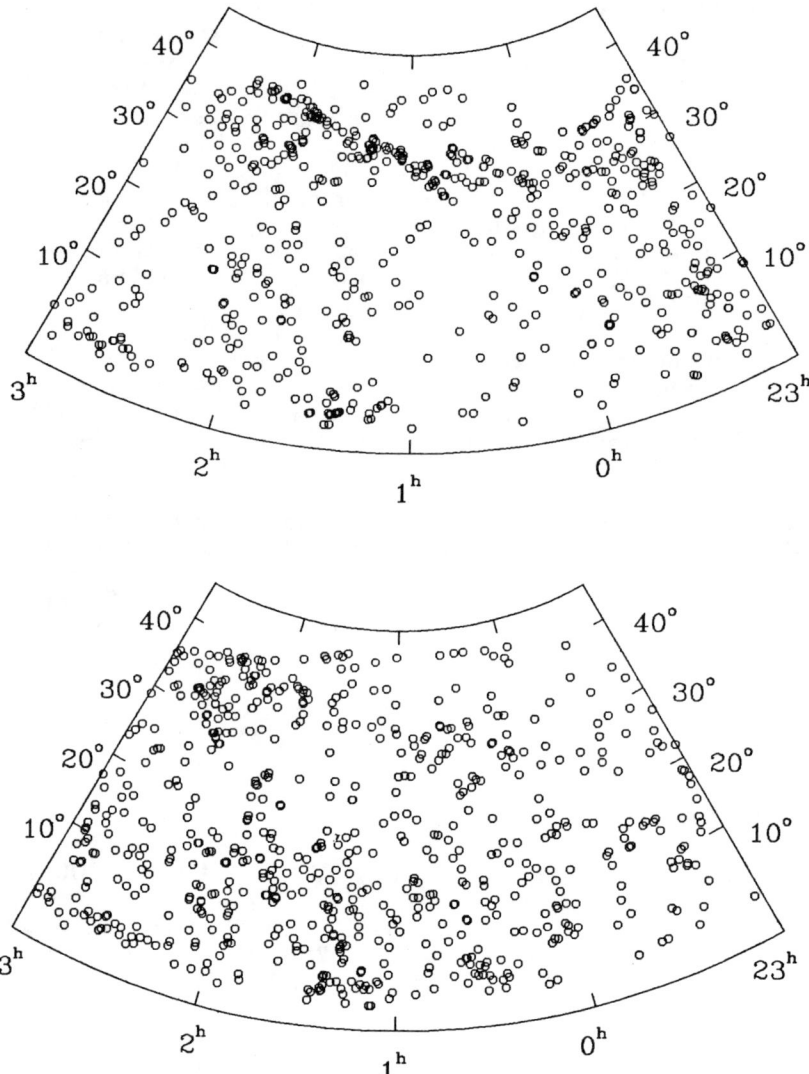

Figure 1. Distributions of galaxies in the Perseus-Pisces region, for the top and the bottom SB quartile galaxies. There is the same number of ellipticals, lenticulars, and spirals in each pannel. Both subsamples have very similar redshift distributions, dominated by the P-P supercluster at $v \sim 5000\ km\ s^{-1}$; thus, the projected clustering differences reflect fairly the "true" three-dimensional clustering differences. High-SB galaxies delineate the well-known filament, and low-SB galaxies fill in the voids. The ratio of three-dimensional clustering amplitudes is about 8.2 .

Index

Abell 194: 117.
Abell 262: 117.
Abell 347: 117.
Abell 426: 117.
Abell 400: 112, 117.
Abell 1060: 101.
Angular Correlation: 149.
Asymptotic Giant Branch Stars: 197.
Baade-Wesselink Method: 11, 61, 180, 285.
Biased Galaxy Formation: 154, 226, 243, 246, 249, 251-254, 255, 256, 276.
B Cas (= SN 1572): 5.
Blast Waves: 271-276.
BL Lac Objects: 211.
Bootes Void: 282.
Bubbles and Bubble Walls: 255, 261, 273, 277, 278.
Cancer Cluster: 101.
Carbon Deflagration: 45, 52.
Carbon Stars: 200.
Centaurus Cluster: 100, 101.
Cepheid Variables: 1, 2, 7-10, 15, 17, 21-24, 29-34, 35, 37, 38, 41, 42, 55-62, 99, 230, 285.
Cetus II Cluster: 101.
Cluster Density Profiles: 163-168.
Cold Dark Matter: 131, 134, 225-228, 243-250, 251-254, 255-264, 265-268.
Color-Magnitude Diagrams: 15, 82, 83, 93-98, 99, 132, 285.
Color-Redshift Diagram: 199.
Coma Cluster: 57, 64, 83, 101, 124, 126.
Convective Overshoot: 17.
Corona Borealis Supercluster: 281, 282.
Correlation Function: 169, 244, 245, 258, 260, 268, 269.
Cosmic Virial Theorem: 151, 154, 244.

Cosmic Background Radiation: 125, 129, 130, 169-176,
    251-254, 286.
Cosmic Strings: 248, 249, 265-268, 278.
Cosmochronology: 177-184, 185-196.
Cosmological Blastwaves: 203-214, 245, 248, 272,
    273-278, 287.
Cosmological Density: 134, 150, 151, 157, 219-223,
    225-228, 229-236, 237-242, 243-246, 255, 256, 266,
276.
DAOPHOT: 16, 22, 37.
DC 2345-32: 124.
DDO 210: 106.
DDO 221: 106.
Dipole Anisotropy: 107, 132, 149, 151, 169-176, 215,
    216, 271, 273.
Distance Indicators: 1, 21, 29, 45-54, 55, 77, 123, 135.
Distance Scale: 1, 7, 21-24, 29-34, 43, 45, 55.
Dwarfs and Irregular Galaxies: 21, 22, 112, 159-162,
    246, 249.
Dynamical Friction: 163-167.
EINSTEIN Observatory: 211.
Elliptical Galaxies: 123-130, 135-138, 197-202.
Environmental Effects: 137, 155, 159-162, 243.
Eridanus Cluster: 101.
ESO 594-G 04: 106.
Faber-Jackson Relation: 3, 5, 99, 123, 135, 267, 285.
Field Galaxies: 76, 81-86, 105-110.
Filaments (Cosmological): 119, 247, 261.
Fornax Cluster: 100, 101, 125.
Fragmentation: 210, 275.
Galactic Cirrus: 148.
Galactic Extinction: 1, 2, 4, 21, 23, 29, 33.
Galaxy Formation: 137, 154, 202, 203, 207, 225, 244,
    246, 251-254, 259, 273- 278.
Globular Clusters: 177-184, 185.
Grus Cluster: 101, 102.
Groups -
    G 0507: 101.
    G 1209: 101.
    G 5077: 101.
    G 5846: 101.
    G 7144: 101.
    G 7196: 101.

Gunn-Peterson Effect: 204.
HEAO 1 : 215, 217.
Hercules Supercluster: 3, 4, 5, 77, 79, 101, 281, 282.
HI Deficiency: 161, 162.
HI Line Width: 3, 127.
Hierarchical Clustering: 246.
Hodge 4: 11, 17.
Hodge 11: 11.
Horizontal Branch: 61, 197.
Hubble Constant: 1, 4, 5, 6, 41, 55-62, 64, 66, 73, 77, 99, 105, 108, 109, 112, 129, 198, 200, 283-288, 286, 287.
Hubble Space Telescope: 23, 41, 52, 61, 285.
Hubble Time: 177-184, 185-196, 200, 206, 286, 287.
Hyades: 8, 17, 61.
Hydra I Cluster: 64, 108.
Hydra-Centaurus Supercluster: 99, 102-104, 109, 113, 114, 125, 133, 139, 142.
IC 10: 106.
IC 1613: 21, 22, 29, 32, 33, 106.
IC 4947 Group: 101.
IC 5152: 106.
"Inflation": 243, 268.
Inner Rings: 4, 5.
Intergalactic Medium: 210, 211, 212.
International Ultraviolet Explorer (IUE) : 10.
IRAS: 1, 113, 132, 147-150, 151, 173, 215, 222, 223, 251, 286.
Isochrone Fitting: 17, 178-180.
Isotopic Ratios: 185-196.
Large Magellanic Cloud: 1, 2, 3-14, 15-18, 25, 29, 30, 33, 61, 62, 106.
Large Scale Structure: 117-122, 169-176, 215-218, 288.
Local Group: 2, 5, 29-34, 35-40, 58, 63, 102, 105, 111-116, 118, 125, 126, 131, 139, 173, 215.
Local Supercluster: 56, 93, 105, 111-113, 121, 133, 139-142, 143-146, 150, 151, 217, 237, 238, 243, 244, 271, 279-282.
Long Period Variables: 1, 2, 10.
Low Density Models: 225, 227, 271.
Luminosity Index: 3, 77.
"Ly-α Forest": 206.
Magnitude-Redshift Diagram: 199.

Main Sequence Fitting: 12, 15-18, 61.
Malmquist Bias: 5, 57, 65-68, 71, 73-76, 77, 78, 112, 124, 127, 132.
Mass-to-Light Ratio: 163, 166, 167.
Mass Density of the Universe: 134, 143, 145, 150, 151, 157, 219-223, 225, 229-236, 237-242, 243-246, 256, 266, 276.
Mass Segregation: 161, 163-168.
Metallicity Effects: 2, 8, 17, 21, 33.
Metric Distance Indicators: 4.
Microwave Dipole Anisotropy: 108, 112, 125, 133, 139, 147, 151, 231, 246, 251, 255, 258.
Microwave Background Fluctuations: 215-218, 258, 266, 273.
Mira Variables: 2, 10, 12.
M3: 180, 183.
M15: 11.
M31 (NGC 221): 2, 4, 5, 18, 21-23, 29, 31, 33, 35-40, 41, 42, 44, 56, 61, 62, 106, 107, 182, 285.
M33 (NGC 598): 2, 18, 21, 22, 29, 31, 33, 58, 61, 62, 106.
M81: 43, 58.
M81 Group: 58, 60.
M87: 43, 44, 56.
M101: 23, 59, 60, 61, 285.
Molecular Hydrogen: 207-210.
N-body Simulations: 144, 154, 159, 219-224, 237-242, 244, 256, 258, 267, 269.
NGC 55: 3.
NGC 147: 107.
NGC 221 (M31): 2, 5, 18, 21-23, 29, 31, 33, 35-40, 41, 42, 44, 56, 61, 106, 107, 182, 285.
NGC 247: 3, 58.
NGC 253: 3, 58.
NGC 300: 3, 21, 22.
NGC 383: 119.
NGC 598 (M33): 2, 18, 21, 22, 29, 31, 33, 60, 106, 107.
NGC 1600-1700 Groups: 129.
NGC 1866: 9, 12.
NGC 2158: 17.
NGC 2162: 11, 12, 16.
NGC 2190: 11, 12.

INDEX

NGC 2210: 10, 11, 12.
NGC 2213: 17.
NGC 2403: 21, 22.
NGC 2420: 17.
NGC 2477: 17.
NGC 2506: 17.
NGC 4214: 23.
NGC 5585: 60.
NGC 6067: 10.
NGC 6093: 43.
NGC 6752: 11.
NGC 6769 Group: 101.
NGC 6822: 18, 21, 22, 29, 32, 33, 106.
NGC 7789: 17.
NGC 7790: 9.
NGC 7793: 3, 58.
Neutrinos: 225, 244, 245, 247, 262, 266, 275.
Novae: 1, 38, 41, 56.
Nucleocosmochronology: 177-184, 185-196, 245.
Null Regions: 60.
Numerical Models: 159, 225-228.
Oosterhoff Type/Effect: 10, 35.
"Pancakes": 207, 221, 244, 247, 261, 268, 275.
Pal 3 (= UGCA 435): 107.
Pavo-Indus Cluster: 125.
Pegasus Dwarf: 106, 107.
Pegasus I Cluster: 100, 101, 117.
Period-Luminosity Relation: 8, 21-23, 29-34, 99.
Pisces Dwarf: 106.
Pisces-Perseus Supercluster: 57, 89, 112, 117-122, 125, 129.
PKS 2155-304: 210, 211, 212.
Pleiades: 8.
Pointless Mass: 243.
Population III Stars: 186, 204, 212.
Primary Distance Indicators: 1.
Prognos 9: 173.
Quadrupole Anisotropy: 139-142, 158, 171.
Radio Galaxies: 197-202.
Radioactive Decay: 47, 185-196.
Ram Pressure Stripping: 162.
RICHFLD: 16.
Rotating Clusters of Galaxies: 163-168.

R-Process Elements: 185-196.
Rubin-Ford Effect: 127, 129, 131.
RR Lyrae Variables: 1, 2, 10, 11, 17, 18, 35-40, 41, 42, 56, 181, 182, 230, 285.
Sculptor Group: 3, 23, 58, 60.
Sersic 129-1: 101.
Sextans A: 21, 22.
Sextans B: 21, 22.
Shells (Cosmological): 127, 271-276.
Shock Waves: 203-214, 271-276, 273-278.
Small Magellanic Cloud: 2, 3-14, 25, 29, 30, 33, 106.
Solar Motion: 73, 102, 105, 108, 172.
"Sosies": 4, 5, 73, 91.
Spin Parameter: 164.
Streaming Velocities: 113, 123-130, 131-134, 151, 203, 251, 262, 271, 267, 273, 288.
Strömgren Radius: 204, 206.
Subdwarf Main Sequence: 61, 180, 181.
Sunyaev-Zel'dovich Effect: 216.
Superclusters: 225, 261, 279-282.
Supergiants Stars: 1, 25-27, 56.
Supergalactic Plane / Supergalactic Co-ordinates: 5.
Supernovae: 3, 4, 5, 45-54, 56, 285.
Surface Brightness: 1, 81, 89, 123, 136, 285.
Surface Photometry: 89.
S Andromedae (SN 1885): 5.
Tidal Interactions: 69, 139-142, 143-146, 158, 163, 230.
T Scorpii: 43.
Tully-Fisher Relation: 3, 5, 55-62, 63-64, 65-72, 73-76, 81, 87-92, 93-96, 99, 105, 106, 112, 118, 124, 125, 127, 131, 132, 135, 137, 160, 229, 267, 285, 286.
UGC 5364: 106.
UHURU: 217.
Ultraviolet Deficit: 46-48.
Ursa Major-Virgo Complex: 127, 129.
Velocity Dispersion: 3, 105-110, 123, 136, 232, 233.
Velocity Field: 5, 73-76, 99-104, 123-130, 219-224.
Virgocentric Flow: 57, 60, 66, 70, 73-76, 99-104, 105-110, 112, 129, 137, 138, 139, 140, 143-146, 151, 157, 215, 219, 220, 225, 229-236, 237, 251, 271, 272, 287.

Virgo Cluster: 3, 4, 41, 57, 60, 63, 66, 67, 70, 73, 83, 77-80, 87, 93, 95, 96, 100, 101, 105, 114, 124, 125, 126, 147, 160, 169, 172, 173.
Virgo-S Cloud: 3, 77, 78, 79.
Virgo-S' Cloud: 3, 78.
Virgo-M Cloud: 3, 78.
Virgo-W Cloud: 3, 78.
Virgo-X Cloud: 3, 78.
Virial Theorem: 151, 154, 232, 244.
Voids: 118, 227, 244, 246, 248, 249, 261, 272, 275, 282.
VY Serpentis: 11.
White Dwarfs: 180.
Wolf-Lundmark-Melotte dwarf galaxy: 21, 22.
ω *Cen*: 10.
X Arietis: 11.
X-Ray Background: 173, 215-218.
Zel'dovich Power Spectrum: 226, 249, 252, 256, 265, 267, 268.
Zwicky Cluster, *Zw 74-23*: 101.
3C 10: 5.